Studies in Big Data

Volume 41

Series editor

Janusz Kacprzyk, Polish Academy of Sciences, Warsaw, Poland
e-mail: kacprzyk@ibspan.waw.pl

The series "Studies in Big Data" (SBD) publishes new developments and advances in the various areas of Big Data- quickly and with a high quality. The intent is to cover the theory, research, development, and applications of Big Data, as embedded in the fields of engineering, computer science, physics, economics and life sciences. The books of the series refer to the analysis and understanding of large, complex, and/or distributed data sets generated from recent digital sources coming from sensors or other physical instruments as well as simulations, crowd sourcing, social networks or other internet transactions, such as emails or video click streams and others. The series contains monographs, lecture notes and edited volumes in Big Data spanning the areas of computational intelligence including neural networks, evolutionary computation, soft computing, fuzzy systems, as well as artificial intelligence, data mining, modern statistics and operations research, as well as self-organizing systems. Of particular value to both the contributors and the readership are the short publication timeframe and the world-wide distribution, which enable both wide and rapid dissemination of research output.

More information about this series at http://www.springer.com/series/11970

Moamar Sayed-Mouchaweh
Editor

Learning from Data Streams in Evolving Environments

Methods and Applications

 Springer

Editor
Moamar Sayed-Mouchaweh
Institute Mines-Telecom Lille Douai
Douai, France

ISSN 2197-6503   ISSN 2197-6511   (electronic)
Studies in Big Data
ISBN 978-3-030-07862-1   ISBN 978-3-319-89803-2   (eBook)
https://doi.org/10.1007/978-3-319-89803-2

Printed on acid-free paper

This Springer imprint is published by the registered company Springer Nature Switzerland AG
The registered company address is: Gewerbestrasse 11, 6330 Cham, Switzerland

Preface

The volume of data is rapidly increasing due to the development of the technology of information and communication. This data comes mostly in the form of streams. Learning from this ever-growing amount of data requires flexible learning models that self-adapt over time. In addition, these models must take into account many constraints: (pseudo) real-time processing, high-velocity, and dynamic multi-form change such as concept drift and novelty. Consequently, learning from streams of evolving and unbounded data requires developing new algorithms and methods able to learn under the following constraints: (1) random access to observations is not feasible or it has high costs, (2) memory is small with respect to the size of data, (3) data distribution or phenomena generating the data may evolve over time, which is known as *concept drift* and (4) the number of classes may evolve overtime, which is known as *concept evolution*. Therefore, efficient data streams processing requires particular drivers and learning techniques able to perform:

- Incremental learning in order to integrate the information carried out by each new arriving data;
- Decremental learning in order to forget or unlearn the data samples which are no more useful;
- Novelty detection in order to learn new concepts.

This edited Springer book presents and discusses recent advanced techniques, methods and tools treating the problem of learning from data streams generated by evolving and non-stationary phenomena. These methods address the different challenges (with concept drift, with concept evolution, with both concept drift and concept evolution) of learning from multidimensional data streams using classification or clustering techniques, mono model or ensemble, online or semi-online, centralized processing or distributed computing, instance based or window based, and incremental decremental with or without transfer learning for different applications (social networks, Twitter data analysis, stream trends and dynamics visualization, user query and preference evaluation, gene network, customer relationship management, electricity price prediction, taxi traffic management, etc.).

Finally, the editor is very grateful to all authors and reviewers for their very valuable contribution allowing enriching the research and publication history of learning from data streams in non-stationary environments. I would like also to acknowledge Mrs. Mary E. James for establishing the contract with Springer and supporting the editor in any organizational aspects. I hope that this volume will be a useful basis for further fruitful investigations and fresh ideas for researcher and engineers as well as a motivation and inspiration for newcomers to address the problems related to this very important and promising field of research.

Douai, France   Moamar Sayed-Mouchaweh

Contents

Introduction

Moamar Sayed-Mouchaweh

Abstract This introductory chapter intends to present the challenges related to the problem of learning from data streams in nonstationary environments. It focuses on the major challenges related to the learning with concept drift, learning with concept evolution, and learning with both concept drift and concept evolution. Then, it classifies the different methods and techniques of the state of the art that are used to address these challenges. This categorization is achieved according to how the learning is performed, how the data streams are processed, and how the changes are detected and integrated into the model. Finally, this chapter ends with a compact summary of the contents of this book by providing a paragraph about each of the contributions and how the learning process from data streams is performed (single learner or ensemble learners, centralized processing or distributed computing, classification, regression or clustering, window-based or sequential-based, applications targeted, etc.).

1 Learning from Data Streams

In the increasing number of real-world applications [1–6], for example, network traffic, web mining, social networking, network monitoring, sensor networks, telecommunication networks, manufacturing systems, electrical energy generation, transmission and distribution grids, data samples arrive continuously online through unlimited streams or flows often at high speed. In addition, the increasing attention on Internet of Things (IoT), web applications, and mobile devices increase the production and proliferation of streaming data.

The generated data streams [7, 8] by these applications possess the 5Vs characteristics of Big Data (i.e., volume, velocity, value, variety, and veracity). Indeed, data streams have unlimited size, so the volume of generated data is massive

M. Sayed-Mouchaweh (✉)
Institute Mines-Telecom Lille Douai, Douai, France
e-mail: moamar.sayed-mouchaweh@imt-lille-douai.fr

© Springer International Publishing AG, part of Springer Nature 2019
M. Sayed-Mouchaweh (ed.), *Learning from Data Streams in Evolving Environments*, Studies in Big Data 41, https://doi.org/10.1007/978-3-319-89803-2_1

and continuously growing at high speed. These data samples are generated from different sources (people and different devices) and types (text, measurements, images, videos, etc.). Finally, the veracity of these data is impacted by noises, outliers, bias, missing information, environment changing, novelties, etc.

Traditional one-shot memory-based learning methods trained offline from a fixed size of historic data sets are not adapted to learn from data streams. This is because, first, it is not feasible to register all the data samples over time and, second, the generated models become quickly obsolete due to the occurrence of changes in their internal dynamics (e.g., because of aging or fault occurrence) or their environments. In traditional machine learning and data mining approaches, the current observed data and the future data are assumed to be sampled independently and from an identical probability distribution (iid). However, the phenomena generating data streams may evolve over time. In this case, the environment in which the system or the phenomenon generated the data is considered to be dynamic, evolving, or nonstationary.

Therefore, methods used to learn from data streams generated from phenomena evolving in dynamic environments are facing several challenges. The first is caused by the huge amount of data that are continuously generated over time. The second challenge is due to the high speed of arrival of data streams. The third challenge may occur when the joint probability distribution of the data samples evolves over time. This is unknown as *concept drift* [2, 4, 9, 10]. The fourth challenge is related to the fact that the number of classes or modes is often unknown in advance. Therefore, new classes, or modes, may appear any time and they must be detected and the model structure must be updated. This is known as *concept evolution* [11–13]. Finally, data streams may be generated and treated by and across a set of distributed sources, sites, or users (i.e., data grid). In this case, processing data using one central model (central computing) may not be efficient, especially if the infrastructure is large and complex (e.g., network). Hence, a decentralized modeling structure based on the use of several local models (distributed computing [14]) is more suitable and efficient to process data grid.

Hence, learning from streams of evolving and unbounded data requires developing new algorithms and methods able to learn under the following constraints: (1) random access to observations is not feasible or it has high costs, (2) memory is small with respect to the size of data, (3) data distribution or phenomena generating the data may evolve over time (concept drift), and (4) the number of classes (occurrence of new classes or disappearance of existing classes) may evolve over time (concept evolution).

2 General Classification of Methods to Learn from Data Streams

There are several machine learning and data mining approaches that are developed to learn from data streams in the presence of concept drift and/or concept evolution. These approaches can be classified according to

- How concept drift and/or concept evolution is handled into informed and blind methods
- How learning is performed into single learner and ensemble learners
- How concept drift and/or concept evolution is monitored into supervised and unsupervised change detection
- How data streams are processed into central and distributed
- How data samples are processed into sequential and windowing approaches

Informed methods explicitly detect changes (concept drift) using triggering or change detection mechanisms such as the performance of a learner [2, 4], or the characteristics of data distributions in the feature space. When a change is detected, the informed methods relearn the model using a recent selection of data samples through a time window or a register. The informed methods are based on three steps: (1) detecting the drift using change or drift-monitoring indicators, (2) deciding the data samples to be used to update the learner (which data samples to keep and which ones to forget), and (3) deciding how to react or update the learner using the collected data samples.

Learning from data streams with concept evolution (abnormal and normal new concepts or classes) is based on the use of clustering techniques [8, 11] in order to evaluate the cohesion between the data samples (registered in a buffer or short time memory) that do not belong to any of the existing classes. In general, the cohesion between data samples in the buffer is evaluated using a distance measure. It is worth mentioning that learning from data streams with both concept drift and concept evolution [15, 16] is a very challenging task since it is necessary to distinguish between the drifting of existing concepts and the occurrence of a new concept.

The blind methods [4] implicitly adapt the learner to the current concept at regular time intervals without any drift detection. They discard old concepts at a constant speed independently of whether changes occurred or not.

Ensemble learners are based on the use of several classifiers, models, or learners built by one learning method with different configurations or by different learning methods in order to achieve the classification or the prediction of a new incoming pattern. The aim of using ensemble methods is to achieve more accurate prediction on training data set and better generalization on unseen data samples. They provide a natural way for adapting to changes by either modifying their structure (combination of the individual learners' decisions, selection of one individual learner's decision, etc.) or updating the combination rules used to fuse the learners' individual decisions into one global decision. In single-learner approaches, only one model is used preferences allows to avoid producing empty results by providing the most relevant results to the user query. However, since the provided results could be too large such that making decision could be complex or impossible, the proposed approach summarizes these results by assigning the similar ones to the same cluster. The preference on data streams is evaluated using the Stream-Based Lattice Skyline algorithm (SLS) which has the advantage to avoid the continuous object to object comparison. Avoiding the object-to-object comparison, required to perform preference evaluation on continuous data streams, allows to reduce the runtime complexity from exponential to linear in the number of input objects. This advantage is essential in order to update continuously the Best Matches Only (BMO) when new data streams arrive since they can contain better objects according to the user preferences. The clustering is based on the use of Borda social choice voting rule. The latter allows the assignment of each object to only one cluster based on an equal importance of the voters (considered dimensions). The vote is based on the calculated distance between the considered object and the cluster centroid according to each dimension. The proposed approach is evaluated (runtime and the number of required iterations to find the BMO) on artificial and real data sets. The latter is data set from Twitter with different input streams (dimensions). The results showed that the proposed approach can be applied for real-time preference evaluation on data streams.

3.4   Chapter 5

This chapter discusses the problem of minimizing the approximation error of streaming data. Indeed, there are several techniques, such as Piecewise Linear Representation (PLR), histograms, and wavelet-based methods, which are used to represent data streams in proper and compact forms. These compact forms allow to perform queries and operations on data streams and to describe the data dynamics as trends, patterns, and outliers. The chapter focuses on the use of PLR because of its efficiency and simplicity. PLR represents data streams with a number of simple line segments, so that the data streams can be efficiently archived, and a query on the stream can be approximately answered by a query on the line segments. In addition, the line segments constructed from PLR provide striking visual outlines of stream trends and can be more efficiently processed and represented in the database. The number of line segments and their parameters are determined based on the extreme slopes, maximum error bound, and convex hulls. These parameters are updated over time with the arrival of new data streams. This chapter studies and compares two PLR approaches: Both OptimalPLR and ParaOptimal achieve optimal representation with linear complexity but process the data points in different spaces, that is, time-value stream space and slope-offset parameter space. The chapter proves formally that the two algorithms are equivalent. However, they may have different performance (efficiency in both memory and time cost) in practice, due to the different ways of recording intermediate data during the processing, which results from the space they are based on.

3.5 Chapter 6

This chapter discusses the problem of choosing the adequate ensemble method to learn from data streams in nonstationary environments where concept drift may occur. It presents the different structures and dynamics of ensemble methods and analyzes their behavior and performances in response to the occurrence of different concept drifts and environments' conditions. Indeed, an ensemble method reacts to a change in its environments' conditions or to a concept drift by adding a new classifier in fixed or variable time, by removing a classifier when the ensemble size is reached, by observing when a classifier's performance is below a specific threshold or when a drift is detected, or by updating the weights/rank of classifiers or by training them using a new data set. Each of these mechanisms, called "dynamics" in the chapter, has its advantages according to the application context and conditions. Therefore, this chapter compares the performances of these ensemble "dynamics" using two academic (Hyperplane and SEA generator) and two real (Forest Cover-type and Electricity prices) data sets. Well-known methods (Adaptive Boosting (Aboost), Dynamic Weighted Majority (DWM), Tracking Recurrent Ensemble (TRE), Adwin Bagging (AdwinBag), Recurrent Concept Drift (RCD), and Online Accuracy Update Ensemble (OAUE)) using these mechanisms are evaluated and compared using different evaluation criteria including classification accuracy, training time, memory usage, average adaptation time to concept drifts, and average accuracy drop upon concept drifts. Each evaluation run in these experiments involves passing one of the chosen data sets through a specific algorithm in a form of data stream with a specified number of instances per interval. The obtained results allow to help in choosing the adequate mechanism or mix of mechanisms in response to the application context and conditions (drift type, nature, speed, etc.).

3.6 Chapter 7

This chapter studies the problem of analyzing online and under one-pass constraints the users' preferences changes in social networks. In the latter, the users' relationships reveal their behavior and preferences. These relationships are continuously evolving over time after each incoming object (i.e., tweets or retweets). In order to update the typology or structure of the network (nodes representing the users and their interactions or relationships), a change detection model is used based on the observation of the deviation of nodes positions. This deviation represents a change in the user's preference over predefined themes or topics. To detect this deviation, three node centralities are used. The first centrality is the degree centrality of a node that measures the number of edges adjacent to it. The second centrality is the betweenness centrality that measures the number of shortest paths passing by this node. The third centrality is the closeness centrality that is the inverse of the average shortest path length between a node and all the other nodes. Three detection models are used: Moving Window Average (MWA), Weighted Moving Window Average (WMWA), and Page Hinckley Test (PH). Then, a change point scoring function is computed based on the centralities and the values of the detection models and then compared to a threshold to decide a change. The accuracy of change detection of users' preferences in a social network (tweets) using each of the three detection modes MWA, WMWA, and PH is compared. This comparison is achieved using homogeneous and bipartite networks. Homogeneous network is based on retweets where nodes are Twitter users and two nodes have a direct edge if the first node retweeted the second node at certain time. Nodes in bipartite network are Twitter users or topics. Topics represent the main themes that users are tweeting about and have been extracted using the LDA (latent Dirichlet allocation) model.

3.7   Chapter 8

This chapter presents a platform, called Apache SAMOA (Scalable Advanced Massive Online Analysis), for mining big data streams. It combines streaming algorithms and distributed computing in order to perform distributed stream processing. It is an open-source gathering framework and library. The framework allows an algorithm developer to reuse his code and run it on several distributed stream processing engines as Storm, Flink, Samza, and Apex. The library allows the implementation of state-of-the-art algorithms for distributed machine learning on streams. It includes three types of algorithms performing basic machine learning functionalities such as classification via a decision tree, Vertical Hoeffding Tree (VHT), clustering (CluStream), and regression rules (HAMR). It also includes adaptive implementations of ensemble methods such as bagging and boosting. These methods include state-of-the-art change detectors such as ADWIN, DDM, EDDM, and Page-Hinckley. The performance of this platform (accuracy and processing time) of its different classification, regression, clustering, and ensemble methods is evaluated using three real data sets (Electricity prices, Particle Physics, and CovertypeNorm). The goal is to compare the performance between local implementation and distributed implementation. The analysis of the obtained results shows the interest of the use of this platform for mining big data streams in distributed fashion.

3.8   Chapter 9

This chapter discusses the problem of building a social network that represents different kinds of relations between resources (employees) and customers (requests, activities). The analysis of this social network allows to improve the quality of "Customer Relationship Management" process. In order to extract the social network from saved sequential event logs, this chapter proposes to use a scheme based on the following steps. First, the process model is built using the Alpha algorithm. This model is complicated since the Alpha algorithm does not consider the activities' frequencies. Therefore in the second step, the infrequent (considered as noises) activities are removed using the heuristics miner algorithm. In order to guarantee the production of a sound process model, the evolutionary tree mine algorithm is applied which reduces the search space, so the unsound models will not be considered. Then, the produced model will be converted into Petri net allowing to code the activities by letters in order to facilitate their readability. The time consumed in each transition of the Petri net model is integrated in order to allow discovering which activity takes much time to be performed. Finally, the social network is built from this Petri net model. Only most relevant activities with a significant impact on the model quality are kept. The proposed scheme is applied to extract a social network using the CRM event logs recorded during the year 2015 between clients and employees of the National Institute of Statistics in Portugal. The extracted social network allows to discover useful information about the behavior of employees (e.g., some employees act as a customer service and internal department at the same time).

3.9   Chapter 10

This chapter studies the problem of clustering in graphs with structures that evolve over time. Indeed in many real-world applications, the number of clusters increases when a cluster divides or decreases when two clusters merge. The evolution of a gene network as an example, which consists of genes and their interactions, produces a graph sequence when genes are added, deleted, or mutated. This chapter analyzes the performances of an algorithm, called O2I, to perform the clustering of evolving graphs. O2I uses spectral clustering and relies on applying the k partition problem to a graph constructed from a graph sequence. It uses a forgetting rate in order to smooth the clusters between timestamps separated by a predefined distance. The performances of O2I are evaluated using academic and real data sets and compared to the ones of the conventional online graph sequence clustering algorithm Preserving Clustering Membership (PCM). In the academic data sets, the sets of points are generated following a Gaussian distribution and move toward and away from the origin as time advances. The real data set is the Enron e-mail data set which is divided onto several periods according to timestamps of e-mails. A unique ID is assigned to the e-mail address for each person participating in the communication, and an edge is assigned to a pair of individuals if they communicate via e-mail within each period. The obtained results show that O2I is a solution for mining evolving graphs in which the connections, number of clusters, and number of vertices evolve over time.

3.10   Chapter 11

This chapter proposes an approach that estimates online the Probability Density Function (PDF) of evolving data streams. The proposed approach, called KDE-Track, is based on the use of the Kernel Density Estimation (KDE) method. However in order to reduce its quadratic time complexity into linear time complexity with respect to the stream size, the proposed approach uses an adaptive resampling strategy in order to control the number of resampling points, that is, more points are resampled in the areas where the PDF has a larger curvature, while less number of points are resampled in the areas where the function is approximately linear. In order to timely track the evolving density, a sliding window is used to estimate the density using the most recent data points. In order to improve the accuracy of the density estimation, the proposed approach uses different bandwidth values for each dimension allowing to capture the spread of the data on each dimension. The accurate bandwidth is selected by minimizing the deviation between the true and the estimated densities. The proposed approach is evaluated using synthetic and real data sets. The real data sets represent two application problems, Taxi traffic real-time visualization and unsupervised online change detection. For instance, the estimated PDF is used for the taxi traffic problem to help in placing taxicabs in places with high pickup rate at a certain period of time. The performances (accuracy, time complexity) of the proposed approach are compared to several well-known density estimation and change detection methods such as the traditional KDE, the FFT-KDE, which deploys Fast Fourier Transform (FFT) to convolve a very fine histogram of the data with a kernel function to produce a continuous density function; the Cluster Kernels, which maintains a specific number of kernels by merging similar kernels; and SOMKE, which employs Self-Organizing Maps (SOM) to cluster the data into a specific number of clusters and uses the centroids of the clusters as the set of kernels.

3.11   Chapter 12

This chapter discusses the use of the classification method Support Vector Machine (SVM) for online learning from data streams. It focuses on the incremental version of SVM that enables it to update the classifier after the arrival of a new data sample without having to rebuild the classifier from scratch. This incremental learning is particularly useful for learning from data streams since the classifier update is performed using limited memory and processing resources. In addition, adding also an unlearning mechanism into the incremental framework allows the classifier to be updated using only useful knowledge and forgetting the useless or obsolete knowledge. Hence, the classifier will be adapted to handle concept drift in the case of evolving data streams. The chapter starts by classifying the major incremental SVM methods in the state of the art into two groups: online and semi-online. The online methods process the data stream one sample at a time and ensure that the Karush–Kuhn–Tucker (KKT) optimality conditions are maintained on all previously seen samples while updating the classifier. The semi-online methods process the samples in batches and while updating the classifier, they discard previously seen samples except those identified as support vectors. Then, this chapter discusses the advantages and drawbacks of these two groups according to their ability to handle concept drift, to learn with unlabeled data samples, and to learn from large scale data sets. The chapter also discusses some of the major applications of these methods and focuses on their performances in the case of data streams.

3.12   Chapter 13

This chapter studies the problem of data streams clustering in the presence of concept drift and concept evolution. Concept drifts occur whenever the data distribution changes, while concept evolutions refer to the appearance or disappearance of clusters. The chapter starts by highlighting the challenges related to the problem of data stream clustering. Indeed, clustering algorithms must (1) detect concept drifts and adapt its clusters accordingly; (2) detect concept evolutions and create/delete clusters independently from user intervention; (3) discern between seeds of new clusters and noisy data; and finally (4) not rely on too much of parameters. The chapter analyzes four selected major data stream clustering methods because of the availability of their code, the high performances of their results in different application domains, and their number of citations. These methods are CluStream, ClusTree, DenStream, and HAStream. Then, the chapter analyzes the social network theory and its use for building social networks from data streams. It studies and compares the performances of three social network-based data stream clustering algorithms: CNDenStream, SNCStream, and SNCStream+. Then, the performances (clustering quality, processing time, and memory usage) of these methods are compared to the ones of the four selected data stream clustering methods using academic and public real data sets. The obtained results show the efficiency of social network-based clustering methods against data stream clustering methods.

References

1. Abdallah, Z.S., Gaber, M.M., Srinivasan, B., Krishnaswamy, S.: Adaptive mobile activity recognition system with evolving data streams. Neurocomputing. **150**, 304–317 (2015)
2. Khamassi, I., Sayed-Mouchaweh, M., Hammami, M., Ghédira, K.: Self-adaptive windowing approach for handling complex concept drift. Cogn. Comput. **7**(6), 772–790 (2015)
3. Rashidi, P., Cook, D.J.: Keeping the resident in the loop: adapting the smart home to the user. IEEE Trans. Syst. Man Cybern. Syst. Hum. **39**(5), 949–959 (2009)
4. Sayed-Mouchaweh, M.: Learning from Data Streams in Dynamic Environments. Springer, Cham (2016)
5. Mouchaweh, M.S.: Diagnosis in real time for evolutionary processes in using pattern recognition and possibility theory. Int. J. Comput. Cogn. **2**(1), 79–112 (2004)

6. Toubakh, H., Sayed-Mouchaweh, M.: Hybrid dynamic data-driven approach for drift-like fault detection in wind turbines. Evol. Syst. **6**(2), 115–129 (2015)
7. Hulten, G., Spencer, L., Domingos, P.: Mining time-changing data streams. In: Proceedings of the Seventh ACM SIGKDD International Conference on Knowledge Discovery and Data Mining, pp. 97–106. ACM, New York (2001)
8. Guha, S., Mishra, N.: Clustering data streams. In: Data Stream Management, pp. 169–187. Springer, Berlin (2016)
9. Hartert, L., Sayed-Mouchaweh, M.: Dynamic supervised classification method for online monitoring in non-stationary environments. Neurocomputing. **126**, 118–131 (2014)
10. Mohamad, S., Bouchachia, A., Sayed-Mouchaweh, M.: A bi-criteria active learning algorithm for dynamic data streams. IEEE Trans. Neural Netw. Learn. Syst. **29**, 74 (2018)
11. Abdallah, Z.S., Gaber, M.M., Srinivasan, B., Krishnaswamy, S.: Anynovel: detection of novel concepts in evolving data streams. Evol. Syst. **7**(2), 73–93 (2016)
12. Faria, E.R., Gonçalves, I.J., de Carvalho, A.C., Gama, J.: Novelty detection in data streams. Artif. Intell. Rev. **45**(2), 235–269 (2016)
13. Patcha, A., Park, J.M.: An overview of anomaly detection techniques: existing solutions and latest technological trends. Comput. Netw. **51**(12), 3448–3470 (2007)
14. Beringer, J., Hüllermeier, E.: Online clustering of parallel data streams. Data Knowl. Eng. **58**(2), 180–204 (2006)
15. Masud, M., Gao, J., Khan, L., Han, J., Thuraisingham, B.M.: Classification and novel class detection in concept-drifting data streams under time constraints. IEEE Trans. Knowl. Data Eng. **23**(6), 859–874 (2011)
16. Mohamad, S., Sayed-Mouchaweh, M., Bouchachia, A.: Active learning for classifying data streams with unknown number of classes. Neural Netw. **98**, 1–15 (2018)

Transfer Learning in Non-stationary Environments

Leandro L. Minku

Abstract The fields of transfer learning and learning in non-stationary environments are closely related. Both look into the problem of training and test data that come from different probability distributions. However, these two fields have evolved separately. Transfer learning enables knowledge to be transferred between different domains or tasks in order to improve predictive performance in a target domain and task. It has no notion of continuing time. Learning in non-stationary environments concerns with updating learning models over time in such a way to deal with changes that the underlying probability distribution of the problem may suffer. It assumes that training examples arrive in the form of data streams. Very little work has investigated the connections between these two fields. This chapter provides a discussion of such connections and explains two existing approaches that perform online transfer learning in non-stationary environments. A brief summary of the results achieved by these approaches in the literature is presented, highlighting the benefits of integrating these two fields. As the first work to provide a detailed discussion of the relationship between transfer learning and learning in non-stationary environments, this chapter opens up the path for future research in the emerging area of transfer learning in non-stationary environments.

1 Introduction

Individuals, organisations and systems have been producing large amounts of data. Such data can be used to gain insights into various areas of interest, such as businesses and processes. In particular, machine learning can be used to create models able to give insights into the form of predictions. Examples of problems involving predictions include spam detection, credit card approval, network intru-

L. L. Minku (✉)
School of Computer Science, University of Birmingham, Birmingham, UK
e-mail: L.L.Minku@cs.bham.ac.uk

© Springer International Publishing AG, part of Springer Nature 2019
M. Sayed-Mouchaweh (ed.), *Learning from Data Streams in Evolving Environments*, Studies in Big Data 41, https://doi.org/10.1007/978-3-319-89803-2_2

sion detection, speech recognition, among many others. This chapter concentrates on machine learning for building predictive models.

In supervised learning, predictive models are created based on existing training examples of the format $(x_i, y_i) \in \mathcal{X} \times \mathcal{Y}$, where x_i is example i's vector of input attributes, y_i is example i's vector of output attributes, $\mathcal{X}$ represents the input space and $\mathcal{Y}$ represents the output space. For example, in the problem of predicting the effort required to develop a software project, each example i could correspond to a software project described by $(x_i, y_i) \in \mathcal{X} \times \mathcal{Y}$, where $\mathcal{X}$ is the four-dimensional space of all possible team expertises, programming languages, types of development and estimated sizes, and $\mathcal{Y}$ is the one-dimensional space of all possible software required efforts in person-hours. In the problem of predicting whether a credit card customer will default their payments, each example i could correspond to a customer described by $(x_i, y_i) \in \mathcal{X} \times \mathcal{Y}$, where $\mathcal{X}$ is the 5-dimensional space of all possible ages, genders, salaries, types of bank accounts and numbers of years consecutively holding a bank account, and $\mathcal{Y}$ is the one-dimensional space of payment categories (*pay* and *default*).[1] When $\mathcal{Y}$ is a one-dimensional space, y_i can be written as y_i. This work concentrates on one-dimensional output spaces, but the ideas discussed herein could be extended to multidimensional output spaces.

Most supervised learning algorithms are offline learning algorithms, defined as follows:

Definition 1 (Offline Learning) Consider a fixed training set $\mathcal{S} = \{(x_i, y_i)\}_{i=1}^m$ $\sim_{\text{iid}}$ $p(x, y)$, where $(x_i, y_i) \in \mathcal{X} \times \mathcal{Y}$, and $p(x, y)$ is the joint probability distribution of the problem. Offline learning consists in using the predefined training set $\mathcal{S}$ to build a model $f : \mathcal{X} \rightarrow \mathcal{Y}$ able to generalise to unseen examples $(x_i, y_i) \sim p(x, y), i > m$.

Given Definition 1, a model f created based on offline learning is appropriate for predicting the output attributes of new instances from the same joint probability distribution $p(x, y)$ as the one underlying the training set $\mathcal{S}$. However, many real-world applications involve in making predictions for a target task in a given domain based on examples coming from different source tasks or domains. In such cases, the probability distributions underlying the target and sources may differ.

The need for using examples from different sources arises from the high cost or even impossibility of collecting training examples from the target task and domain. For example, when building a model to predict the effort required to develop a software project in a given company, it is typically expensive to collect labelled examples describing projects from within this company [14]. Examples of projects developed by other companies are available in existing software project repositories [13, 19], but they may come from different underlying probability distributions.

Machine learning algorithms operating in this type of scenario must be able to tackle the different source and target probability distributions. Specifically, they

---

[1]The software required effort and credit card payment problems will be used as illustrative examples for the concepts explained in this chapter.

need to transfer knowledge from the source task/domain to the target task/domain. Algorithms designed to achieve such knowledge transfer are referred to as *Transfer Learning* (TL) algorithms.

Many real-world applications pose yet another challenge to machine learning algorithms. Instead of providing a fixed training set, they provide a potentially infinite sequence of training sets $S = \langle S_1, S_2, \ldots \rangle$, where $S_t = \{(x_t^{(i)}, y_t^{(i)})\}_{i=1}^{m_t} \sim_{\text{iid}} p_t(x, y)$; $t > 0$ is a time step, that is the sequential identifier of a time moment when a new training set was received; and $m_t > 0$ is the size of the training set. Such sequence is referred to as a *data stream*. Its unbounded nature provides a clear notion of continuing time. For example, in the problem of predicting whether a credit card customer will default their payments, new examples describing the behaviour of additional customers may be received over time [22].

Data streams may suffer changes in their underlying probability distributions over time, that is examples drawn at different time steps may belong to different probability distributions. Such changes are referred to as *concept drifts* [10]. They can be seen as the result of (1) actual changes in the underlying data-generating process or (2) insufficient, unknown or unobservable input attributes, which result in the probability distributions underlying observable attributes to change, despite the true data-generating process being stationary [7]. For example, customers' defaulting behaviour may be affected by the beginning of an economic crisis. If there are no input attributes describing the presence of a crisis, the beginning of the economic crisis would be perceived as a change in the joint probability distribution underlying the observable attributes.

Machine learning algorithms for Non-stationary Environments (NSE) must be able to adapt predictive models to concept drift. They need to maintain up-to-date predictive models reflecting the current joint probability distribution, even considering that part of the past training examples may belong to a different joint probability distribution.

As we can see from the above, both TL and learning in NSE involve training and test data that potentially come from different probability distributions. However, these two fields have evolved separately, and little work has investigated the connections between them. A discussion of the relationship between these two fields could greatly benefit future work, as TL approaches could inspire new approaches to overcome problems in learning algorithms for NSE, and vice-versa.

This chapter provides a novel discussion of the relationship between the fields of TL and learning in NSE, and of the potential benefit of using one field to improve the other. It explains two existing approaches that combine the strengths of TL and learning in NSE: Diversity for Dealing with Drifts (DDD) [22] and Dynamic Cross-company Mapped Model Learning (Dycom) [23]. The results achieved by these two approaches in the literature are briefly explained, highlighting the benefits of transferring knowledge in NSE.

This chapter is further organised as follows. Sections 2 and 3 explain TL and learning in NSE, respectively. They are not intended to provide an extensive literature review of these two fields, which can be found elsewhere [7, 11, 17, 29].

Instead, they are intended to provide definitions and examples of representative algorithms to inform and enable the discussion of the relationship between these two fields, which is provided in Sect. 4. After discussing this relationship, Sect. 5 discusses the potential of combining these two types of approach to improve their weaknesses. It includes an explanation of the approaches DDD [22] and Dycom [23] for TL in NSE, and of the results they have achieved in the literature. Finally, Sect. 6 concludes this work.

2   Transfer Learning (TL)

TL has been studied by the machine learning and data mining communities for many years [28]. When defining TL, several authors rely on the distinction between the terms "domain" and "task" [28]:

Definition 2 (Domain) *Domain* is a pair $\langle \mathcal{X}, p(x) \rangle$, where $\mathcal{X}$ is the input space; $p(x)$ is the unconditional probability distribution function (pdf); and $x \in \mathcal{X}$.

Definition 3 (Task) Given a *Domain* $= \langle \mathcal{X}, p(x) \rangle$, *Task* is a pair $\langle \mathcal{Y}, p(y|x) \rangle$, where $\mathcal{Y}$ is the output space; $p(y|x)$ represents the posterior probabilities of the output attributes; $x \in \mathcal{X}$; and $y \in \mathcal{Y}$.

Given existing TL approaches [29], we can consider existing work to be adopting the following definition of TL:

Definition 4 (Transfer Learning) Consider N source tasks, $Task_i = \langle \mathcal{Y}_i, p_i(y|x) \rangle$, $1 \leq i \leq N$, and their corresponding source domains, $Domain_i = \langle \mathcal{X}_i, p_i(x) \rangle$. Consider also a target task, $Task' = \langle \mathcal{Y}', p'(y|x) \rangle$, and its corresponding target domain, $Domain' = \langle \mathcal{X}', p'(x) \rangle$, where $\forall i$, [$Domain_i \neq Domain'$ or $Task_i \neq Task'$]. The sources are associated to fixed (labelled or unlabelled) data sets $\mathcal{S}_i$. The target is associated to a fixed (labelled or unlabelled) data set $\mathcal{S}'$. TL consists in using both the source and target data sets to build a target model $f' : \mathcal{X}' \rightarrow \mathcal{Y}'$ able to generalise to unseen examples of $p'(x, y) = p'(y|x)p'(x)$. Its aim is to improve learning in comparison with algorithms that use only $\mathcal{S}'$.

The use of the source data sets $\mathcal{S}_i$, $1 \leq i \leq N$, may be direct or indirect. In the former case, relevant source examples can be selected to train the target model f'. In the latter, source data sets can be used to learn parameters of models or representations, which are in turn used to help building f'.

TL is typically useful when there is not enough target data to produce a good target predictive model. The lack of target data may result from the high cost or even impossibility of collecting target training examples, as explained in Sect. 1. However, it is worth noting that TL can only be beneficial if the sources and target share some similarities. If they are too dissimilar, the use of source examples could even be detrimental to the performance of the target predictive model, depending on the TL algorithm being used [31]. This phenomenon is called *negative transfer*.

Depending on whether sources and target differ in terms of their domains or tasks, TL approaches can be categorised into two types—transductive (Sect. 2.1) and inductive (Sect. 2.2).

2.1 Transductive TL

Given Definition 4 (TL), transductive TL can be further defined as follows:

Definition 5 (Transductive Transfer Learning) Transductive transfer learning consists in transferring knowledge between different domains that share the same task. More formally, $\forall i$, $[\mathcal{X}_i \neq \mathcal{X}' \text{ or } p_i(x) \neq p'(x)]$ and $[\mathcal{Y}_i = \mathcal{Y}' \text{ and } p_i(y|x) = p'(y|x)]$.

Transductive TL is typically used when we do not have access to labelled target examples, but we do have one or more training sets containing labelled source examples. Let's take the software required effort problem introduced in Sect. 1 as an example. Consider that three software development companies c_1, c_2 and c' can develop software projects whose estimated size varies from small to large. However, company c_1 is more often involved with large software projects, company c_2 is more often involved with medium software projects and company c' is more often involved with small software projects. So, $p_1(x) \neq p_2(x) \neq p'(x)$. Consider also that the three companies adopt largely the same practices. So, it is likely that $p_1(y|x) = p_2(y|x) = p'(y|x)$. Companies c_1 and c_2 have collected several examples of their completed software projects, including information on their required efforts. Such data was donated to the International Software Benchmarking Standards Group [13]. Company c', on the other hand, collected only the input attributes of its projects, because required efforts are expensive to collect. Therefore, company c' may wish to use transductive TL to benefit from the data collected by companies c_1 and c_2 to build a model for predicting software required effort.

In transductive TL, as the source and target tasks are the same, labelled source examples can be used to learn the target task. However, as the source and target domains are different, the sources may not cover the same regions of the input space as the target, or may not share the same input space as the target. This needs to be tackled to avoid incorrectly biasing the target predictive model.

The differences between the source and target domains can be addressed by filtering or weighting source examples, so that the ones most relevant to the target domain are emphasised. For example, Turhan et al. [39] filter source examples based on their distance to the target examples in the input space. The source examples that are closest to target examples are used to build the target predictive model. Huang et al. [12] learn weights for the source examples so as to minimise the difference between the means of the source and target examples in a kernel Hilbert space.

These weights can then be used when learning a kernel-based model for the target domain based on the labelled source examples.

Another way to deal with the differences between source and target domains is to transfer parameters that compose models or feature representations. For example, Dai et al. [6] proposed a naïve Bayes transfer classification algorithm based on Expectation-Maximisation. The parameters of a naïve Bayes model are first estimated based on a labelled source data set. Expectation-Maximisation is then used based on an unlabelled target data set to gradually converge the source parameters to the target probability distribution. Pan et al. [29] proposed to learn a transformation of the source and target input attributes into a new space, called the latent space. This transformation is learned so as to minimise the distance between the transformed source and target domains. This idea shares some similarities with Huang et al. [12]'s approach. However, Huang et al. [12] learn weights to be used with the transformed examples, whereas Pan et al. [29] learn the transformation itself. Once the transformation is learnt, the target model can be learnt based on the transformed source examples and their corresponding output attributes.

2.2   Inductive TL

Given Definition 4 (TL), inductive TL can be further defined as follows:

Definition 6 (Inductive Transfer Learning) Inductive transfer learning consists in transferring knowledge between different tasks. The domains may or may not be different. More formally, $\forall i, [\mathcal{Y}_i \neq \mathcal{Y}'$ or $p_i(y|x) \neq p'(y|x)]$.

As the output space or the posterior probabilities of the output attributes are different, the source examples provide no information about the possible outputs or the relationship between inputs and outputs of the target task. Therefore, a few labelled target examples are necessary to learn such information. TL approaches operating in this scenario are thus useful when the cost of acquiring labelled target examples is high, but there are some labelled target examples available. The source examples may or may not need to be labelled, depending on the learning algorithm.

Let's take again the problem of software required effort introduced in Sect. 1 as an example. Consider that a given software development company c_1 donated a data set containing several examples of completed software projects and their required efforts to the International Software Benchmarking Standards Group [13]. Company c', on the other hand, has collected only a few examples of completed software projects with their required efforts, due to the high cost of collecting required efforts. Companies c_1 and c' typically conduct the same type of software projects (i.e. $p_1(x) = p'(x)$). However, the underlying function mapping $\mathcal{X}$ to $\mathcal{Y}$ may differ between these two companies (i.e. $p_1(y|x) = p'(y|x)$) because they adopt different practices and such practices have not been collected as input attributes. In this case, company c' may wish to perform inductive TL based on c_1's projects in order to improve its software required effort predictions.

As with transductive TL, one way to perform inductive TL is to filter or weigh source examples based on how well they are believed to match the target probability distributions. Such examples can then be used to help training the target predictive model. A very popular approach in this category is TrAdaBoost [5]. This approach extends the well-known AdaBoost [33] ensemble learning algorithm to perform inductive TL in scenarios where both domains and tasks may differ between sources and target. Base models of the ensemble are trained sequentially, as in the original Adaboost. Labelled target examples are weighted based on AdaBoost's original weighting rule, that is target training examples have their weights increased/decreased if they are incorrectly/correctly classified by former base models. In this way, misclassified target examples are emphasised, encouraging the ensemble to learn how to classify them correctly. For source training examples, which are also required to be labelled examples, this strategy is inverted. Source training examples correctly classified/misclassified by former base models have their weights increased/decreased, because they may match the target probability distributions better/worse.

Other inductive TL approaches transfer parameters that compose models or feature representations, which can then be used with the target model. For example, at the same time as training source and target predictive models, Argyriou et al. [1] learn a feature representation that is common to the source and target domains. Learning consists in concurrently determining (1) target predictive model parameters, (2) source predictive model parameters and (3) a transformation of the input space, so that the regularised error of the source and target predictive models is minimised. The error of the source/target predictive models is calculated based on the source/target training examples. For that, both source and target training examples need to be labelled. This approach works in scenarios where both tasks and domains may differ between sources and target, but a common feature space exists among them. Different from Argyriou et al. [1], Raina et al. [30] learn the feature representation and target model separately, so that unlabelled source examples can be used. Oquab et al. [26] use the internal layers of a convolutional neural network as a generic extractor of higher-level features from the source domain. For that, it requires labelled source examples. The neural network parameters representing such higher-level features are then reused by the target predictive model. Their corresponding internal layers are followed by an adaptation layer, which enables the target task to be learnt. This approach considers that both tasks and domains may differ between sources and target but relies on domain-specific preprocessing of the source input attributes to produce a domain more similar to the target one.

Some inductive TL approaches also exist to transfer knowledge between relational domains, where data can be represented by multiple relationships, such as social networks [28]. Relational domains are out of the scope of this chapter.

3   Learning in Non-stationary Environments (NSE)

NSE are environments where training examples arrive in the form of data streams which may suffer concept drift. Concept drift can be defined as follows [17]:

Definition 7 (Concept Drift) Let $p_t(x, y) = p_t(y|x)p_t(x)$ be the joint probability distribution (concept) underlying a machine learning problem at time step t. Concept drift occurs when the joint probability distribution changes over time. More formally, if, for any time steps t and $t + \Delta$, $p_t(x, y) \neq p_{t+\Delta}(x, y)$, then concept drift has occurred.

Concept drifts can be either the result of changes in the actual data-generating process, or simply perceived, rather than actual changes. As explained by Ditzer et al. [7], the latter case can be "caused by insufficient, unknown, or unobservable attributes, a phenomenon known as 'hidden context' [42]". In this case, there is a stationary data-generating process, but it is hidden from the machine learning algorithm, which will perceive it as non-stationary. Therefore, this work will refer to learning in both cases as learning in NSE.

From Definition 7, we can see that concept drift may involve changes in $p(y|x)$, $p(x)$ or both. This leads to the following widely used additional definitions:

Definition 8 (Real Concept Drift) A concept drift is referred to as a *real* concept drift if it involves changes in $p(y|x)$. More formally, if, for any time steps t and $t + \Delta$, $p_t(y|x) \neq p_{t+\Delta}(y|x)$, then a real concept drift has occurred.

For example, in the credit card payment problem introduced in Sect. 1, an economic crisis may cause customers that used to pay their bills in the past to start defaulting their payments, representing a change in $p(y|x)$.

Definition 9 (Virtual Concept Drift) A concept drift is referred to as a *virtual* concept drift if it only involves changes in $p(x)$. More formally, let t and $t + \Delta$ be two time steps where $p_t(x) \neq p_{t+\Delta}(x)$. If $p_t(y|x) = p_{t+\Delta}(y|x)$, then the differences in the probability distributions between t and $t + \Delta$ represent a virtual concept drift.

For example, in the credit card payment problem introduced in Sect. 1, a given credit card company may start receiving and accepting more credit card applications from younger customers, leading to a change in the problem's $p(x)$.

Concept drifts are also frequently categorised with respect to their speed (number of time steps taken for a change to complete), severity (how large the changes in the probability distributions are), recurrence (whether the concept drift takes us to a previously seen concept) and periodicity (whether concept drifts occur periodically) [7, 17, 24, 38].

In particular, it is worth noting that $p(x, y)$ may suffer several small changes between a number of consecutive time steps before becoming stable. Some authors refer to that as a single gradual concept drift [38], whereas others refer to that as a sequence of concept drifts of low severity [24]. This is distinguished from an abrupt or sudden concept drift, which is an isolated concept drift that takes a single time step to complete. The term "gradual concept drift" can also be used to describe a single concept drift that takes several time steps to complete because the old and new joint probability distributions are active concurrently for a given period of time [2, 24]. In this scenario, the chances of an example being drawn from the old/new joint probability distribution gradually reduce/increase, until the new joint probability distribution completely replaces the old one. Certain data streams may also continuously suffer concept drifts, that is they may not have any significant period of complete stability.

Based on existing work on learning in NSE [7, 11, 17], we can consider existing approaches for learning in NSE to be adopting the following definition:

Definition 10 (Learning Algorithms for Non-stationary Environments) Consider a process generating a data stream $S = \langle S_1, S_2, \cdots \rangle$, where $S_t = \{(x_t^{(i)}, y_t^{(i)})\}_{i=1}^{m_t} \sim_{\text{iid}} p_t(x, y)$; $m_t > 0$ is the size of the training set received at time step t; $(x_t^{(i)}, y_t^{(i)}) \in \mathcal{X} \times \mathcal{Y}$; and $p_t(x, y)$ is the joint probability distribution of the problem at time step t. Consider that at a current time step t, we are given access to a model $f_{t-1} : \mathcal{X} \rightarrow \mathcal{Y}$ created based on past examples from S, a new training set $S_t \in S$ and possibly a set S_{past} containing a limited number of past examples from S. Learning in non-stationary environments aims at creating an updated model $f_t : \mathcal{X} \rightarrow \mathcal{Y}$ able to generalise to unseen examples of $p_t(x, y)$, based on the given information.

The following observations must be made when using Definition 10:

- The explicit index t in the probability distributions distinguishes these algorithms from algorithms for stationary environments [7], as it takes the possibility of concept drift into account.
- A model f_t may be an ensemble model, possibly composed of several predictive models created with data from previous training sets.
- The size of a training set can be one, that is the training set may consist of a single example.
- Many algorithms discard past training sets once they are processed. However, some approaches make use of a limited number of examples from past training sets, as will be explained in Sects. 3.1 and 3.2. When used, the number of past examples must be limited, given that data streams have potentially infinite size. The unbounded nature of data streams means that it is infeasible to always store all past examples for future access.

Learning algorithms for NSE may process data streams chunk-by-chunk (Sect. 3.1) or example-by-example (Sect. 3.2). Most of these algorithms are prepared to deal with changes that affect the suitability of the learnt decision boundaries. Different from real concept drifts, virtual concept drifts do not affect the true decision boundary of the problem. However, they may still affect the suitability of the learnt decision boundary [41]. Therefore, most existing algorithms are applicable to data streams with both types of concept drifts, despite having different strengths and weaknesses depending on the context.

It is also worth noting that, even though in theory each training example from a given data stream could potentially come from a completely different probability distribution, it would be impossible for learning algorithms to build well-performing predictive models in such scenario. In practice, it would be rather unlikely that a given learning problem continuously suffers very large changes. For instance, in the problem of predicting whether credit card customers will default their payments, it would be rather unlikely that the defaulting behaviour of customers erratically changes all the time. Therefore, most learning algorithms for NSE implicitly assume that there will be some periods of relative stability, or that there are very frequent/continuous changes, but such changes are frequently small.

3.1 Chunk-by-Chunk Approaches

Given Definition 10, chunk-based learning algorithms for NSE can be defined as follows:

Definition 11 (Chunk-Based Learning for Non-stationary Environments) Chunk-based learning algorithms for non-stationary environments are algorithms that perform learning in non-stationary environments by processing the data stream chunk-by-chunk, where the chunk size is larger than 1. These algorithms need to wait for a whole chunk of examples to become available before learning it.

Intuitively, a chunk would be equivalent to a training set S_t provided by the data stream. However, it is also possible to set the chunk size in such a way that they are not equivalent. For instance, a chunk may contain more than one training set within it, or a given training set may be broken down into different chunks. Even though the size of each chunk could potentially be very small, it is typically implicitly assumed that the size is set to be large enough for a predictive model trained only on it to be better than random guess.

Most chunk-based learning algorithms for NSE are ensemble learning algorithms whose predictions are the weighted average or weighted majority vote among the predictions of their base models. The weights enable the ensemble to emphasise the base models most appropriate to the current concept. These ensembles typically perform learning as follows [9, 17, 35, 40]:

1. Train an initial base model using the first chunk of training examples.
2. For each new chunk, do:

    (a) Use this chunk to evaluate each predictive model that composes the ensemble, based on a given performance measure.
    (b) Assign a weight to each predictive model based on its performance calculated above.
    (c) Create a new predictive model using this chunk.

(d) Add the new predictive model to the ensemble if the maximum ensemble size has not been reached; otherwise, replace an existing predictive model by the new one.
(e) Discard the current chunk.

One of the problems of such approaches is that they are sensitive to the chunk size. A too small chunk size means that there are not enough examples to learn a good predictive model. A too large chunk means that a single chunk may contain examples from different joint probability distributions, resulting in the inability to deal with concept drifts adequately.

Some chunk-based approaches try to reduce sensitivity to the chunk size. For example, Scholz and Klinkenberg [34] allow a new chunk to be used to update an existing predictive model, rather than always creating a new model for each new chunk. This enables chunk sizes to be small enough without necessarily hurting predictive performance. Some chunk-based approaches also enable a limited number of examples from past chunks to be reprocessed. For example, Chen and He [4] preserve certain minority class examples seen in past chunks in order to deal with class imbalanced problems.

3.2 Example-by-Example Approaches

Given Definition 10, example-by-example learning algorithms for NSE can be defined as follows:

Definition 12 (Example-by-Example Learning for Non-stationary Environments) Example-by-example learning algorithms for non-stationary environments are algorithms that perform learning in non-stationary environments by processing the data stream example-by-example. These algorithms can update the predictive model whenever a new training example is received.

The simplest type of example-by-example algorithms are algorithms that maintain a sliding window over the data stream [18]. They build a new classifier to replace the old one whenever the window slides, by making use of the examples within the window. Similar to most chunk-based algorithms, sliding window algorithms are also sensitive to the window size. They assume that the size must be large enough to produce a well-performing predictive model, but small enough not to delay adaptation to concept drifts due to examples belonging to past concepts being within the window.

Another type of example-by-example algorithms are *online learning algorithms*, which process each training example separately upon arrival and then immediately discard it. The fact that each training example is immediately discarded leads to significant differences between the mechanisms that typically underlie these algorithms and other example-by-example or chunk-by-chunk algorithms. Chunk-based or sliding window algorithms typically use offline learning algorithms to learn each new chunk or window of examples. In many cases, this involves training a new model using solely the examples within the new chunk or window [17]. This would lead to poor predictive performance if each chunk or window contained a single training example. Moreover, offline learning algorithms often require iterating through training examples several times. Therefore, there is frequently an implicit assumption that training examples from a chunk or window can be reprocessed several times before the chunk or window is discarded. Online learning algorithms are much stricter—each training example must be discarded before a new training example is used for training. They are therefore more suitable for applications with very tight time and/or memory constraints, such as applications where the rate of incoming data is very large or certain embedded systems.

Many of the online learning algorithms for NSE use concept drift detection methods to actively detect concept drift. This is typically done by monitoring an online learning model for stationary environments, for example naive Bayes [3] or Hoeffding tree [8]. An example of well-known concept drift detection method is Gama et al. [10]'s. This method tracks the error of an online learning model over time. If this error significantly increases, a concept drift detection is triggered. Other authors proposed different concept drift detection methods by monitoring different quantities. For instance, Baena-Garcia et al. [2] monitor the distance between misclassifications over time, whereas Ross et al. [32] monitor the exponentially weighted average of the errors. A typical way to deal with concept drifts once they are detected is to reset the online learning model, so that it can start learning the new concept from scratch [2, 10, 32]. However, this strategy is sensitive to false positive drift detections (a.k.a., false alarms). Strategies such as creating a new online learning model upon concept drift detection, but maintaining old online learning models in case they turn out to be still useful, can help to improve robustness to false positive drift detections [22, 25].

Some online learning algorithms deal with concept drift passively, that is they do not use any concept drift detection method. A well-known example is the Dynamic Weighted Majority (DWM) algorithm [16]. This algorithm maintains an ensemble of online learning models, each associated to a weight. For classification problems, the prediction given by the ensemble is the weighted majority vote among the predictions given by the base learners. When a new training example becomes available, each online learning model is asked to predict the output attribute of the example before learning it. If a given online learning model misclassifies the training example, its weight is reduced. In this way, the ensemble can automatically emphasise the online learning models most suitable to the current concept. Online learning models whose weight is below a certain threshold are deemed outdated and are thus eliminated. New online learning models can also be created when the ensemble as a whole misclassifies a training example. In this way, new models can be created to learn new concepts from scratch.

4   The Relationship Between TL and Learning in NSE

This section discusses the similarities (Sect. 4.1) and differences (Sect. 4.2) between TL and learning in NSE.

4.1   Similarities

As we can see from Definitions 4 and 10, TL and learning in NSE both involve training and test data that potentially come from different probability distributions. In TL, training examples coming from different sources may have different domains and tasks than the target test data. In learning in NSE, past training examples may come from a different joint probability distribution from that underlying the current test data.

In particular, transductive TL is concerned with sources and targets that share the same task but have different domains. This means that sources and targets differ in terms of their unconditional pdf $p(x)$. This is similar to learning under virtual concept drifts, which also represent changes in $p(x)$. In inductive TL, sources and targets have different tasks, that is they differ in terms of the posterior probabilities of the classes $p(y|x)$. This is similar to learning under real concept drifts, which also represent changes in $p(y|x)$. Moreover, sources and targets may or may not differ in terms of $p(x)$ in inductive TL. This is similar to real concept drifts, which may or may not involve changes in $p(x)$.

The similarities above translate into similarities in the approaches proposed to perform TL and learning in NSE. As explained in Sects. 3.1 and 3.2, many approaches for learning in NSE are ensemble approaches that maintain weighted predictive models trained on data from different periods of time. So, each predictive model could potentially represent a different joint probability distribution. In TL terms, they could be seen as representing different sources. These approaches could arguably be seen as a form of inductive TL. They allow knowledge from different (source) joint probability distributions to possibly help making predictions for a given (target) concept. Moreover, the fact that predictive models are weighted based on how well they match the current concept has strong resemblance to approaches such as TrAdaBoost [5], which weigh examples from different sources based on how well they match the target joint probability distribution.

TL approaches could also potentially be seen as transferring knowledge from the past to the present. This is because different periods of time of a given data stream could be seen as different sources, which may have different domains and/or tasks from the present (target) data. Therefore, both TL and learning in NSE could be seen as trying to create good predictive models to a given present time.

Table 1 summarises the similarities between TL and learning in NSE.

Table 1 Similarities between TL and learning in NSE

TL	Learning in NSE
Test data come from a different joint probability distribution from that underlying (part of) the training data.	Test data may come from a different joint probability distribution from that underlying (part of) the past training data.
Transductive TL deals with changes in $p(x)$.	Virtual concept drifts consist in changes in $p(x)$.
Inductive TL deals with changes in $p(y\|x)$.	Real concept drifts consist in changes in $p(y\|x)$.
Inductive TL may deal with changes in $p(x)$.	Real concept drifts may involve changes in $p(x)$.
TL tries to use data from different sources to build a target model.	NSE approaches that use past predictive models could be seen as using knowledge from different sources to build a target model.
TL could be used to transfer knowledge from the past in order to perform well in the present.	Learning in NSE consists in creating up-to-date predictive models that perform well in the present.

4.2 Differences

Despite the strong similarities between TL and learning in NSE presented in Sect. 4.1, there are also significant differences. The main difference is that TL has no notion of continuing time, as explained by Ditzler et al. [7] and elucidated by Definition 4 (TL). In particular, TL approaches currently assume that there are pre-existing source and target data sets coming from fixed joint probability distributions. Even if the sources are used to represent data from different past periods of time and the target is used to represent a given present period of time, this would still capture only a fixed snapshot of the environment. The *continuing* nature of time captured by data streams, which is a fundamental aspect of learning in NSE (Definition 10), is not considered.

The consequence of that is that TL approaches are not designed to process additional data over time. Therefore, they cannot automatically cope with concept drifts that may affect the present and cause the current target model to become obsolete. In order to transfer knowledge across time, these approaches require us to know beforehand which past and present periods of time represent a given source/target. Therefore, concept drifts resulting in a change in target (with the previous target becoming a source) would need to be manually identified and the whole TL approach re-run from scratch.

TL approaches also do not make provision for processing data with gradual concept drifts, where probability distributions slowly change until they become stable, or where there are two different probability distributions concurrently active before the concept drift completes. As each source and target should come from a fixed joint probability distribution in TL, it is unclear what to do with transitional periods when trying to transfer knowledge from the past to the present. Examples produced during such periods may need to be manually discarded. This issue becomes even more significant for real-world data streams presenting continuous changes, that is whose underlying joint probability distribution is always changing from one time step to another. TL approaches are not prepared to deal with this type of problem.

Moreover, TL typically requires past sources to be reprocessed several times. For example, TrAdaBoost [5] creates predictive models for its ensemble sequentially, and requires iterating over all source and target examples again for each new predictive model being created. The process is similar to AdaBoost [33], which is an offline ensemble learning algorithm. Argyriou et al. [1]'s feature learning requires iterating over all sources and target several times until a convergence criterion is reached. Given the unbounded nature of data streams (they are potentially infinite), this is infeasible for learning in NSE.

It is also worth mentioning that the time order between sources representing different past periods of time would not be taken into account by TL approaches, even if they were trying to transfer knowledge across time. This is because they do not distinguish between the moment in time where different past sources have been produced. However, this is a less significant issue in the context of NSE than the absent notion of continuing time itself. Even though some NSE approaches take the age of past predictive models into account (e.g. by eliminating older models), this is not necessarily a good strategy to learn in NSE [15].

Another difference between TL and learning in NSE is that TL is explicitly concerned with using different sources to create a better target model than one produced using only the target data. Even though the sources come from different probability distributions from that of the target, they are expected to be useful and are exploited. For example, many TL approaches try to transform the input space of the source and target into a feature space where they become more similar [1, 29, 30], as discussed in Sect. 2. Others try to find out which source examples match the target probability distribution well, even though the source as a whole is known to follow a different probability distribution from that of the target [5].

Learning in NSE, on the other hand, is concerned with creating an appropriate predictive model for the current concept. Even though models representing the past can be used for that, attempting to use past knowledge to help learning a new concept is not a requirement. For example, many NSE approaches delete old predictive models once a concept drift is detected [2, 10, 32], without even trying to check whether such past models could be somehow helpful for building a new model. Even when models representing different periods of time are used by ensemble approaches for NSE [9, 17, 34, 35, 40], the ultimate goal of these approaches is to identify when the past models represent concepts that match the current concept well, so that they can be used in the present. Although past models representing somewhat different concepts may end up being used and result in possible benefits, this is different from deliberately trying to make use of source models/data when we know that they do come from different probability distributions, as done by TL approaches.

In addition, the sources used by TL do not necessarily need to come from the same data-generating process. For instance, in the problem of software effort

Table 2 Differences between TL and learning in NSE

TL	Learning in NSE
No notion of continuing time.	Continuing time is a fundamental aspect.
Not designed to automatically process incoming data over time.	Designed to automatically process incoming data over time.
Unable to automatically cope with changes in the present.	Designed to automatically cope with changes in the present.
No provision for processing examples from transitional periods between different joint probability distributions.	Can process examples from transitional periods between different joint probability distributions.
Typically has to reprocess examples several times, being unsuitable for potentially infinite data streams.	Only requires repeated access to a limited number of past examples, being feasible for potentially infinite data streams.
Unable to distinguish between the time order of different past sources.	Can potentially take the age of different past models into account.
Aims to use sources with different probability distributions to improve target model.	Aims to create an up-to-date predictive model, without necessarily using past data or models to help learning a new concept.
Sources may come from different data-generating processes.	All training examples come from the same data-generating process.
Explicitly considers sources and targets with different input and output spaces.	Changes in the input and output spaces are usually not explicitly considered.

estimation, training examples may be acquired from different companies than the target one. Learning in NSE assumes that all training examples come from the same data-generating process, even though there may be concept drift. Therefore, learning in NSE approaches are not prepared to benefit from different data-generating process. In particular, they can process a single data stream over time.

Yet another difference is that TL explicitly considers different input and output spaces, as illustrated by Definition 4. Learning in NSE could potentially involve changes in the input and output spaces, as they are intrinsically related to changes in $p(x)$ and $p(y)$, which compose the joint probability distribution of a problem. For instance, Sun et al. [36] proposed a NSE approach that explicitly takes class evolution (appearance, disappearance and reoccurrence of classes) into account. However, most existing NSE work does not explicitly tackle changes in the input and output space.

Table 2 summarises the differences between TL and learning in NSE.

5 The Potential of Transfer Learning in NSE

Sections 4.1 and 4.2 show that, even though there are strong similarities between current work on TL and learning in NSE, there are also significant differences. These differences lead to limitations that prevent these approaches to effectively deal with certain types of problem, or that cause these approaches to potentially miss useful knowledge that could lead to better predictive performance.

In particular, TL is not designed to automatically process incoming data over time and is typically unable to deal with potentially infinite data streams. It cannot automatically cope with changes to the present and has no provision to process examples from transitional periods between concepts. However, several real-world problems that could potentially benefit from TL provide data streams rather than fixed training sets. Let's take the example of the software required effort problem mentioned in Sect. 1. Software development companies develop additional software projects over time, which could be provided as training examples in the form of data streams. Such data streams are likely to present concept drift, given that, for example the practices adopted by a software company and the type of projects that it develops may change over time. So, a TL approach able to deal with data streams would be desirable.

Meanwhile, NSE approaches are potentially wasting useful knowledge from past concepts that could be used to help learning a new concept. As concept drifts could lead to new concepts that share some similarities with respect to old concepts, it would be desirable to have NSE approaches able to transfer knowledge from old concepts to better learn new concepts. Moreover, NSE approaches cannot benefit from examples coming from different data-generating processes. However, several learning problems that operate in NSE could benefit from data coming from different data-generating processes. In particular, as explained in the previous paragraph, the software required effort problem introduced in Sect. 1 is actually a problem that both operates in NSEs and could benefit from data coming from different data-generating processes (i.e. different companies).

By combining TL and learning for NSE, the individual limitations of these approaches could be overcome. For instance, two approaches called Dynamic Cross-company Learning (DCL) [21] and Dynamic Cross-company Mapped Model Learning (Dycom) [23] make use of data from different source data-generating processes in order to improve predictive performance in a target non-stationary data stream. The former can still only benefit from knowledge from a different data-generating process when it matches the current target concept well. However, the latter can transform knowledge from sources with different tasks and domains into useful knowledge to learn a new concept in NSE. This approach enables TL to process data streams in NSE. It can be seen as using ideas from learning in NSE to make TL aware of continuing time.

Another online learning approach called Diversity for Dealing with Drifts (DDD) [22] attempts to use knowledge acquired from a past concept's $p(x)$ and $p(y|x)$ to aid the learning of a new concept. This helps it to improve predictive performance in the presence of gradual or not severe concept drifts. A recent chunk-based approach [37] called Diversity and Transfer based Ensemble Learning (DTEL) uses knowledge acquired from past concepts' $p(x)$ to aid the learning of a new concept. These approaches can be seen as getting inspiration from TL to improve learning in NSE.

The combination of TL and learning for NSE environments could lead to a whole new range of machine learning approaches, forming a new combined area of TL in NSE that overcomes the limitations of these areas in isolation. Sections 5.1 and 5.2 explain Dycom [23] and DDD [22], as well as the results achieved by these approaches in the literature [22, 23], highlighting the potential benefit of TL in NSE.

5.1 Dynamic Cross-company Mapped Model Learning (Dycom)

Dycom [23] is a regression approach for online inductive transfer learning in NSE where both tasks and domains may differ between the sources and target. It assumes that a good number of labelled source training examples are available from different data-generating processes, but that the specific source generating a given example is unknown. It considers that collecting labelled training examples from the target data-generating process is expensive, and that only very few of such examples can be acquired over time. Therefore, transferring knowledge from different sources may help to learn a better target predictive model. Such TL should be able to tackle concept drift, given that the target is non-stationary.

The approach is illustrated in Fig. 1. It separates the set of source training examples into M partitions according to their similarities. This could be done based on a clustering approach [20] or on prior knowledge [23]. Each of the M partitions is used to create a source predictive model (source models 1 to M in Fig. 1). In this version of Dycom, the source models are trained based on offline learning. However, this approach could be extended to use source data streams.

A target data stream is used to train a target predictive model (target model 0 in Fig. 1) based on an online supervised learning algorithm. This predictive model is expected to be weak and perform poorly, due to the small number of labelled target

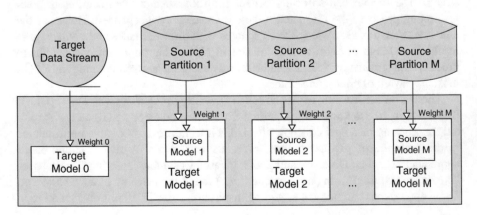

Fig. 1 Dycom approach for TL in NSE. Arrows represent flow of information

training examples. However, depending on how long the periods of stability are, it may happen that, over time, it will be trained with enough examples to perform well. Therefore, it is worth building and maintaining this model.

The target data stream is also used to learn M functions $g_t^{(i)} : \mathcal{Y} \to \mathcal{Y}$ able to map the predictions made by each of the M source models i into the target concept at time t. These functions thus compose M target models (target models 1 to M in Fig. 1). Dycom assumes that, as the source training examples have been separated into similar partitions, there exists a reasonably linear relationship between each of the source models and the target concept, which can be learnt based on a few labelled target training examples. This relationship is described as follows:

$$g_t^{(i)}(f^{(i)}(x)) = f^{(i)}(x) \cdot b_t^{(i)},$$

where $g_t^{(i)}$ is the function to map source model $f^{(i)}$'s predictions to the target concept, $b_t^{(i)}$ is a learnt parameter, and $x \in \mathcal{X}$ are the input attributes of the target example being predicted.

Parameter $b_t^{(i)}$ is learnt in an online manner based on the following equation:

$$b_t^{(i)} = \begin{cases} 1, & \text{if no target training example} \\ & \text{has been received yet;} \\ \dfrac{y_t}{f^{(i)}(x_t)}, & \text{if}(x_t, y_t) \text{ is the first} \\ & \text{target training example;} \\ lr \cdot \dfrac{y_t}{f^{(i)}(x_t)} + (1 - lr) \cdot b_{t-1}^{(i)}, & \text{otherwise,} \end{cases} \quad (1)$$

where (x_t, y_t) is the current target training example, and $lr \in (0, 1)$ is a predefined smoothing factor.

As explained in [23], the mapping function performs a direct mapping $b_t^{(i)} = 1$ if no target training example has been received yet. When the first target training example is received, $b_t^{(i)}$ is set to the value $y_t/f^{(i)}(x_t)$. This gives a perfect mapping of $f^{(i)}(x_t)$ to the target concept for the example being learnt, as $f^{(i)}(x_t) \cdot b_t^{(i)} = f^{(i)}(x_t) \cdot y_t/f^{(i)}(x_t) = y_t$. For all other target training examples received, exponentially decayed weighted average with smoothing factor lr is used to set $b_t^{(i)}$. Higher lr will cause more emphasis on the most recent target training examples and higher adaptability to changing environments, whereas lower lr will lead to a more stable mapping function. So, the weighted average allows learning mapping functions that provide good mappings based on previous target training examples, while allowing adaptability to changes that may affect the target.

Dycom's predictions for new target examples are based on the weighted average of all target models. The weights are initialised to 1 and are updated whenever a new target training example becomes available as follows. The winner model is set to be the target model whose prediction for the current target training example is the most accurate. All other models are the looser models, and have their weights multiplied by a predefined factor $\beta \in (0, 1]$. All weights are then normalised so that they sum up to 1. The weights of the target models thus allow more emphasis to be placed on the models that are currently more accurate. In particular, if the target model that does not use source knowledge (target model 0) is inaccurate, its corresponding weight will be low, so that its predictions will not hinder Dycom's predictive performance.

This approach has been evaluated in the context of software effort estimation, where we are interested in predicting the effort for projects from a given target company. In this problem, we have access to training examples from other source companies, despite the fact that acquiring labelled training examples from within the target company is expensive. The problem of software effort estimation gave the name Dynamic *Cross-company* Mapped Model Learning to this approach, but Dycom is also applicable to other problems, as we can see from its description above.

Experiments on five databases containing software development projects [23] using regression trees as the base learner show that Dycom achieved similar or better predictive performance while requiring *10 times less target training examples* than a target model created using only target (no source) training examples. This highlights the benefits of using TL in NSE when the cost of acquiring labelled target training examples is high. Moreover, the experiments showed that Dycom's mapping functions can be visualised so that project managers can see how the relationship between the efforts required by their company and other companies changes over time. This could be potentially used to aid the development of strategies to improve a company's productivity. Therefore, the insights provided by TL in NSE could go beyond mere predictions.

5.2 Diversity for Dealing with Drifts (DDD)

DDD is a classification ensemble approach for online learning in NSE. It was not originally described as an approach for TL in NSE. However, this is essentially what it does. This approach is based on two key findings [22]:

- Knowledge from a past concept can be useful when learning in the presence of concept drifts that occur gradually or are not severe. When a gradual concept drift occurs, knowledge from the past concept remains useful for a certain period of time, until the drift completes. And, more importantly, if there is a non-severe concept drift (i.e. if the old and new concepts share some similarities), knowledge from the old concept could be useful to help learning the new concept. This is in line with the fact that TL approaches can be beneficial if the source and target share some similarities, as explained in Sect. 2. On the other hand, if the concept drift occurs very fast and is very severe, the old and new concepts will not share

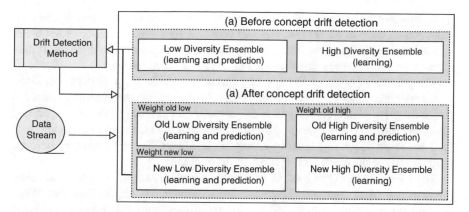

Fig. 2 DDD approach for TL in NSE. Arrows represent flow of information

enough similarities for knowledge from the old concept to help learning the new concept.

- Learning a given concept using very highly diverse ensembles will cause them to perform poorly on this concept. However, such weak learning enables these ensembles to quickly adapt to a new concept, if this new concept shares similarities with the given concept. In essence, this enables knowledge of a given concept to be transferred to a new concept in order to improve predictive performance in NSE.

Therefore, DDD maintains online learning ensembles with different diversity levels in order to achieve robustness to different data stream conditions (i.e. to different types of concept drift or periods of stability). Figure 2 illustrates its behaviour. Before a concept drift is detected, a low-diversity ensemble is used both for learning incoming training examples and for making predictions. A very highly diverse ensemble is used for learning, but not for predictions. This is because this ensemble is expected to be weak and perform poorly in the current concept, but it may become helpful if there are gradual or not severe concept drifts, based on the findings above.

The low-diversity ensemble is monitored by a drift detection method. If a drift is detected, DDD switches to the mode "after concept drift detection". In this mode, the previous low- and high-diversity ensembles are kept as old ensembles, and both are activated for learning and predictions. Low diversity is enforced into the learning procedure of the old high-diversity ensemble, so that it can strongly learn the new concept while transferring knowledge from the old concept. These ensembles may be beneficial if the concept drift is gradual or not severe. They may also be beneficial in case the drift detection was a false positive drift detection (false alarm), in which case the concept did not change and the old ensembles remain representative of the current concept. A new low-diversity ensemble is created to start learning the new concept from scratch. It may be useful if the concept drift is very fast and severe.

This ensemble is therefore activated for predictions. It is also monitored by the drift detection method to detect new concept drifts. A new high-diversity ensemble is created to weakly learn the new concept, and is not active for predictions. It may become useful if there is a new gradual or not severe concept drift.

The prediction given by the system in the mode "after concept drift detection" is the weighted majority vote of the predictions given by the ensembles that are active for predictions. The weight is the normalised accuracy of the corresponding ensemble since the last concept drift detection, and is calculated in an online way [22] based on incoming training examples. It allows the right ensembles to be emphasised for the given concept drift (or false alarm).

The approach switches back to the mode "before concept drift detection" once the accuracy of the new low-diversity ensemble becomes higher than that of the old ensembles, or the accuracy of the old high-diversity ensemble becomes higher than that of the low-diversity ensembles, with a certain margin. The ensemble which became more accurate than the others and the new high-diversity ensemble become the low- and high-diversity ensembles in the mode "before concept drift detection".

Any online ensemble learning algorithm and method to encourage high or low diversity could potentially be used. Online bagging ensembles [27] were used in the paper that proposed DDD [22], and different levels of diversity were achieved by using different sampling rates based on the parameter lambda of the Poisson distribution used by online bagging. In particular, the lambda value was kept with the original value of one used by the online bagging algorithm in order to create low-diversity ensembles. Lower lambda values (less than one) lead to higher diversity, and were used to produce the high-diversity ensembles.

Experiments were performed to evaluate DDD based on several synthetic data streams containing different types of concept drift and on real-world data streams from the areas of credit card approval, electricity price prediction and network intrusion detection [22]. The results show that DDD was usually able to maintain or improve predictive performance in comparison with other approaches for learning in NSE (DWM [16] and Baena-Garcia et al. [2]'s approach) and online bagging without mechanisms to deal with concept drifts. Predictive performance was improved specially for gradual and non-severe drifts, which are exactly the cases for which TL via the the old ensembles had been found to be helpful (see first bullet point in the beginning of this subsection).

Overall, DDD and its achieved results further illustrate the benefits of combining TL and learning in NSE. In particular, they show that knowledge from old concepts can be used to help learning the new concept, improving learning in NSE especially when the old and new concepts share some similarities.

6   Conclusions

This chapter provided background in the fields of TL and learning in NSE to enable understanding the relationship between these two fields. It gave definitions that existing work in these fields can be considered to be adopting, and examples of several representative approaches. Based on that, a discussion of the similarities and differences between these two fields was provided. This discussion reveals that approaches in these fields have limitations that could be overcome through a better integration between them. For instance, NSE approaches are potentially wasting useful knowledge from past concepts that could be helpful for learning a new concept. Moreover, they cannot benefit from examples coming from different data-generating processes. TL, on the other hand, has no notion of continuing time. It is not designed to automatically process incoming data over time and is typically unable to deal with potentially infinite data streams. It cannot automatically cope with concept drifts affecting the present and has no provision to process examples from transitional periods between concepts.

Therefore, this chapter discussed the potential benefits of better integrating the fields of TL and learning in NSE. In particular, two existing approaches for TL in NSE called Dycom and DDD were explained. The positive results obtained by these approaches in the literature highlight the benefit of TL in NSE. Dycom shows how ideas from learning in NSE can be used to make TL aware of continuing time, enabling it to deal with data streams and automatically cope with concept drift. Conversely, DDD shows how ideas from TL can be used to inspire better algorithms for learning in NSE, by enabling knowledge from past concepts to aid the learning of new concepts.

As the first work to provide a detailed discussion of the relationship between TL and NSE, this chapter opens up the path for future research in the emerging area of TL in NSE. It encourages the research communities in these fields to work together, so that improved algorithms can be proposed to tackle challenging real-world problems.

Acknowledgements This work was partially funded by EPSRC Grant No. EP/R006660/1.

References

1. Argyriou, A., Evgeniou, T., Pontil, M.: Multi-task feature learning. In: Proceedings of the 19th Annual Conference on Neural Information Processing Systems, pp. 41–48 (2007)
2. Baena-García, M., Del Campo-Ávila, J., Fidalgo, R., Bifet, A.: Early drift detection method. In: Proceedings of the 4th International Workshop on Knowledge Discovery from DataStreams, Berlin, pp. 77–86 (2006)
3. Bishop, C.: Pattern Recognition and Machine Learning. Springer, Singapore (2006)
4. Chen, S., He, H.: SERA: selectively recursive approach towards nonstationary imbalanced stream data mining. In: Proceedings of the 2009 International Joint Conference on Neural Networks (IJCNN), pp. 522–529 (2009)
5. Dai, W., Yang, Q., Xue, G., Yu, Y.: Boosting for transfer learning. In: Proceedings of the 24th International Conference on Machine Learning, pp. 193–200 (2007)
6. Dai, W., Xue, G.R., Yang, Q., Yu, Y.: Transferring Naive Bayes classifiers for text classification. In: Proceedings of the 22nd National Conference on Artificial Intelligence (2007)

7. Ditzler, G., Roveri, M., Alippi, C., Polikar, R.: Learning in nonstationary environments: a survey. IEEE Comput. Intell. Mag. **10**(4), 12–25 (2015)
8. Domingos, P., Hulten, G.: Mining high-speed data streams. In: Proceedings of the 6th ACM SIGKDD International Conference on Knowledge Discovery and Data Mining (KDD), pp. 71–80 (2000)
9. Elwell, R., Polikar, R.: Incremental learning of concept drift in nonstationary environments. IEEE Trans. Neural Netw. Learn. Syst. **22**(10), 1517–1531 (2011)
10. Gama, J., Medas, P., Castillo, G., Rodrigues, P.: Learning with drift detection. In: Proceedings of the 7th Brazilian Symposium on Artificial Intelligence (SBIA), São Luiz do Maranhão. Lecture Notes in Computer Science, vol. 3171, pp. 286–295. Springer, Berlin (2004)
11. Gama, J., Zliobaite, I., Bifet, A., Pechenizkiy, M., Bouchachia, A.: A survey on concept drift adaptation. ACM Comput. Surv. **46**(4), 44:1–44:44 (2015)
12. Huang, J., Smola, A., Gretton, A., Borgwardt, K., Scholkopf, B.: Correcting sample selection bias by unlabeled data. In: Proceedings of the 19th Annual Conference on Neural Information Processing Systems (2007)
13. ISBSG. The International Software Benchmarking Standards Group. (2011) http://www.isbsg.org
14. Kitchenham, B., Mendes, E., Travassos, G.: Cross versus within-company cost estimation studies: a systematic review. IEEE Trans. Softw. Eng. **33**(5), 316–329 (2007)
15. Kolter, J.Z., Maloof, M.A.: Using additive expert ensembles to cope with concept drift. In: Proceedings of the 22nd International Conference on Machine Learning (ICML), Bonn, pp. 449–456 (2005)
16. Kolter, J.Z., Maloof, M.A.: Dynamic weighted majority: an ensemble method for drifting concepts. J. Mach. Learn. Res. **8**, 2755–2790 (2007)
17. Krawczyk, B., Minku, L., Gama, J., Stefanowski, J., Wozniak, M.: Ensemble learning for data stream analysis: a survey. Inform. Fusion **37**, 132–156 (2017)
18. Kuncheva, L., Zliobaite, I.: On the window size for classification in changing environments. Intell. Data Anal. **13**(6), 861–872 (2009)
19. Menzies, T., Caglayan, B., He, Z., Kocaguneli, E., Krall, J., Peters, F., Turhan, B.: The promise repository of empirical software engineering data (2012). http://promisedata.googlecode.com
20. Minku, L., Hou, S.: Clustering dycom: an online cross-company software effort estimation study. In: Proceedings of the 13th International Conference on Predictive Models and Data Analytics for Software Engineering (PROMISE), pp. 12–21 (2017)
21. Minku, L., Yao, X.: Can cross-company data improve performance in software effort estimation? In: Proceedings of the 8th International Conference on Predictive Models in Software Engineering (PROMISE), Lund, pp. 69–78 (2012)
22. Minku, L.L., Yao, X.: DDD: a new ensemble approach for dealing with concept drift. IEEE Trans. Knowl. Data Eng. **24**(4), 619–633 (2012)
23. Minku, L., Yao, X.: How to make best use of cross-company data in software effort estimation? In: Proceedings of the 36th International Conference on Software Engineering (ICSE), pp. 446–456 (2014)
24. Minku, L.L., White, A., Yao, X.: The impact of diversity on on-line ensemble learning in the presence of concept drift. IEEE Trans. Knowl. Data Eng. **22**(5), 730–742 (2010)
25. Nishida, K.: Learning and detecting concept drift. PhD thesis, Hokkaido University (2008)
26. Oquab, M., Bottou, L., Laptev, I., Sivic, J.: Learning and transferring mid-level image representations using convolutional neural networks. In: Proceedings of the 2014 IEEE Conference on Computer Vision and Pattern Recognition (CVPR), pp. 1717–1724 (2014)
27. Oza, N.C., Russell, S.: Online bagging and boosting. In: Proceedings of the 2005 IEEE International Conference on Systems, Man and Cybernetics, New Jersey, vol. 3, pp. 2340–2345. Institute for Electrical and Electronics Engineers (2005)
28. Pan, S., Yang, Q.: A survey on transfer learning. IEEE Trans. Knowl. Data Eng. **22**(10), 1345–1359 (2010)
29. Pan, S.J., Tsang, I.W., Kwok, J.T., Yang, Q.: Domain adaptation via transfer component analysis. IEEE Trans. Neural Netw. **22**(2), 199–210 (2011)

30. Raina, R., Battle, A., Lee, H., Packer, B., Ng, A.: Self-taught learning: transfer learning from unlabeled data. In: Proceedings of the 24th International Conference on Machine Learning (ICML), pp. 759–766 (2007)
31. Rosenstein, M., Marx, Z., Kaelbling, L.: To transfer or not to transfer. In: Proceedings of the Conference on Neural Information Processing Systems (NIPS) Workshop Inductive Transfer: 10 Years Later (2005)
32. Ross, G., Adams, N., Tasoulis, D., Hand, D.: Exponentially weighted moving average charts for detecting concept drift. Pattern Recogn. Lett. **33**, 191–198 (2012)
33. Schapire, R.E., Singer, Y.: Improved boosting algorithms using confidence-rated predictions. Mach. Learn. **37**(3), 297–336 (1999)
34. Scholz, M., Klinkenberg, R.: Boosting classifiers for drifting concepts. Intell. Data Anal. **11**(1), 3–28 (2007)
35. Street, W., Kim, Y.: A streaming ensemble algorithm (SEA) for large-scale classification. In: Proceedings of the 7th ACM SIGKDD International Conference on Knowledge Discovery and Data Mining (KDD), pp. 377–382 (2001)
36. Sun, Y., Tang, K., Minku, L.L., Wang, S., Yao, X.: Online ensemble learning of data streams with gradually evolved classes. IEEE Trans. Knowl. Data Eng. **28**(6), 1532–1545 (2016)
37. Sun, Y., Tang, K., Zhu, Z., Yao, X.: Concept drift adaptation by exploiting historical knowledge (2017). ArXiv https://arxiv.org/abs/1702.03500
38. Tsymbal, A.: The problem of concept drift: definitions and related work. Technical Report 106, Computer Science Department, Trinity College, Dublin (2004)
39. Turhan, B., Menzies, T., Bener, A., Di Stefano, J.: On the relative value of cross-company and within-company data for defect prediction. Empir. Softw. Eng. **14**(5), 540–578 (2009)
40. Wang, H., Fan, W., Yu, P.S., Han, J.: Mining concept-drifting data streams using ensemble classifiers. In: Proceedings of the 9th ACM SIGKDD International Conference on Knowledge Discovery and Data Mining (KDD), pp. 26–235 (2003)
41. Wang, S., Minku, L., Yao, X.: A systematic study of online class imbalance learning with concept drift. IEEE Trans. Neural Netw. Learn. Syst., 20 pp (2018). https://doi.org/10.1109/TNNLS.2017.2771290
42. Widmer, G., Kubat, M.: Learning in the presence of concept drift and hidden contexts. Mach. Learn. **23**(1), 69–101 (1996)

A New Combination of Diversity Techniques in Ensemble Classifiers for Handling Complex Concept Drift

Imen Khamassi, Moamar Sayed-Mouchaweh, Moez Hammami, and Khaled Ghédira

Abstract Recent advances in Computational Intelligent Systems have focused on addressing complex problems related to the dynamicity of the environments. Generally in dynamic environments, data are presented as streams that may evolve over time and this is known by *concept drift*. Handling concept drift through ensemble classifiers has received a great interest in last decades. The success of these ensemble methods relies on their *diversity*. Accordingly, various diversity techniques can be used like *block-based data*, *weighting-data* or *filtering-data*. Each of these diversity techniques is efficient to handle certain characteristics of drift. However, when the drift is complex, they fail to efficiently handle it. *Complex drifts* may present a mixture of several characteristics (speed, severity, influence zones in the feature space, etc.) which may vary over time. In this case, drift handling is more complicated and requires new detection and updating tools. For this purpose, a new ensemble approach, namely EnsembleEDIST2, is presented. It combines the three diversity techniques in order to take benefit from their advantages and outperform their limits. Additionally, it makes use of EDIST2, as drift detection mechanism, in order to monitor the ensemble's performance and detect changes. EnsembleEDIST2 was tested through different scenarios of complex drift generated from synthetic and real datasets. This diversity combination allows EnsembleEDIST2 to outperform similar ensemble approaches in terms of accuracy rate, and present stable behaviors in handling different scenarios of complex drift.

I. Khamassi (✉) · M. Hammami · K. Ghédira
Université de Tunis, Institut Supérieur de Gestion de Tunis, Tunis, Tunisia
e-mail: imen.khamassi@isg.rnu.tn; moez.hammami@isg.rnu.tn; khaled.ghedira@isg.rnu.tn

M. Sayed-Mouchaweh
Institute Mines-Telecom Lille Douai, Douai, France
e-mail: moamar.sayed-mouchaweh@imt-lille-douai.fr

© Springer International Publishing AG, part of Springer Nature 2019
M. Sayed-Mouchaweh (ed.), *Learning from Data Streams in Evolving Environments*, Studies in Big Data 41, https://doi.org/10.1007/978-3-319-89803-2_3

1 Introduction

Learning from evolving data stream has received a great attention. It addresses the state of data being non-stationary over time, which is known by *concept drift*. The term *concept* refers to data distribution, represented by the joint distribution $p(x, y)$, where x represents the *n-dimensional* feature vector and y represents its class label. The term *concept drift* refers to a change in the underlying distribution of new incoming data. For example, in intrusion detection application, the behavior of an intruder may evolve in order to confuse the system protection rules. Hence, it is essential to consider these changes for updating the system in order to preserve its performance.

Ensemble classifiers appear to be promising approaches for tracking evolving data streams. The success of the ensemble methods, according to single classifier, relies on their *diversity* [17, 21, 22]. *Diversity* can be achieved according to three main strategies[15]: *block-based data*, *weighting-data*, or *filtering-data*. In *block-based ensembles* [5, 16, 20], the training set is presented as blocks or chunks of data at a time. Generally, these blocks are of equal size and the evaluation of base learners is done when all instances from a new block are available. In *weighting-data ensembles*[2, 13, 18], the instances are weighted according to some weighting process. For example in Online Bagging [19], the weighting process is based on reusing instances for training individual learners. Finally, *filtering-data ensembles* [1] are based on selecting data from the training set according to a specific criterion, for example, similarity in feature space.

In many real-life applications, the concept drift may be *complex* in the sense that it presents time-varying characteristics. For instance, a drift can present different characteristics according to its speed (*abrupt or gradual*), nature (*continuous or probabilistic*), and severity (*local or global*). Accordingly, *complex drift* can present a mixture of all these characteristics over time. It is worth underlining that each characteristic presents its own challenges. Hence, a mixture of these different characteristics may accentuate the challenge issues and complicate the drift handling.

In this paper, the goal is to underline the complementarity of the diversity techniques (*block-based data, weighting-data*, and *filtering-data*) for handling different scenarios of complex drift. For this purpose, a new ensemble approach, namely EnsembleEDIST2, is proposed. The intuition is to combine these three diversity techniques in order to efficiently handle different scenarios of complex drift. Firstly, EnsembleEDIST2 defines a data-block with variable size for updating the ensemble's members, thus it can avoid the problem of tuning off size of the data-block. Secondly, it defines a new filtering criterion for selecting the most representative data of the new concept. Thirdly, it applies a new weighting process in order to create diversified ensemble's members. Finally, it makes use of EDIST2 [11, 14], as drift detection mechanism, in order to monitor the ensemble's performance and detect changes.

EnsembleEDIST2 has been tested through different scenarios of complex drifts generated from synthetic and real datasets. This diversity combination allows EnsembleEDIST2 to outperform similar ensemble approaches in terms of accuracy rate, and present a stable behavior in handling different scenarios of complex drift.

The remainder of the paper is organized as follows. In Sect. 2, the challenges of complex concept drift are exposed. In Sect. 3, the advantages and the limits of each diversity technique are studied. In Sect. 4, the proposed approach, namely EnsembleEDIST2, is detailed. Section 5, the experimental setup and the obtained results are presented. Finally, in Sect. 6, the conclusion and some future research directions are exposed.

2   Complex Concept Drift Characteristics and Challenges

A drift occurs when a new concept replaces an old one, and it may by characterized by its speed (*abrupt, gradual*), nature (*continuous, probabilistic*), or severity (*local, global*).

2.1   Speed

Speed refers to how long the drift lasts. Hence, the drift can be categorized as

- *Abrupt drift* which occurs when the new concept suddenly replaces the old one. This drift is challenging because it immediately deteriorates the learner performance and causes a rapid accuracy decrease.
- *Gradual drift* which occurs when the drifting time is relatively large. As a result, there are two types of gradual drift:

    - *Gradual probabilistic drift* which refers to a period when both new and old concepts are active. As time passes, the probability of sampling from the old concept decreases, whereas the probability of sampling from new concept increases, until the new concept totally replaces the old one (see Fig. 1). Namely, this drift is challenging, because it creates a period of uncertainty, where both new and old concepts are active at the same time.
    - *Gradual continuous drift* occurs when the concept itself continuously changes from the old to the new concept, by suffering small modifications at every time step (see Fig. 2). Notice that these changes are so small that they are only noticed during a long time period; and this may lead to a delay of detection.

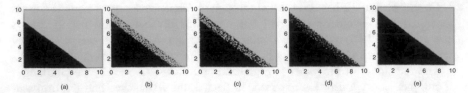

Fig. 1 Gradual probabilistic local drift: SEA dataset. (**a**) concept1: 100%; concept2: 0%. (**b**) concept1: 75%; concept2: 25%. (**c**) concept1: 50%; concept2: 50%. (**d**) concept1: 25%; concept2: 75%. (**e**) concept1: 0%; concept2: 100%

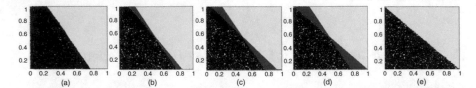

Fig. 2 Gradual continuous local drift: (**a**) concept1, (**b**)–(**d**) instance space affected by the drift and (**e**) concept2

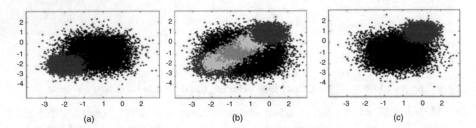

Fig. 3 Gradual continuous global drift: (**a**) concept1, (**b**) concept evolution and (**c**) concept2

2.2 Severity

Severity refers to the amount of change caused by the drift. Accordingly, the drift can be categorized as

- *Local drift* where changes only occur in some regions of the instance space. Hence, when looking at the overall instance space, we notice that only some subsets are affected by the drift (see Figs. 1 and 2). Namely, the time until local concept drift is detected can be arbitrarily long. This is due to the rarity of data representing the drift, which in turn makes it difficult to confirm the presence of drift.
- *Global drift* where changes affect the overall instance space. In such a case, the difference between the old and the new concept is more noticeable and the drift can be earlier detected (see Fig. 3). Namely, handling this drift is also challenging, because the performance's decrease of the learner is more important than the other types of drift.

2.3   Complex Concept Drift

In many real-life applications, the concept drift may be *complex* in the sense that it presents time-varying characteristics. Let us take the example of a drift with three different characteristics according to its speed *(gradual or abrupt)*, nature *(continuous or probabilistic)*, and severity *(local or global)*. It is worth underlining that each characteristic presents its own challenges (as stated in Sects. 2.1 and 2.2). As a result, a mixture of these different characteristics may accentuate the challenge issues and complicate the drift handling.

For instance, we can consider the drift depicted in Fig. 2 as complex drift as it simulates a Gradual Continuous Local Drift, in the sense that the hyperplane class boundary is gradually rotating during the drifting phase and continuously presenting changes with each instance in local regions. Namely, the time until this complex drift is detected can be arbitrarily long. This is due to the rarity of data source representing the drift, which in turn makes it difficult to confirm the presence of drift. Moreover, in some cases, this drift can be considered as noise by confusion, which makes the model unstable. Hence, to overcome the instability, the model has to (1) effectively differentiate between local changes and noises, and (2) deal with the scarcity of instances that represent the drift in order to effectively update the learner.

Another interesting complex drift represents the Gradual Continuous Global Drift (see Fig. 3). During this drift, the concept is gradually changing and continuously presenting modifications with each instance. Namely, during the transition phase, the drift evolves and presents several intermediate concepts until the emergence of the final concept (see Fig. 3b). Hence, the challenging issue is to efficiently decide the end time of the old concept and detect the start time of the new concept. The objective is to update the learner with the data that represent the final concept (see Fig. 3c) and not with data collected during the concept evolution (see Fig. 3b). Moreover, this drift is considered as global because it is affecting all the instances of the drifting class. Namely, handling this complex drift is also challenging, because the performance's decrease of the learner is more pronounced than the other types of drifts.

3   Related Work

The *diversity* [15] among the ensemble can be fulfilled by applying various techniques such as *block-based data*, *weighting-data* or *filtering data*, in order to differently train base learners (see Fig. 4). Hence, the objective in this investigation is to highlight the advantages and drawbacks of each diversity technique in handling complex drift (see Table 1).

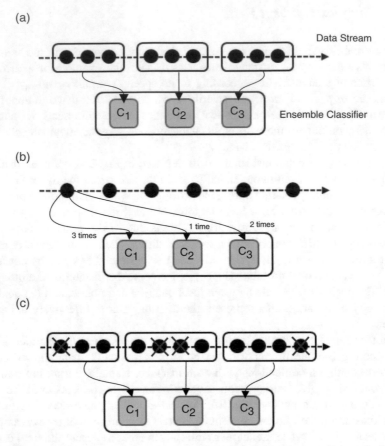

Fig. 4 Different diversity techniques among the ensemble. (**a**) Block-based (**b**) Weighting-data (**c**) Filtering-data

Table 1 Summary of the advantages (+) and drawbacks (−) of diversity techniques for handling complex drift

Complex drift	Gradual continuous		Gradual probabilistic		Abrupt	
	Local	Global	Local	Global	Local	Global
Block-based	+	+	+	+	−	−
Weighting-data	+	−	+	−	+	−
Filtering-data	−	+	−	+	−	+

3.1 Block-Based Technique

According to the *block-based technique*, the training set is presented as blocks or chunks of data at a time. Generally, these blocks are of equal size and the construction, evaluation, or updating of base learners is done when all instances from a new block are available. Very often, ensemble learners periodically evaluate their components and substitute the weakest one with a new (candidate) learner after each data-block [4, 16, 20]. This technique preserves the adaptability of the ensemble in such way that learners, which were trained in recent blocks, are the most suitable for representing the current concept.

The block-based ensembles are suitable for handling gradual drifts. Generally, during these drifts, the change between consecutive data-blocks is not quite pronounced; thus, it can be only noticeable in long periods. The interesting point in the block-based ensembles is that they can enclose different learners that are trained in different periods of time. Hence, by aggregating the outputs of these base classifiers, the ensemble can offer accurate reactions to such gradual drifts.

In contrast, the main drawback of block-based ensembles is the difficulty of tuning off the block size to offer a compromise between fast reactions to drifts and high accuracy. If the block size is too large, they may slowly react to abrupt drift; whereas small size can damage the performance of the ensemble in stable periods.

3.2 Weighting-Data Technique

In this technique, the base learners are trained according to weighted instances from the training set. A popular instance weighting process is presented in the Online Bagging ensemble [19]. For ease of understanding, the weighting process is based on reusing instances for training individual classifiers. Namely, if we consider that each base classifier C_i is trained from a subset M_i from the global training set, then the instance$_i$ will be presented k times in M_i; where the weight k is drawn from a Poisson(1) distribution.

Online Bagging has inspired many researchers in the field of drift tracking [2, 13, 17]. This approach can be of great interest for:

– Class imbalance: where some classes are severely underrepresented in the dataset
– Local drift: where changes occur in only some regions of the instance space.

Generally, the weighting process intensifies the reuse of underrepresented class data and helps to deal with the scarcity of instances that represent the local drift. However, the instance duplication may impact the ability of the ensemble in handling global drift. During global drift, the change affects a large amount of data; thus, when reusing data for constructing base classifiers, the performance's decrease is accentuated and the recovery from the drift may be delayed.

3.3 Filtering-Data Technique

This technique is based on selecting data from the training set according to a specific criterion, for example similarity in the feature space. Such technique allows to select subsets of attributes that provide partitions of the training set containing maximally similar instances, that is, instances belonging to the same regions of feature space. Thanks to this technique, base learners are trained according to different subspaces to get benefit from different characteristics of the overall feature space.

In contrast with conventional approaches which detect drift in the overall distribution without specifying which feature has changed, ensemble learners based on filtered data can exactly specify the drifting feature. This is a desired property for detecting novel class emergence or existing class fusion in unlabeled data. However, these approaches may present difficulty in handling local drifts if they do not define an efficient filtering criterion. It is worth underlining that during local drift, only some regions of the feature space are affected by the drift. Hence, only the base classifier which is trained on changing region is the most accurate to handle the drift. However, when aggregating the final decision of this classifier with the remained classifiers, trained from unchanged regions, the performance recovery may be delayed [12].

4 The Proposed Approach

The intuition behind EnsembleEDIST2 is to combine the three diversity techniques (*Block-based*, *Weighting-data*, and *Filtering data*) in order to take benefit from their advantages and avoid their drawbacks.

The contributions of EnsembleEDIST2 for efficiently handling complex concept drifts are as follows, it:

- Explicitly handles drift through a drift detection method EDIST2 [14] (Sect. 4.1)
- Makes use of data-block with variable size for updating the ensemble's members (Sect. 4.2)
- Defines a new filtering criterion for selecting the most representative data of the new concept (Sect. 4.3)
- Applies a new weighting process in order to create diversified ensemble's members (Sect. 4.4)

4.1 Drift Monitoring Process in EnsembleEDIST2

EnsembleEDIST2 is an ensemble classifier designed to explicitly handle drifts. It makes use of EDIST2 [14], as drift detection mechanism, in order to monitor the ensemble's performance and detect changes (see Fig. 5).

EDIST2 monitors the prediction feedback provided by the ensemble. More precisely, EDIST2 studies the distance between two consecutive errors of classification. Notice that the distance is represented by the number of instances between

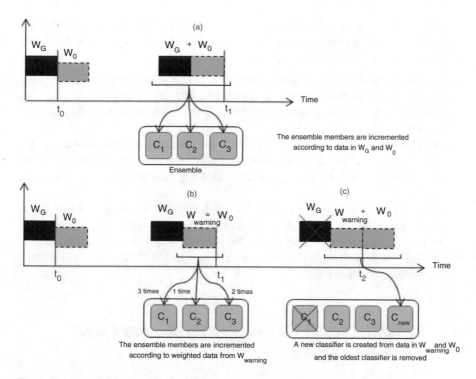

Fig. 5 EnsembleEDIST2's adapting process according to the three detection levels: (**a**) In-control, (**b**) Warning and (**c**) Drift

two consecutive errors of classification. Accordingly, when the data distribution becomes nonstationary, the ensemble will commit much more errors and the distance between these errors will decrease.

In EDIST2, the concept drift is tracked through two data windows, a 'global' one and a 'current' one. The global window W_G is a self-adaptive window which is continuously incremented if no drift occurs and decremented otherwise; and the current window W_0 which represents the batch of current collected instances.

In EDIST2, we want to estimate the error distance distribution of W_G and W_0 and make a comparison between the averages of their error distance distributions in order to check a difference. As stated before, a significant decrease in the error distance implies a change in the data distribution and suggests that the learning model is no longer appropriate.

EDIST2 makes use of a statistical hypothesis test in order to compare W_G and W_0 error distance distributions and checks whether the averages differ by more than the threshold ϵ. It is worth underlining that there is no a priori definition of the threshold ϵ, in the sense that it does not require any a priori adjusting related to the expected speed or severity of the change. ϵ is autonomously adapted according to a statistical hypothesis test (for more details please refer to [14]).

The intuition behind EDIST2 is to monitor μ_d which represents difference between W_G and W_0 averages and accordingly three thresholds are defined (see EnsembleEDIST2 Algorithm):

- *In-Control level*: $\mu_d \leq \epsilon$; within this level, we confirm that there is no change between the two distributions, so we enlarge W_G by adding W_0 's instances. Consequently, all the ensemble members are incremented according to data samples in W_G and W_0.
- *Warning level*: $\mu_d > \epsilon$; within this level, the instances are stored in a warning chunk W_{warning}. Hence, all the ensemble members are incremented according to weighted data from W_{warning}. (The weighting process will be explained in *Sect. 4.4*)
- *Drift level*: $\mu_d > \epsilon + \sigma_d$; within this level, the drift is confirmed and W_G is decremented by only containing the instances stored since the warning level, that is, in W_{warning}. Additionally, a new base classifier is created from scratch and trained according to data samples in W_{warning}, then the oldest classifier is removed from the ensemble.

4.2   EnsembleEDIST2's Diversity by Variable-Sized Block Technique

In EnsembleEDIST2, the size of data-block is not defined according to the number of instances, as it is the case of conventional *block-based ensembles*, but according to the number of errors committed during the learning process. More precisely, the data-block W_0, in EnsembleEDIST2, is constructed by collecting the instances that exist between N_0 errors (see *CollectInstances* Procedure in EnsembleEDIST2 Algorithm).

As depicted in Fig. 6, when the drift is abrupt, the ensemble commits N_0 errors in short drifting time. However, when the drift is gradual, the ensemble commits N_0 errors in relatively longer drifting time. Hence, according to this strategy, the block size is variable and adjusted according to drift characteristics.

It is worth underlining that EnsembleEDIST2 can offer a compromise between fast reactions to abrupt drift and stable behavior regarding gradual drift. This is a desirable property for handling complex drift which may present different characteristics at the same time, and accordingly EnsembleEDIST2 can avoid the problem of tuning off the size of data-block as it is the case of most block-based approaches.

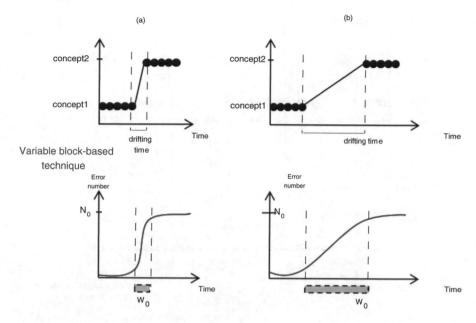

Fig. 6 Variable data-block technique in EnsembleEDIST2. (**a**) Abrupt drift (**b**) Gradual drift

4.3 EnsembleEDIST2's Diversity by New Filtering-Data Criterion

Different from conventional filtering-data ensembles, which filter data according to similarity in the feature space, EnsembleEDIST2 defines a new filtering criterion. It filters the instances that trigger the warning level. More precisely, each time the ensemble reaches the warning level, the instances are gathered in a warning chunk W_{warning} in order to reuse them for training the ensemble's members (see Fig. 7a). This is an interesting point when dealing with local drift because drifting data are scarce and not continuously provided. It is possible that a certain amount of drifting data can be found in zones (1), (2), (3) and (4) but not quite sufficient to reach the drift level. Hence, by considering these data for updating the ensemble's members, EnsembleEDIST2 can ensure a rapid recovery from local drift.

In contrast, conventional filtering-data ensembles are unable to define in which zone the drift has occurred; thus, they may update the ensemble's members with data filtered from unchanged feature space, which in turn may delay the performance correctness.

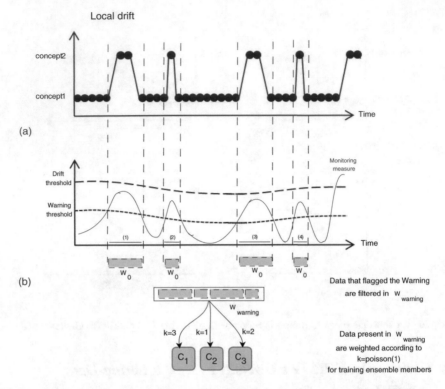

Fig. 7 (**a**) Filtering-data technique and (**b**) Weighting-data technique in EnsembleEDIST2

4.4 EnsembleEDIST2's Diversity by New Weighting-Data Process

The focus in EnsembleEDIST2 is to maximize the use of data present in W_{warning} for accurately updating the ensemble. More precisely, the data in W_{warning} are weighted according to the same weighting process used in Online bagging [19]. Namely, each $instance_i$ from W_{warning} is reused k times for training the base classifier C_i, where the weight k is drawn from a $Poisson(1)$ distribution (see *WeightingDataProcess* Procedure in *EnsembleEDIST2* Algorithm).

Generally, the weighting process in EnsembleEDIST2 offers twofold advantages. First, it intensifies the reuse of underrepresented class data and helps to deal with scarcity of instances that represent the local drift. Second, it permits faster recovery from global drift than conventional weighting-data ensembles. As it is known, during global drift, the change affects a large amount of data. Hence, different from conventional weighting-data ensembles, which apply the weighting process to all the datasets; EnsembleEDIST2 only weights the instances present in W_{warning} (see Fig. 7b). In consequence, it can avoid to accentuate the decrease of the ensemble's performance during global drift, and ensure a fast recovery.

Algorithm *EnsembleEDIST2*

Input: (x, y): Data Stream

        N_0: number of error to construct the window

        m: number of base classifier

Output: Trained ensemble classifier E

1.   **for** each base classifier C_i from E

2. $InitializeClassifier(C_i)$

3.   **end for**

4. $W_G \leftarrow CollectInstances(E, N_0)$

5. $W_{warning} \leftarrow \emptyset$

6.   **repeat**

7. $W_0 \leftarrow CollectInstances(E, N_0)$

8. $Level \leftarrow DetectedLevel(W_G, W_0)$

9.   **switch** (Level)

10.   **case 1:** *Incontrol*

11. $W_G \leftarrow W_G \cup W_0$

12. $UpdateParameters(W_G, W_0)$

13.   Increment all ensemble's members of E according to instances in W_G

14.   **end case 1**

         case 2: *Warning*

15. $W_{warning} \leftarrow W_{warning} \cup W_0$

16. $UpdateParameters(W_{warning}, W_0)$

17. $WeightingDataProcess(E, W_{warning})$

18.   **end case 2**

         case 3: *Drift*

19.   Create a new base classifier C_{new} trained on instances in $W_{warning}$

20. $E \leftarrow E \cup C_{new}$

21.   Remove the oldest classifier from E

22. $W_G \leftarrow W_{warning}$

23. $W_{warning} \leftarrow \emptyset$

24.   **end case 3**

25. **end switch**

26.**until** The end of the data streams

Algorithm *DetectedLevel(* W_G, W_0 *)*

Input: W_G: Global data window characterized by:

        N_G: error number

        μ_G: error distance mean

        σ_G: error distance standard deviation

      W_0: Current data window characterized by:

        N_0: error number,

        μ_0: error distance mean,

        σ_0: error distance standard deviation

Output: *Level*: detection level

1. $\mu_d \leftarrow \mu_G - \mu_0$

2. $\sigma_d \leftarrow \sqrt{\dfrac{\sigma_G^2}{N_G} + \dfrac{\sigma_0^2}{N_0}}$

3. $\epsilon \leftarrow t_{1-\alpha} * \sigma_d$

4.   **if** $(\mu_d > \epsilon + \sigma_d)$

5. $Level \leftarrow Drift$

6.   **else if** $(\mu_d > \epsilon)$

7. $Level \leftarrow Warning$

8.   **else** $Level \leftarrow Incontrol$

9.   **end if**

10. **end if**

11. **return** (Level)

Algorithm *CollectInstances(E, N_0)*

Input: (x, y): Data Stream

           N_0: number of error to construct the window

           C: trained ensemble classifier E

Output: W: Data window characterized by:

           N: error number

           μ: error distance mean

           σ: error distance standard deviation

1. $W \leftarrow \emptyset$

2. $N \leftarrow 0$

3. $\mu \leftarrow 0$

4. $\sigma \leftarrow 0$

5.   **repeat** for each instance x_i

6. $Prediction \leftarrow unweightedMajorityVote(E, x_i)$

7.   **if** $(Prediction = false)$

8. $d_i \leftarrow computeDistance()$

9. $\mu \leftarrow \dfrac{N}{N+1}\mu + \dfrac{d_i}{N+1}$

10. $\sigma \leftarrow \sqrt{\dfrac{N-1}{N}\sigma^2 + \dfrac{(d_i-\mu)^2}{N+1}}$

11. $N \leftarrow N + 1$

12.   **end if**

13. $W \leftarrow W \cup \{x_i\}$

14. **until** $(N = N_0)$

15. **return** (W)

Algorithm *UpdateParameters(W_G, W_0)*

Input: W_G: Global data window characterized by:

           N_G: error number

           μ_G: error distance mean

           σ_G: error distance standard deviation

W_0: Current data window characterized by:

N_0: error number,

μ_0: error distance mean,

σ_0:error distance standard deviation

Output: Updated parameters of W_G

1. $\mu_G \leftarrow \frac{1}{N_G+N_0}(N_G.\mu_G+N_0.\mu_0)$ $\sigma_G \leftarrow \sqrt{\frac{N_G\sigma_G^2+N_0\sigma_0^2}{N_G+N_0} + \frac{N_GN_0}{(N_G+N_0)^2}(\mu_G - \mu_0)^2}$

2. $N_G \leftarrow N_G + N_0$

Algorithm *WeightingDataProcess(E, $W_{warning}$)*

Input: E: Ensemble Classifier

   $W_{warning}$: Window of data

Output: E: Updated ensemble classifier

1. **for** each instance x_i from $W_{warning}$
2.   **for** each base classifier C_i from E
3. $k \leftarrow poisson(1)$
4.   **do** k times
5. $TrainClassifier(C_i, x_i)$
6.   **end do**
7.   **end for**
8. **end for**

---

5 Experimental Evaluation

5.1 Synthetic Datasets

All Synthetic datasets contain 100,000 instances and one concept drift where the starting and the ending time are predefined. For gradual drift, the drifting time lasts 30,000 instances (it begins at $t_{start} = 40,000$ and ends at $t_{end} = 70,000$). For abrupt drift, the drift occurs at $t = 50,000$.

(a) *Rotating Hyperplane* Hulten et al. [9] is used to simulate the *Gradual Continuous Local Drift*.

It is based on moving hyperplane which is represented in *d-dimensional* space by: $\sum_{i=1}^{d} w_i x_i = w_0$. Where x_i is the instance, w_i is the corresponding weight to each *attribute$_i$* and w_0 is the total weight. The instances which satisfy $\sum_{i=1}^{d} w_i x_i \geq= w_0$ are labeled as positive, otherwise negative.

In this investigation, the hyperplane is represented in *2-dimentional* space and the concept drift is simulated by slightly rotating the hyperplane with each consecutive instance. This rotation is done by gradually modifying w_i by

0.1 with each instance. As depicted in Fig. 4, the drift is locally affecting the instance space.

(b) *RBF (Radial Basis Function)* Bifet et al. [3] is used to simulate the *Gradual Continuous Global Drift.*

A fixed number of random centroids are generated. Each center has a random position, a single standard deviation, class label, and weight. The instances are generated by selecting a center at random. A random direction is chosen to offset the attribute values from the central point. The length of the displacement is randomly drawn from a Gaussian distribution with standard deviation determined by the chosen centroid.

In this investigation, we have used two centroids which represent two classes. The drift is introduced by gradually moving the center of only one centroid in a linear direction; the center is slowly moving at speed level of 0.001 with each instance. As depicted in Fig. 5, the drift is considered as global because it is affecting all the instances of the drifting centroid.

(c) *SEA Concepts Generator* Street and Kim [24] models three independent real valued attributes in [0, 10], only the first two attributes are relevant for prediction. In original data, the class decision boundary is defined as follows: $x_1 + x_2 \le \theta$ where x_1 and x_2 are the first two attributes and θ is a threshold value. Four functions are defined for generating binary class labels: $f_1 : \theta = 9$, $f_2 : \theta = 8$, $f_3 : \theta = 7$, and $f_4 : \theta = 9.5$.

In this investigation, we have generated SEA Gradual dataset for simulating *Gradual Probabilistic Local Drift*, where the drift is introduced by gradually substituting the class function f_1 by f_4 in a probabilistic way (See Fig. 3). We have also generated another dataset, namely SEA Abrupt, where the drift is introduced by suddenly replacing f_1 by f_4 in order to simulate *Abrupt Local Drift.*

(d) *STAGGER Concepts Generator* Schlimmer and Granger Jr. [23] models three independent categorical attributes: size (small, medium, large), color (red, green, blue), and shape (square, circular, triangular). The classification task is defined according to three decision functions:
$f_1 : size = small$ and $color = red$, $f_2 : color = green$ or $shape = circular$, and $f_3 : size = medium$ or $size = large$.

In this investigation, we have generated STAGGER Gradual dataset for simulating *Gradual Probabilistic Global Drift*, where the drift is introduced by gradually substituting the class function f_1 by f_3 in a probabilistic way. We have also generated another dataset, namely STAGGER Abrupt, where the drift is introduced by suddenly replacing f_1 by f_3 with 10% of noise in order to simulate *Abrupt Local Drift* with noise. In this paper, by noise we refer to class noise, that is, errors artificially introduced to class labels (Table 2).

Table 2 Different types of complex drift handled in this investigation

Complex drift characteristics			
Speed	Nature	Severity	Synthetic datasets
Gradual	Continuous	Local	Hyperplane [9]
		Global	RBF [3]
	Probabilistic	Local	SEA gradual [24]
		Global	STAGGER gradual [23]
Abrupt		Local	SEA abrupt [24]
		Global	STAGGER abrupt [23]

5.2 Real Datasets

(a) *Electricity Dataset* (48,312 instances, 8 attributes, 2 classes) is a real-world dataset from the Australian New South Wales Electricity Market [8]. In this electricity market, the prices are not fixed and may be affected by demand and supply. The dataset covers a period of 2 years and the instances are recorded every half an hour. The classification task is to predict a rise (UP) or a fall (DOWN) in the electricity price. Three numerical features are used to define the feature space: the electricity demand in the current region, the electricity demand in the adjacent regions, and the schedule of electricity transfer between the two regions.

This dataset may present several scenarios of complex drift. For instance, a gradual continuous drift may occur when the users progressively change their consumption habits during a long time period. Likewise, an abrupt drift may occur when the electricity prices suddenly increase due to unexpected events (e.g., political crises or natural disasters). Moreover, the drift can be local if it impacts only one feature (e.g., the electricity demand in the current region), or global if it impacts all the features.

(b) *Spam Dataset* (9,324 instances, 500 attributes, 2 classes) is a real-world dataset containing e-mail messages from the Spam Assassin Collection Project [10]. The classification task is to predict if a mail is spam or legitimate. The dataset contains 20% of spam mailing. The feature space is defined by a set of numerical features such as the number of receptors, textual attributes describing the mail content, and sender characteristics...

This dataset may present several scenarios of complex drift. For instance, a gradual drift may occur when the user progressively changes his preferences. However, an abrupt drift may occur when the spammer rapidly changes the mail content to trick the spam filter rules. In one side, the drift can be continuous when the spammer starts to change the spam content, but the filter continues to correctly detect them. On the other side, the drift can be probabilistic when the spammer starts to change the spam content, but the filter fails in detecting some of them.

5.3   Evaluation Criteria

When dealing with evolving data streams, the objective is to study the evolution of the EnsembleEDIST2 performance over time and see how quick the adaptation to drift is. According to Gama et al. [7], the prequential accuracy is a suitable metric to evaluate the learner performance in presence of concept drift. It proceeds as follows: Each instance is firstly used for testing, then for training. Hence, the accuracy is incrementally updated using the maximum available data; and the model is continuously tested on instances that it has not already seen (for more details please refer to [7]).

5.3.1   Parameter Settings

All the tested approaches were implemented in the java programming language by extending the Massive Online Analysis (MOA) software [3]. MOA is an online learning framework for evolving data streams and supports a collection of machine learning methods.

For comparison, we have selected well-known ensemble approaches according to each category:

- *Block-based ensemble*: AUE (Accuracy Updated Ensemble) [5], AWE (Accuracy Weighted Ensemble) [16], and LearnNSE [20] with block size equal to 500 instances.
- *Weighting-data ensemble*: LeveragingBag [2] and OzaBag [19]
- *Filtering-data ensemble*: LimAttClass [1]

For all these approaches, the ensemble's size was fixed to 10 and the Hoeffding Tree (HT) [6] was used as a base learning algorithm.

It is worth noticing that EnsembleEDIST2 makes use of two parameters: N_0 which is the number of errors in W_0 and m which is the number of base classifiers among the ensemble. In this investigation, we respectively set $N_0 = 30$ and $m = 3$ according to empirically studies done in Sects. 6.1 and 6.2.

6   Comparative Study and Interpretation

6.1   Impact of N_0 on EnsembleEDIST2 Performance

EnsembleEDIST2 makes use of the parameter N_0 in order to define the minimum number of errors occurred in W_0. Recall that W_0 represents the batch of current collected instances. This batch is constructed by collecting the instances that exist between N_0 errors.

Table 3 Prequential accuracy for different values of N_0 in EnsembleEDIST2

Complex drifts	Gradual continuous		Gradual probabilistic		Abrupt	
	Local	Global	Local	Global	Local	Global
Synthetic databases	Hyperplane	RBF	SEA gradual	STAGGER gradual	SEA abrupt	STAGGER abrupt
$N_0 = 30$	**98.6**	**95.9**	**97.2**	91.6	97.9	**99.6**
$N_0 = 60$	98.2	**95.9**	**97.2**	91.5	98.1	**99.6**
$N_0 = 90$	98.2	95.6	97.1	91.6	97.5	**99.6**
$N_0 = 120$	98.3	**95.9**	97.1	91.6	**98.2**	**99.6**
$N_0 = 150$	98.3	95.6	97.1	**91.7**	97.5	**99.6**

It is interesting to study the impact of N_0 on the accuracy according to different scenarios of complex drift. For this purpose, we have done the following experiments: For each scenario of complex drift, the accuracy of EnsembleEDIST2 is presented by varying N_0 values (see Table 3).

Based on these results, we can conclude that the performance of EnsembleEDIST2 in handling different scenarios of complex drifts is weakly sensitive to N_0. Hence, we have decided to use $N_0 = 30$ as it has achieved the best accuracy rate in most cases.

6.2 Impact of Ensemble Size on EnsembleEDIST2 Performance

EnsembleEDIST2 makes use of the parameter m in order to define the number of classifiers in the ensemble. Accordingly, it is interesting to study the impact of m on ensemble's performance according to different scenarios of complex drift.

According to Table 4, it is noticeable that the size of EnsembleEDIST2 does not impact significantly the performance in handling different scenarios of complex drift. Hence, we have decided to use $m = 3$ as it achieved the best accuracy rate in most cases and it allows to limit the computational complexity of the ensemble.

6.3 Accuracy of EnsembleEDIST2 Vs Other Ensembles

Table 5 summarizes the average of prequential accuracy during the drifting phase. The objective of this experiment is to study the ensemble performance in the presence of different scenarios of complex drift. Firstly, it is noticeable that EnsembleEDIST2 has achieved better results than *block-based ensembles* in handling different types of abrupt drift. During abrupt drift (independently of being local of global), the change is rapid; thus AUE, AWE, and LearnNSE present difficulty in

Table 4 Accuracy of EnsembleEDIST2 with a different number of base classifiers

	Gradual continuous		Gradual probabilistic		Abrupt	
Complex drifts	Local	Global	Local	Global	Local	Global
Synthetic databases	Hyperplane	RBF	SEA Gradual	STAGGER Gradual	SEA Abrupt	STAGGER Abrupt
$m = 3$	**98.6**	**95.9**	**97.2**	91.6	97.9	**99.6**
$m = 5$	**98.6**	**95.9**	97.1	91.6	**98**	**99.6**
$m = 10$	98.4	95.8	**97.2**	**91.1**	97.6	**99.6**

Table 5 Accuracy of EnsembleEDIST2 vs. other ensembles in synthetic datasets

		Gradual continuous		Gradual probabilistic		Abrupt	
Complex drifts		Local	Global	Local	Global	Local	Global
Synthetic databases		Hyperplane	RBF	SEA Gradual	STAGGER Gradual	SEA Abrupt	STAGGER Abrupt
EnsembleEDIST2		**98.604**	**95.982**	**97.211**	**91.609**	**98.196**	**99.605**
Block-based	AUE	94.187	95.611	94.547	90.381	95.234	98.367
	AWE	94.054	95.018	94.563	90.551	95.23	98.367
	LearnNSE	96.369	95.44	94.372	85.873	95.079	39.049
Weighting-data	LeveragingBag	98.6	95.8	97.1	89.1	**98.2**	94.3
	OzaBag	98.195	93.533	96.982	69.21	98.132	96.64
Filtering-data	LimAttClass	91.281	94.186	91.126	86.553	91.226	94.893

tuning off the block size to offer a compromise between fast reactions to drift and high accuracy. However, EnsembleEDIST2 is able to autonomously train ensemble members with variable amount of data at each time process, thus it can efficiently handle abrupt drift.

Secondly, it is noticeable that EnsembleEDIST2 outperforms *weighting-data ensembles* in handling different categories of global drift. During global drift (either continuous, probabilistic or abrupt), the change affects a large amount of data; thus when LeveragingBag and OzaBag intensify the reuse of data for training ensemble members, the performance's decrease is accentuated. In contrast, EnsembleEDIST2 duplicates only a set of filtered instances for training the ensemble members, that is why it is more accurate in handling global drift.

Thirdly, it is noticeable that EnsembleEDIST2 outperforms the *filtering-data ensembles* in handling different categories of local drift. During local drift (either continuous, probabilistic or abrupt), the change affects a little amount of data; thus, the choice of the filtering criterion is an essential point for efficiently handling local drift. EnsembleEDIST2 defines a new filtering criterion, which is based on selecting the data that triggered the warning level. These data are the most representative of the new concept; thus, when training the ensemble's members accordingly, it makes it more efficient for handling local drift.

EnsembleEDIST2 has also been tested through real-world datasets which represent different scenarios of drift. It is worth underlining that the size of these datasets is relatively small compared to the synthetic ones. Despite the different features of

Table 6 Accuracy of EnsembleEDIST2 vs. other ensembles in real datasets

Real dataset		Electricity	Spam
EnsembleEDIST2		**84.8**	**89.2**
Block-based	AUE	69.35	79.34
	AWE	72.09	60.25
	LearnNSE	72.07	60.33
Weighting-data	LeveragingBag	83.8	88.2
	OzaBag	82.3	82.7
Filtering-data	LimAttClass	82.6	63.9

each real dataset, encouraging results have been found where EnsembleEDIST2 has achieved the best accuracy in all the datasets (see Table 6).

To sum, it is worth underlining that the combination of the three diversity techniques in EnsembleEDIST2 is beneficial for handling different scenarios of complex drift at the same time.

7   Conclusion

In this paper, we have presented a new study of the role of diversity among the ensemble. More precisely, we have highlighted the advantages and the limits of three widely used diversity techniques (*block-based data*, *weighting-data* and *filtering data*) in handling *complex drift*.

Additionally, we have presented a new ensemble approach, namely EnsembleEDIST2, which combines these three diversity techniques. The intuition behind this approach is to explicitly handle drifts by using the drift detection mechanism EDIST2. Accordingly, the ensemble performance is monitored through a self-adaptive window. Hence, EnsembleEDIST2 can avoid the problem of tuning off the size of the batch data as it is the case of most block-based ensemble approaches, which is a desirable property for handling abrupt drifts. Secondly, it defines a new filtering criterion, which is based on selecting the data that trigger the warning level. Thanks to this property, EnsembleEDIST2 is more efficient for handling local drifts than conventional filtering-data ensembles, which are only based on filtering data according to similarity on feature space. Then, differently from the conventional weighting-data ensembles which apply the weighting process to all the data stream; EnsembleEDIST2 only intensifies the reuse of the most representative data of the new concept, which is a desirable property for handling global drifts.

EnsembleEDIST2 has been tested in different scenarios of complex drift. Encouraging results were found, compared to similar approaches, where EnsembleEDIST2 has achieved the best accuracy rate in all datasets, and presented a stable behavior in handling different scenarios of complex drift.

It worth underlining that in the present investigation, the ensemble size, that is, the number of ensemble members, was fixed. Hence it is interesting, for future work, to perform a strategy for dynamically adapting the ensemble size. The focus is that, during stable periods, the ensemble size is maintained fixed, whereas during the drifting phase the size is autonomously adapted. This may ameliorate the performance and reduce the computational cost among the ensemble.

References

1. Bifet, A., Frank, E., Holmes, G., Pfahringer, B., Sugiyama, M., Yang, Q.: Accurate ensembles for data streams: combining restricted hoeffding trees using stacking. In: 2nd Asian Conference on Machine Learning (ACML2010), pp. 225–240 (2010)
2. Bifet, A., Holmes, G., Pfahringer, B.: Leveraging bagging for evolving data streams. In: Proceedings of the 2010 European Conference on Machine Learning and Knowledge Discovery in Databases: Part I. ECML PKDD'10, pp. 135–150. Springer, Berlin, Heidelberg (2010)
3. Bifet, A., Holmes, G., Kirkby, R., Pfahringer, B.: MOA: massive online analysis. J. Mach. Learn. Res. **11**, 1601–1604 (2010)
4. Brzezinski, D., Stefanowski, J.: Accuracy updated ensemble for data streams with concept drift. In: Corchado, E., Kurzyński, M., Woźniak, M. (eds.) Hybrid Artificial Intelligent Systems. Lecture Notes in Computer Science, vol. 6679, pp. 155–163. Springer, Berlin, Heidelberg (2011)
5. Brzezinski, D., Stefanowski, J.: Reacting to different types of concept drift: the accuracy updated ensemble algorithm. IEEE Trans. Neural Netw. Learn. Syst. **25**(1), 81–94 (2014)
6. Domingos, P., Hulten, G.: Mining high-speed data streams. In: Proceedings of the Sixth ACM SIGKDD International Conference on Knowledge Discovery and Data Mining, pp. 71–80. KDD00. ACM, New York (2000)
7. Gama, J., Sebastião, R., Rodrigues, P.: On evaluating stream learning algorithms. Mach. Learn. **90**(3), 317–346 (2013)
8. Harries, M.: Splice-2 comparative evaluation: electricity pricing. Technical Report, The University of South Wales (1999)
9. Hulten, G., Spencer, L., Domingos, P.: Mining time-changing data streams. In: Proceedings of the Seventh ACM SIGKDD International Conference on Knowledge Discovery and Data Mining, San Francisco, CA, August 26–29, 2001, pp. 97–106 (2001)
10. Katakis, I., Tsoumakas, G., Vlahavas, I.: Tracking recurring contexts using ensemble classifiers: an application to email filtering. Knowl. Inform. Syst. **22**(3), 371–391 (2010)
11. Khamassi, I., Sayed-Mouchaweh, M.: Drift detection and monitoring in non-stationary environments. In: Evolving and Adaptive Intelligent Systems (EAIS), Linz, pp. 1–6 (2014)
12. Khamassi, I., Sayed-Mouchaweh, M.: Self-adaptive ensemble classifier for handling complex concept drift. In: 2nd ECML/PKDD 2017 Workshop on Large-scale Learning from Data Streams in Evolving Environments, Skopje, pp. 52–72 (2017)
13. Khamassi, I., Sayed-Mouchaweh, M., Hammami, M., Ghédira, K.: Ensemble classifiers for drift detection and monitoring in dynamical environments. In: Annual Conference of the Prognostics and Health Management Society, New Orlean (2013)
14. Khamassi, I., Sayed-Mouchaweh, M., Hammami, M., Ghédira, K.: Self-adaptive windowing approach for handling complex concept drift. Cogn. Comput. **7**(6), 772–790 (2015)
15. Khamassi, I., Sayed-Mouchaweh, M., Hammami, M., Ghédira, K.: Discussion and review on evolving data streams and concept drift adapting. Evol. Syst. **9**(1), 1–23 (2018)
16. Kolter, J.Z., Maloof, M.A.: Dynamic weighted majority: an ensemble method for drifting concepts. J. Mach. Learn. Res. **8**, 2755–2790 (2007)
17. Minku, L., White, A., Yao, X.: The impact of diversity on online ensemble learning in the presence of concept drift. IEEE Trans. Knowl. Data Eng. **22**(5), 730–742 (2010)
18. Minku, L., Yao, X.: Ddd: a new ensemble approach for dealing with concept drift. IEEE Trans. Knowl. Data Eng. **24**(4), 619–633 (2012)

19. Oza, N.C., Russell, S.: Online bagging and boosting. In: Artificial Intelligence and Statistics 2001, pp. 105–112. Morgan Kaufmann, Boston (2001)
20. Polikar, R., Upda, L., Upda, S., Honavar, V.: Learn++: an incremental learning algorithm for supervised neural networks. IEEE Trans. Syst. Man Cybern. Part C Appl. Rev. **31**(4), 497–508 (2001)
21. Ren, Y., Zhang, L., Suganthan, P.N.: Ensemble classification and regression-recent developments, applications and future directions. IEEE Comput. Intell. Mag. **11**(1), 41–53 (2016)
22. Sayed-Mouchaweh, M.: Handling Concept Drift. In: Learning from Data Streams in Dynamic Environments, pp. 33–59. Springer International Publishing, Cham (2016)
23. Schlimmer, J.C., Granger Jr., R.H.: Incremental learning from noisy data. Mach. Learn. **1**(3), 317–354 (1986)
24. Street, W.N., Kim, Y.: A streaming ensemble algorithm (sea) for large-scale classification. In: Proceedings of the Seventh ACM SIGKDD International Conference on Knowledge Discovery and Data Mining. KDD01, pp. 377–382. ACM, New York (2001)

Analyzing and Clustering Pareto-Optimal Objects in Data Streams

Markus Endres, Johannes Kastner, and Lena Rudenko

Abstract Stream data analysis is a high relevant topic in various academic and business fields. Users want to analyze data streams to extract information in order to learn from this ever-growing amount of data. Although many approaches exist for effective processing of data streams, learning from streams requires new algorithms and methods to be able to learn under the evolving and unbounded data. In this chapter we focus on the task of *preference-based stream processing and clustering to analyze data streams*. We show that this method is a real alternative to the state-of-the-art approaches.

1 Introduction

Today, data processed by humans as well as computers is very large, rapidly increasing, and often in form of data streams. Many modern applications such as network monitoring, financial analysis, infrastructure manufacturing, sensor networks, meteorological observations, or social networks require query processing over data streams, e.g., [1, 9, 12, 56]. Users want to analyze this data to extract personalized and customized information in order to learn from this ever-growing amount of data, e.g., [16, 26, 38, 58]. However, queries on streams run continuously over a period of time and return different results as new data arrive. Therefore, stream data processing and analyzing can be considered as a high relevant, but difficult and complex task, which is in the focus of current research.

Although many approaches exist for effective processing of data streams, learning from streams requires new algorithms and methods to be able to learn under the evolving and unbounded data. In this chapter we focus on the task of *preference-based query processing and clustering* to analyze data streams. We show that this method is a real alternative to the state-of-the-art approaches. Queries in

M. Endres (✉) · J. Kastner · L. Rudenko
Institute for Computer Science, University of Augsburg, Augsburg, Germany
e-mail: Markus.Endres@informatik.uni-augsburg.de;
Johannes.Kastner@informatik.uni-augsburg.de; Lena.Rudenko@informatik.uni-augsburg.de

© Springer International Publishing AG, part of Springer Nature 2019
M. Sayed-Mouchaweh (ed.), *Learning from Data Streams in Evolving Environments*, Studies in Big Data 41, https://doi.org/10.1007/978-3-319-89803-2_4

Table 1 Example of Twitter data about the Confed Cup

Tweet.ID	Hashtag	user.followers_count	user.status_count	tweet.text
76513	#fifa	32.109	4.430	#RapidReplay #FIFA #ConfedCup #WorldCup Julian Draxler captained Germany to the #ConfedCup title and wins the Gold
81365	#fifa	42.171	2.014	Germany's B team beating everyone else's A team. #ConfedCup #FIFA
65230	#soccer	53.093	1.087	Today, it's @miseleccionmx vs @DFB_Team_EN! Where in #McAllen are you watching the big game? #ConfedCup #Soccer
77514	#fifa	9.316	15.866	Germany top #FIFA World Rankings after #ConfedCup triumph
99142	#football	20.639	6.057	#ConfederationsCupfinal #Germany defeated #Chile 1-0 to win the #ConfedCup for the 1st time! #football
53614	#football	14.006	9.918	#ConfedCup #CHIGER Chile was more aggressive n deserved to win but the more mature team won n consolidated their position in world #football

the context of preferences are soft constraints that should be fulfilled as closely as possible [51, 52]. If exact matches are not available, optimal alternatives are better than nothing.

Example 1 Consider the sample Twitter data[1] about the 2017 FIFA Confederations Cup presented in Table 1.

Assume a person wants to retrieve only *high quality* tweets and comments and therefore specifies that he *prefers* users on Twitter with *at least 10,000 followers* (user.followers_count $\geq$ 10,000), *at least 10,000 tweets* overall (user.status_count $\geq$ 10,000), and a hashtag *in {#fifa, #soccer, #football}*. The person assumes that a tweet having these field values is from a reliable user due to the high number of followers and tweets.

Formulating this query with *hard constraints* might lead to an empty result as on the data in Table 1, since no object in the data stream fulfills all conditions. However, expressing the wishes above as *preferences* and combining them as

---

[1]Note that Twitter provides stream data in the JSON document format. We merged and re-formatted fields from tweets for a better overview.

equally important would lead to a result set which contains all tweets in Table 1, because all objects are Pareto-optimal w.r.t. to the given preferences on the attribute fields. Note that the *Pareto-optimal* objects are those objects which are not dominated by any other object. An object p having d dimensions (attributes) *dominates* an object q, if p is better than q in at least one dimension and not worse than q in all other dimensions for the defined preference. The Pareto-optimal set is also known as *the Skyline* [14]. At the end, the searching person retrieves only personalized and valuable information w.r.t. the given preference.

Personalized stream processing reduces the huge amount of data to high relevant information without producing an empty result. However, during the nature of streams, even the information reduced w.r.t. user's preferences could be too large such that making some decision could be impossible. One typical learning and decision-making task in stream scenarios is clustering to summarize similar data for an overview over the data stream content and to recommend items to a user, e.g., in decision support systems, cp. [57]. Based on the Pareto-frontier a preference query provides, we present a novel clustering method exploiting the Borda social choice voting rule as criterion for the cluster allocation in order to present representatives to the user such that one can make a decision.

In this chapter we propose an approach to *analyze data streams with the support of user preferences*. Our contributions are

- a *preference-based stream processing framework* for analyzing data streams.
- the *preference continuous query language* (PCQL) for effective personalized query formulation on streams.
- a novel *stream-based lattice skyline algorithm* (SLS) for efficient real-time preference evaluation of continuous data streams.
- a groundbreaking *clustering technique* for Pareto-optimal objects based on the *Borda social choice* voting rule.

The remainder of this book chapter is organized as follows: Sect. 2 highlights related work. Section 3 recapitulates essential concepts of the used preference model. In Sect. 4 we describe our preference-based stream analysis framework, the preference continuous query language, and our SLS algorithm. Our clustering approach based on the Borda social choice rule is described in Sect. 5. Section 6 demonstrates an application use case. Experiments in Sect. 7 show the advantages of our approach in comparison to existing methods. Finally, we conclude in Sect. 8 and give an outlook on future work and open challenges.

2   Related Work

When dealing with Pareto-optimal objects and preferences in general, several models play an important role. For example, Kasabov and Song [33] and Dovžan et al. [17] handle preferences with fuzzy values, whereas Boutilier et al. [11] use

Ceteris-Paribus nets (CP-nets) do describe user wishes. Other models like Chomicki [13] and Kießling [35, 38] use strict partial orders to represent preferences in information systems and are often more flexible than other approaches.

Another central aspect in this chapter is stream processing to extract important information from continuous data flows. Babu and Widom [6], e.g., focus primarily on the problem how to define and evaluate continuous queries over data streams. Ribeiro et al. [50] describe an approach for processing data streams according to temporal conditional preferences. In [41], Lee et al. propose a new method for processing multiple continuous Skyline queries over a data stream.

In [5] the authors motivate the need for and research issues arising from stream data processing. Faria et al. [23] describe various applications of novelty detection in data streams, and Krempl et al. [40] discuss challenges for data stream mining such as protecting data privacy, handling incomplete and delayed information, or analysis of complex data. In [39] the authors examine the characteristics of important preference queries (Skyline, top-k and top-k dominating) and review algorithms proposed for the evaluation of continuous preference queries under the sliding window streaming model. However, they do not present any framework for preference-based stream evaluation nor discuss clustering on Pareto-optimal objects.

An important research direction associated with stream processing is data stream clustering. This issue is discussed, e.g., in [7, 8, 15, 25, 27, 29, 30, 42]. Clustering data of social networks, e.g., Twitter, is discussed in [45]. The need for stream processing has also been shown in, e.g., [44], where the authors describe an event notification system that monitors and delivers semantically relevant tweets if these meet the user's information needs. Railean and Moraru [49] address the problem of determining the popularity of social events based on their presence in Twitter. For this they compute an association coefficient for an event-tweet pair and use it to determine the popularity. In [46], Mahardhika et al. present an evolving fuzzy-rule-based classifier on streaming data.

The most related work on preference-based clustering is [24] and [28]. In [24] Ferligoj and Batagelj present a Pareto-efficient clustering approach, where several criteria for clustering are consulted in order to find one and only one Pareto-dominant clustering, which dominates all other clusterings. Huang et al. [28] presented a SkyClustering method based on k-means, which works within a Skyline/Pareto computation in a relational database. This approach explores the diversity of Skylines by compressing large sets of Pareto-optima and finally presents k representative and diverse Skyline objects to the user. Another approach to handle with Skylines was presented in [59], where supervised alternative clusterings are introduced. The main focus of the paper is to find clusterings of good quality starting from given negative clusterings which should be as different as possible and at the same time a Pareto-optimal solution.

In contrast to previous work, our approach is based on preference stream processing which never yields an empty result set and only extracts the most relevant information w.r.t. the user's preferences. Our subsequent clustering of Pareto-optimal objects is done by exploiting the Borda social choice rule to weight distances to clusters according to each object.

3 Background

Preference modeling has been in focus for some time, leading to diverse approaches, e.g., [13, 36]. A preference $P = (A, <_P)$ is a strict partial order (SPO) on the domain of A, $dom(A)$. Thus $<_P$ is irreflexive and transitive. Some values are considered to be better than some others. The term $x <_P y$ can be interpreted as *"y is preferred over x"*. Two values not ordered by the strict partial order $<_P$ are regarded as *indifferent*, i.e. $\neg(x <_P y) \wedge \neg(y <_P x)$. The *maximal objects* of a preference $P = (A, <_P)$ on an input data set D are all objects that are not dominated by any other object w.r.t. the given preference P.

Definition 1 (Best-Matches-Only Set (BMO) [36]) The *Best-Matches-Only* (BMO) result set contains only the best matches w.r.t. the strict partial order of a preference P. These objects are computed by the preference selection operator $\sigma[P](D)$. The query model retrieves exact matches if such objects exist and best alternatives else.

$$\sigma[P](D) := \{o \in D \mid \neg\exists o' \in D : o <_P o'\} \tag{1}$$

3.1 Preference Constructors

To express preferences on data attributes, *preference constructors* were defined in [36–38, 60]. There are preference constructors for single attributes on *categorical*, *numerical*, *temporal*, and *spatial* domains as well as on multiple attributes. Subsequently, we present some selected constructors used in this chapter.

The *categorical base preference* POS(A, POS-set) expresses that a user has a set of preferred values, the POS-set. The preference NEG(A, NEG) is the counterpart to the POS preference. It is possible to combine these preferences to POS/POS or POS/NEG.

The *numerical base preference* AROUND(A, z) favors values close to a preferred numerical target value z. If this is infeasible, values with less deviation from the specified value z are preferred. HIGHEST(A) and LOWEST(A) allow users to express easily their desire for values as high or as low as possible. The preferences AT_LEAST(A, z) and AT_MOST(A, z) prefer values higher and lower than z, resp.

Complex preferences determine the relative importance of preferences and combine base or again complex preference constructors. For example, in a *Pareto preference* $P := P_1 \otimes P_2 = (A_1 \times A_2, <_P)$ all preferences are of equal importance.

Definition 2 (Pareto Preference) Given two Preferences $P_1 = (A_1, <_{P_1})$, $P_2 = (A_2, <_{P_2})$ with $x = (x_1, x_2)$, $y = (y_1, y_2) \in dom(A)$, the Pareto preference constructor $P := P_1 \otimes P_2 = (A_1 \times A_2, <_P)$ is defined as:

$$(x_1, x_2) <_p (y_1, y_2) \Leftrightarrow (x_1 <_{P_1} y_1) \wedge (x_2 <_{P_2} y_2 \vee x_2 = y_2)) \vee$$
$$(x_2 <_{P_2} y_2) \wedge (x_1 <_{P_1} y_1 \vee x_1 = y_1))$$

If we consider only LOWEST and HIGHEST preferences in Pareto, then this definition coincides with the one of *Skyline* queries, cp. [10, 14].

There are more complex preferences, e.g., in a *Prioritization preference* $P :=$ P_1 & P_2 the preference $P_1 = (A_1, <_{P_1})$ is more important than $P_2 = (A_2, <_{P_2})$ and a RANK$_F$ preference allows to construct weighted preferences. For a formal definition and more detailed information, we refer to [36, 38].

Example 2 Remember the introductory Example 1 where a user wants to filter high quality tweets. The mentioned conditions can now be modeled as soft constraints by using base preferences combined to Pareto:

$$P := \text{AT\_LEAST(user.followerscount, 10.000)} \otimes$$
$$\text{AT\_LEAST(user.statuscount, 10.000)} \quad \otimes$$
$$\text{POS(hashtag, \{\#fifa, \#soccer, \#football\})}$$

3.2 PreferenceSQL

The PreferenceSQL query language (cp. [38]) is a declarative extension of SQL by strict partial order preferences, behaving like soft constraints under the BMO query model described in Definition 1. Syntactically, PreferenceSQL extends the SELECT statement of SQL by an optional PREFERRING clause, cp. Fig. 1. The keywords SELECT, FROM, and WHERE are treated as the standard SQL keywords. The PREFERRING clause specifies a preference which is evaluated after the WHERE condition.

Example 3 The Pareto preference in Example 2 can be expressed in PreferenceSQL as follows, where ConfedCup is the data set which contains information about the *FIFA Confederations Cup 2017* in Russia.

```
SELECT * FROM ConfedCup
PREFERRING
    user.followers_count AT LEAST 10.000    PARETO
    user.status_count AT LEAST 10.000       PARETO
    hashtag IN ('#fifa', '#soccer', '#football');
```

Fig. 1 Simplified
PreferenceSQL query block

```
SELECT      ... <projection, aggregation>
FROM        ... <table_reference>
WHERE       ... <hard_conditions>
  PREFERRING ... <soft_conditions>
```

In PreferenceSQL IN expresses a POS preference. The keyword PARETO states a Pareto preference, and PRIOR TO would lead to a Prioritization (not shown).

Note that for the evaluation of preferences special algorithms are needed, cp. [14] for an overview.

4 Preference-Based Stream Processing

In this section we describe our *preference-based stream processing framework*, introduce the *Preference Continuous Query Language* (PCQL), and show how to find the BMO-set by using our *Stream-based Lattice Skyline* (SLS) algorithm.

4.1 The Preference-Based Stream Processing Framework

For the processing of preferences on data streams we exploit the University prototype of PreferenceSQL[2] [38, 55]. PreferenceSQL originally was developed to run queries against bounded data sets that are stored persistently in a relational database and for this it provides several database related evaluation and optimization techniques. Note that we rely on PreferenceSQL, because it is a well-known, approved, and high-performance software prototype. However, for analyzing continuous, unbounded data, it is necessary to extend the PreferenceSQL system in order to process data from streams.

We use *Apache Flink*,[3] an open source platform for scalable stream and batch data processing, to transform continuous data into a PreferenceSQL compatible and processable format. Figure 2 depicts our stream processing framework for preference evaluation.

The incoming data stream is processed by Apache Flink in an ETL (Extract, Transform, Load) process. Since streams are often encoded in various format, e.g., as JSON[4]-objects as in Twitter, data must be transformed into a PreferenceSQL

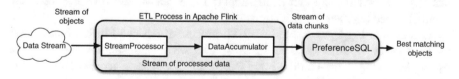

Fig. 2 Streaming architecture for preference analytics

---

[2]PreferenceSQL: http://www.preferencesql.com.

[3]Apache Flink: https://flink.apache.org/.

[4]JSON: http://www.json.org/.

```
{"created-at":"Thu Jul 03 14:44:05 +0000 2017", "id":77514,
"text":"Germany top #FIFA World Rankings after #ConfedCup triump",
"source":"Twitter for iPhone", ..., "user": {"id":228586199,
"name":"Mario Gomez , "screen-name":"mariohonduran10", "location":"South
Carolina","url":null , "description":"My biggest passion ,  soccer  ? the love
 of my life! #HalaMadrid #Honduras #Olimpia. San Pedro Sula, Honduras/SC,
USA #ACL surgery on both knees", "protected":false , "verified":false ,
"followers-count":9316, "friends-count":5, "listed-count":22,
"favourites-count":5916, "statuses-count":15866},...}
```

Fig. 3 Simplified JSON object of a tweet

readable format. For this one has to implement the mapping of a stream object to a relational structure inside the `StreamProcessor`, which provides a corresponding interface implementation. The data types of the fields can be extracted from the stream objects, e.g., by using the Twitter API. The `DataAccumulator` builds (finite) chunks of objects, which can then be processed by PreferenceSQL to find the best-matches-only set w.r.t. the preference specified by a user. This grouping can be *size* (how many objects are in one chunk) or *time* (the number of objects per chunk is determined by the time) based. For more details and a prototype implementation, we refer to [54, 55].

Example 4 Figure 3 shows a simplified JSON object of a twitter message. There are several different *fields* which describe a tweet in detail, e.g., *created_at*, *id*, *text*, *name*, *followers_count*, *status_count*, and many more. Our `StreamProcessor` converts such objects into more structured data, e.g., as shown in Table 1.

4.2   The Preference Continuous Query Language (PCQL)

In this section we propose a language for preference-based stream analysis. Our *Preference Continuous Query Language* (PCQL) is based on the PreferenceSQL query language described in Sect. 3.2 and in [54], but additionally has the STREAM keyword to emphasize that we connect or query a data stream. Note that there are also other approaches to stream analysis and querying, e.g., [2, 3, 32], each of them has SQL-like syntax and enhanced support for windows and ordering, but they cannot be used together with preferences.

As shown in Fig. 2, we exploit the ETL process of Apache Flink, which allows us to connect to different data streams. The *stream connection* can be done by implementing a pre-defined interface, which leads to a generic and flexible framework for arbitrary stream processing. In our current University prototype we provide connectors for Twitter data (`TwitterStream`), for the DAX stock market index (`StockStream`) as well as data from Flickr (`FlickrStream`).

By using a pre-defined stream connector or implementing one, a user connects to a stream as depicted in the syntax diagram in Fig. 4. The keywords CREATE STREAM introduce a stream connection with a user defined stream

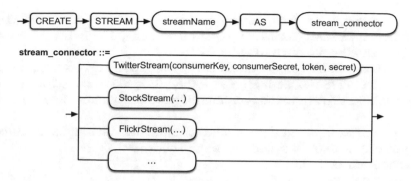

Fig. 4 Syntax diagram for the connection of data streams

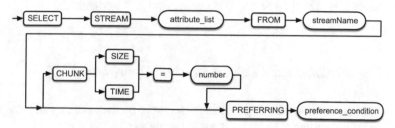

Fig. 5 Syntax diagram for preference stream queries

connection name streamName. The value of stream_connector corresponds to the name of a pre-implemented connector which, e.g., gets the user-specific login credentials as shown in the TwitterStream invocation in Fig. 4.

When connected to a stream, the data can be queried with the SELECT STREAM clause as depicted in Fig. 5. After SELECT STREAM one can specify a list of attributes attribute_list, where each attribute corresponds to a field in a stream object. The stream name streamName corresponds to the user defined name given in the CREATE STREAM statement. The kind and the quantity of the chunks, that are built from the endless stream, can be defined by the user. Thereby, the chunks can be size or time based. When specifying a time (in seconds), the DataAccumulator in the ETL process builds chunks based on the data retrieved within the given time slot. By specifying a size the ETL processor combines *size* objects to one chunk. PREFERRING introduces the preference conditions as in PreferenceSQL which will be evaluated on the chunks of the data stream.

Example 5 Remember Example 3 with the preference query on the Twitter stream. If we want to directly connect and query the stream data, we use the following statements, where the chunks are constructed time based in intervals of 60 s.

```
-- Connect to Twitter by using (invalid) credentials
CREATE STREAM ConfedCupTwitterStream AS TwitterStream
('NQEk0KbszVbaAcjCWLksbodkN',
 'HDKxlp2REOHvuq59oKrZZsdFovItwG6upOGJuSN4btr6npp2c3',
 '2400192752-DQSedtepr68SerQVyjHLzpHhMitcwJQfbvwxnLi',
 'BAk0krCYq77W4p45UwyuAuNnpR3nrv9WofO9PNL46YFch');

-- Preference query on the chunks of the data stream
SELECT * FROM ConfedCupTwitterStream
CHUNK TIME = 60
PREFERRING
   user.followers_count AT LEAST 10.000   PARETO
   user.status_count AT LEAST 10.000      PARETO
   hashtag IN ('#fifa', '#soccer', '#football');
```

4.3   The Stream-Based Lattice Skyline Algorithm (SLS)

In this section we show how to evaluate preferences on data streams and suggest a real-time lattice skyline algorithm for stream analysis.

4.3.1   Finding the BMO-Set of a Data Stream

Preference processing requires efficient evaluation algorithms, especially on time-oriented data streams. In addition, result computation must be adapted to stream properties since the dataflow is continuous and there is no "final" result after some data of the stream is processed. The result must be calculated and adjusted as soon as new data arrive, since new stream objects received later can match the user preferences better than the objects already recognized in previously computed BMO-sets.

To the best of our knowledge only Block-Nested-Loop style algorithms (e.g., BNL [10]) can be adapted to preference evaluation on continuous stream data. These algorithms are based on an object-to-object comparison approach and have a worst-case runtime complexity of $O(n^2)$, where n is the number of input objects. New objects have to be compared with all objects from the current BMO-set.

More detailed: Let $c_1, c_2, \ldots$ be the chunks provided by the ETL processor. A BNL-style algorithm evaluates the user preference P on the first chunk, i.e., $\sigma[P](c_1)$, cp. Eq. (1). Since c_2 could contain better objects we also have to compare the new objects from c_2 with the current BMO-set, i.e., we have to compute

$$\sigma[P](\sigma[P](c_1) \cup c_2), \tag{2}$$

and so on. However, this leads to a computational overhead if c_2 is large, which is the usual case. Therefore, this continuous comparing process is the most expensive operation of preference-based stream evaluation.

4.3.2 The SLS Algorithm

For preference stream evaluation we implemented the *Stream-based Lattice Skyline* algorithm (SLS), which avoids the annoying object-to-object comparison of BNL. SLS is based on the algorithms *Hexagon* [48] and *Lattice Skyline* [43], can be parallelized as shown in [18, 19], and has a linear runtime complexity. Our algorithm exploits the *lattice* induced by a Pareto preference over discrete domains to compute the best objects. Visualization of such lattices is often done using *Better-Than-Graphs* (BTG) (similar to Hasse diagrams), graphs in which edges state dominance. The nodes in the BTG represent *equivalence classes*. Each equivalence class contains the objects mapped the same feature vector constructed by a preference. All values in the same class are considered substitutable.

An example of a BTG is shown in Fig. 6. We write [2,4] to describe a two-dimensional domain as well as the maximal possible values of the feature vector representing preference values. For example, the BTG could present a Pareto preference on the *activity status* of a user ({active, non-active, unknown}) (values 0, 1, and 2) and the *hashtag*, which might be an element of {#fifa, #soccer, #football, #confed_cup, #others} (values 0, ... 4). The arrows show the dominance relationship between nodes of the lattice. The node (0,0) presents the *best node*, whereas (2,4) is the *worst node*. All gray nodes are occupied with an element of the data stream (in this case with an object of the first chunk), white nodes are empty.

The elements of the data stream that compose the (temporary) BMO set are those in the BTG that have *no path leading to them from another non-empty node*. In Fig. 6 these are the nodes (0,1) and (2,0). All other nodes have direct or transitive edges from these both nodes, and therefore are dominated. We exploited these

Fig. 6 Data stream processing with SLS

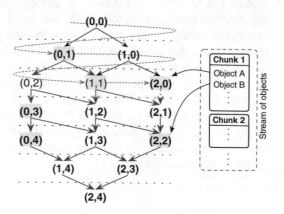

observations to develop an algorithm for efficient stream processing. Our approach in general works as follows:

1. *The* construction phase *initializes the data structure (the lattice) which depends on the Pareto preference and domain size (see [18] for details). Initially, all nodes of the lattice are marked as* empty.
2. *Adding phase*

   (a.) *Read the next chunk from the data stream.*
   (b.) *Iterate through the objects of the chunk. Each object will be mapped to one node in the lattice and this node is marked as* non-empty.

3. Removal phase: *After all objects in the chunk have been processed, the nodes of the lattice that are marked as* non-empty *and are not reachable by the transitive dominance relationship from any other* non-empty *node of the lattice represent the (temporary) BMO-set. From an algorithmic point of view this is done by a combination of* breadth-first traversal *(BFT) and* depth-first traversal *(DFT).*

   (a.) *Start a BFT at node* (0, 0) *to find non-empty and non-dominated nodes*
   (b.) *For each non-empty and non-dominated node start a DFT to mark nodes as* dominated, *if a transitive better-than relationship exists.*
   (c.) *The remaining nodes contain the temporary BMO-set and can be presented to the user*

4. *Go to Step 2.*

Note that the BMO-set computation in Step 3 can be done after an arbitrary number of processed chunks or after a pre-defined time. Therefore, our algorithm can be used for *real-time* preference evaluation. It is also possible to parallelize this approach in the sense of [18, 19]: After adding an object, directly start a DFT to mark nodes as *dominated*. In this case it is not necessary to "add" objects to the lattice if nodes are already marked as *dominated*.

Example 6 As mentioned, data streams are processed in finite chunks, as can be seen in Fig. 6. After constructing the lattice, which only must be done once, all objects of the current chunk (chunk 1) are mapped to the corresponding lattice nodes in a consecutive way. Assume all gray nodes in Fig. 6 are occupied with data from the first chunk. Afterwards we run a BFT to find the non-empty nodes (blue dashed line in Fig. 6). The first non-empty node is $(0, 1)$ and we start a DFT (red arrows) to mark all transitive dominated nodes as *dominated*. If the DFT reaches the bottom node $(2, 4)$ (or an already dominated node), it will recursively follow all other edges.

Afterwards the BFT continues with node $(1, 0)$, which is empty. The next non-empty node is $(1, 1)$, but dominated. Continue with $(2, 0)$. Since all other nodes are marked as dominated, the remaining nodes, $(0, 1)$ and $(2, 0)$, present the temporary BMO-set for the current chunk.

Now consider the next chunk (chunk 2). Read all objects form chunk 2, add them to the lattice, and perform a BFT and DFT. Again, the remaining nodes (maybe including the objects from the previous computation) contain the temporary best objects. Continue with chunk 3, and so on.

Note that all lattice based algorithms suffer from two restrictions: First, they are restricted to low-cardinality domains, and second, in general they only work with Pareto preferences which construct a lattice. To overcome the first restriction a method was developed by Morse et al. [43], Endres and Kießling [19] to remove one unrestricted domain in the BMO computation. The second restriction can be discarded by using the approach presented in [21] which allows the embedding of any strict partial order into a lattice structure.

5 Clustering of Pareto-Optimal Objects

In this section we show how to cluster Pareto-optimal objects on continuous data streams after a pre-defined time for each PCQL query. Before we explain our clustering framework, we introduce the most important basics and background knowledge of our approach. Afterwards we present our Borda social choice approach for clustering multi-dimensional objects.

5.1 Clustering Background

The most relevant work for our approach is the k-means clustering algorithm [31], which is an iterative partitioning algorithm with a convergence criterion. Hereby each object gets allocated to one of k clusters iteration by iteration, until a stable configuration is found. The k-means clustering approach is described in Algorithm 1 and Function 1 and generally works as follows:

1. *Find an initial partition for the cluster centroids by choosing a random point of set X for each of the k centroids in line 2.*
2. *Calculate for each point the distances to all centroids by Function 1. The point is being allocated to the closest centroid in line 9 of Algorithm 1, by using, e.g., the Euclidean distance (line 5 in Function 1).*
3. *Recalculate the centroids by averaging the contained points in line 14 of Algorithm 1.*
4. *Proceed with step 5.1 until two succeeding clusterings are stable, which means that all clusters from the last iteration contain the equal set of points as in the current iteration (Algorithm 1, line 6).*

Algorithm 1 k-means-clustering

Input: Set $X = \{x_i \mid i = 1, ..., n\}$ of d-dimensional points $x_i = (x_{i1}, ..., x_{id})$ of size n, set of k cluster centroids $C = \{c_j \mid j = 1, ..., k\}$.
Output: Sets $E_k = \{e_j \mid j = 0, ..., m\}$ of clustered d-dim. points $e_i = (e_{i1}, ..., e_{id})$.

  1: **function** K-MEANS(X, C)
  2:    $C \leftarrow random(X, k)$ / Random initial partition for k centroids
  3:    $E \leftarrow \emptyset$ / Current clustered set
  4:    $E' \leftarrow \emptyset$ / Clustered set from last iteration
  5:    $equals \leftarrow false$ / Variable for stop-criterion
  6:    **while** !$equals$ **do**   / Termination-criterion: Old Set equals current set
  7:        **for** $x_i \in X$ **do**   / Iterate through all points
  8:           $id \leftarrow getClosestCentroidId(x_i, C)$ / Get id of closest cluster
  9:           $E_{id} \leftarrow E_{id} \cup \{x_i\}$ / Add the current point to the closest centroid
10:        **end for**
11:        $equals \leftarrow checkClusterings(E, E')$ / Check if clusters are equals
12:        $E' \leftarrow E$ / Save the current clustering for next iteration
13:        **for** $E_i \in E$ **do**
14:           $c_i \leftarrow recalculateCentroid(E_i)$ / Recalculate centroids of each cluster
15:        **end for**
16:    **end while**
17:    **return** E
18: **end function**

Function 1 Closest centroid for traditional distances

Input: d-dim. point $x_i = (x_{i1}, ..., x_{id})$, set of k centroids $C = \{c_j \mid j = 1, ..., k\}$.
Output: id of the closest centroid, the point gets allocated to.

  1: **function** GETCLOSESTCENTROIDID(x_i, C)
  2:    $dist \leftarrow maxDistance$ / Distance to closest centroid
  3:    $id \leftarrow 0$
  4:    **for** $c_j \in C$ **do**   / Iterate through all clusters
  5:        $d \leftarrow distance(x_i, c_j)$ / Current distance between point and cluster
  6:        **if** $d < dist$ **then**
  7:           $dist \leftarrow d$ / Replace closest distance if current distance is closer
  8:           $id \leftarrow j$ / Replace id if current id has closer distance
  9:        **end if**
10:    **end for**
11:    **return** id
12: **end function**

5.2 The Borda Social Choice Voting Rule for Clustering

In [34] we presented an approach for clustering Pareto-optimal objects which was based on the Pareto-dominance criterion. However, this only works well for two-dimensional use cases. Since objects extracted from data streams contain several different dimensions, we need a more suitable approach for allocating objects to clusters than our former method.

We use the Borda social choice voting rule, because it represents a suitable decision criterion for the allocation of each point to one and only one cluster with taking account of equal importance of the considered dimensions as required by Pareto preferences. In our approach each dimension is considered self-contained and has the same influence in the form of a weighting on the cluster allocation. This overcomes the crucial influence on the clustering process of dimensions with extensive domains as in traditional distance-measures like the Euclidean distance.

5.2.1 The Borda Social Choice Voting Rule

In the Borda social choice rule each candidate receives equal weighted votes from each voter, as it is defined in [53].

Definition 3 (Borda Social Choice Voting Rule) Given m candidates C_i, $i = 1 \ldots m$ and d voters V_j, $j = 1 \ldots d$ where every voter votes for each candidate. Each voter V_j has to allocate the votings v_{jk} from a pairwise distinct set of $k = 0 \ldots m-1$.

After all voters assigned their votes, the votes for each candidate are summed up as it can be seen in Eq. (3), while the Borda winner is determined as mentioned in Eq. (4).

$$\text{bordasum}_{C_m} = \sum_{j=1}^{d} v_{jm} \tag{3}$$

$$\text{bordawinner}_{C_m} = \max\{\text{bordasum}_{C_m}\} \tag{4}$$

If we apply this approach to our clustering-framework, we replace the candidates with the available clusters and the voters with the d-dimensional stream objects, which should be allocated to one and only one cluster. Finally, each stream object votes for each cluster. After the voting, Eq. (3) determines the sum of all votes for each cluster, and Eq. (4) defines a cluster as the Borda winner, which got the most overall votes. Finally, the Borda social choice voting rule ensures that each dimension receives equal weighted votes, whether they have a large or a small domain. So small domains have a higher influence on the clustering using Borda social choice compared to traditional distance measures like Euclidean.

Note that we use the Jaccard similarity measure when clustering categorical domains in order to get a numerical representation, since the Borda social choice rule only works on numerical values.

5.2.2 Cluster Allocation

In order to realize a clustering approach based on the Borda social choice rule as decision criterion for the cluster allocation, we modify the k-means algorithm.

For this we replace line 8 of Algorithm 1 by Function 2. In Function 2 the Borda votes for each point x_i to each cluster is assigned as explained in Eq. (3). In order to sum up the votes for the clusters, an array is created in line 2. Thus the distances in each dimension between the considered point and the cluster centroid are calculated, which is performed in Function 3. For each dimension these distances are calculated and saved together with the *id* in an object-based data-structure in lines 6 and 7. After that the object is set on the current last position of the array in line 8 and sorted with *insertion sort* between lines 10 and 22 to the right position according to the distances (line 12), in order to avoid a further sorting if all objects would be added unsorted to the array.

While the last distance is sorted to the right position in the array, the votes for each cluster are determined (line 17) and summed up in the array (line 19) for the absolute votes, in order to avoid another iteration through the array afterwards. After the array with the summed up votes is returned from Function 3, it is sorted with *insertion sort* in line 3 of Function 2 in order to find the Borda winner as explained in Eq. (4). If there is only one Borda winner (line 5), return the *id* of the centroid.

Function 2 Closest centroid for Borda social choice voting rule

---

Input: d-dim. point x_i, clusterset C, cluster-id last iteration id_{last}.
Output: id of the closest cluster for point x_i.

```
 1: function GETBESTCLUSTERID(x_i, C, id_Last)
 2:    bordaVals[] ← calculateBordaVals(x_i, C)              / calculate Borda values
 3:    insertion_sort(bordaVals[])                            / Sort Borda values
 4:    counter ← 1
 5:    if bordaVals[0].val > bordaVals[1] then               / Only one Borda winner
 6:       return bordaVals[0].id                             / Return id of first element
 7:    end if
 8:    if bordaVals[0].val == bordaVals[1] then              / More than one Borda winner
 9:       while bordaVals[counter].val == bordaVals[0].val do
10:          if bordaVals[counter].id == id_last then        / Check id_last
11:             return id_last                               / Return id of element from last iteration
12:          end if
13:          counter ← counter + 1
14:          if counter >= bordaVals[].Size() then           / All centroids Borda winner
15:             return bordaVals[random(0, counter − 1)].id  / Return random id
16:          end if
17:       end while
18:    end if
19:    return bordaVals[random(0, counter − 1)].id           / Return random id
20: end function
```

We decided to add a specific decision criterion for the cluster allocation at the appearance of more than one Borda winner. Our approach looks back to the last iteration, in order to check if the centroid was allocated to one of the Borda winners before (Function 2, line 10). If not, one of the centroids *id*'s, which is represented in the Borda winners, is randomly chosen in line 19. Finally the k-means clustering continues in line 9 of Algorithm 1.

Function 3 Calculate the votes for each centroid

Input: d-dim. point x_i, clusterset C.
Output: bordaVals for each centroid.

```
 1: function CALCULATEBORDAVALS(x_i, C)
 2:     bordaVals[] ← bordaVals[C.size()]
 3:     for i = 0; x < x_i.getDims().size(); i ← i + 1 do
 4:         bordaObj[] ← bordaObj[C.size()]                          / Array for voters
 5:         for k = 0; k < C.size(); k ← k + 1 do                    / For each voter(centroid)
 6:             val =| x_i.getDims()[i] − c_j.getDims()[i] |         / Calculate distance
 7:             obj ← newbordaObj(val, c_j.id)                        / Init. object for voter
 8:             bordaObj[k] = obj                            / Set object on last available position
 9:             voteDiff ← k + 1                                     / Difference for Votes
10:             for j = k; j >= 0; j ← j − 1 do                      / Start insertion_sort
11:                 if j > 0 then                            / Check if not last item is sorted
12:                     if obj.val < bordaObj[j − 1].val then        / Compare obj. pairwise
13:                         swap(bordaObj[j − 1], bordaObj[j])        / Swap objetcs
14:                     end if
15:                 end if
16:                 if k == C.size() − 1 then                        / Last Element is sorted
17:                     val ← bordaObj[].size() − voteDiff           / Determine Votes
18:                     id ← bordaObj[j].id                          / Fetch id of Object to sum up
19:                     bordaVals[id].val ← bordaVals[id].val + val   / Sum up the value
20:                     voteDiff ← voteDiff − 1               / Reduce voteDiff for growing values
21:                 end if
22:             end for
23:         end for
24:     end for
25:     return bordaVals[]
26: end function
```

5.2.3 Complexity and Convergence

Our algorithm reaches a complexity of $\mathcal{O}(n \cdot (d \cdot c \cdot c^2 + c^2 + c))$ where n is the number of d-dimensional points that should be clustered in c clusters. For each point n the distances of each dimension d for each cluster c is calculated in $\mathcal{O}(d \cdot c)$. These distances are sorted by *insertion sort* with a complexity of $\mathcal{O}(c^2)$ for the worst case, while the list of the summed up votes is sorted with *insertion sort* in $\mathcal{O}(c^2)$, too. Finally the list is iterated one more time in order to check if there is more than one Borda winner in $\mathcal{O}(c)$. Hence we get a complexity of $\mathcal{O}(n \cdot d \cdot c^3)$.

After some preliminary tests, we found out that especially for growing numbers of dimensions, our approach is not terminating, which leads to alternating between two or more clusters. Therefore, we decided to adjust Step 1 of the k-means clustering algorithm (Algorithm 1, line 2) by using k-means++, in order to start the clustering process with a better starting position. The seeding technique of k-means++ speeds up the clustering process as published in [4] and reaches a higher accuracy as well. The feature of using the seeding technique of k-means++ ensures that our approach terminates in higher dimensions likewise, which will be documented and discussed in Sect. 7.

Another well-known problem in clustering is the appearance of empty clusters, e.g., if the initial partition of the cluster centroids are populated with very similar objects in the first step of k-means clustering. With k-means++ the probability of choosing objects which are duplicates or direct neighbors is minimized. In order to prevent the phenomenon of empty clusters, k-medoids is a very suitable approach.

6 Application Use Case

In this section we want to exemplify a whole run through our analyzing and clustering approach of Pareto-optimal objects in data streams.

Assume that the preference query in Example 1 exactly retrieves the Pareto-optimal objects shown in Table 1. We now want to apply our Borda approach to cluster these objects in order to learn some kind of "similarity" between these Pareto-optimal objects. For this, we apply k-means++ with Borda, use k=3 initial clusters, and exploit the lookback to the last iteration at the appearance of more than one Borda winner as explained in Sect. 5.

First of all we initialize the random centroids of k-means++ using the Borda rule as follows:

1. The Stream object 65230 is chosen randomly as centroid C_1. After that the distances D for each point to the current centroid C_1 are calculated as explained in Sect. 5. These computations can be seen in Table 2.
2. Then the next stream object is set as new centroid C_2. A determined random value of 65.026 is exceeded of the summed up squared distances at object 99142 while iterating through the object set.
3. For C_3 the stream object with ID 77514 is determined likewise C_2 after updating the distances to the previously chosen centroid shown in Table 2. If a distance is now smaller than a previous one, the current is set and the probability to be chosen decreases.

After the initialization we allocate objects to clusters using the Borda social choice rule as follows:

4. We show the allocation for the stream objects with IDs 81356 and 53614 to the initialized centroids C_1, C_2, and C_3, as it can be seen in Table 3 in the first iteration of the k-means clustering algorithm.

Table 2 Creating the initial cluster partition with kmeans++

Tweet.ID	$D(x)^2$	$\sum$	Tweet.ID	$D(x)^2$	$\sum$
76513	8.0360	8.0360	77514	50.0025	60.0410
81365	2.0025	10.0385	**99142**	18.0360	**78.0770**
65230	–	–	53614	32.0000	110.0770

Table 3 Cluster allocation for some stream-objects

	ID 81365			ID 53614		
	$D(C_1)$	$D(C_2)$	$D(C_3)$	$D(C_1)$	$D(C_2)$	$D(C_3)$
followers_count	1 (2)	2 (1)	4 (0)	4 (0)	1 (1)	1 (2)
status_count	1 (2)	2 (1)	4 (0)	4 (0)	1 (2)	1 (1)
Hashtag	0.05 (1)	0.11 (0)	0.00 (2)	0.00 (2)	0.06 (0)	1.05 (1)
$\sum$	**5**	2	2	2	3	**4**

5. For the stream object with ID of 81365 the distances D to centroid C_1 are the closest for both the dimensions of the followers count and the status count. Thus centroid C_1 receives a maximum possible vote of 2 for both distances. In the third dimension concerning the hashtags, C_1 is only the second closest centroid and thus receives only a voting of 1. Nevertheless C_1 receives a higher sum of five votes overall, while both C_2 and C_3 only get a total of 2 votes.

6. For the stream object with the ID of 53614 each dimension has a different closest centroid. Thus the second closest centroids are crucial for the overall voting. C_3 receives both a second highest voting of 1 from the dimensions of the status count and the hashtags and finally wins the Borda voting with a concise advance of 4 against C_2, which receives a total of three votes.

7. If there would be a draw in the summed up votings between two or more centroids in one iteration, i.e, we have more than one Borda winner, the cluster ID of the previous iteration would tip the balance to one of these Borda winner centroids if the ID of the previous Borda winner is represented in the current set of the Borda winners. Otherwise this object will be allocated randomized to one of the centroids.

8. The algorithm continues with the recalculation of the centroids and the check of the termination criterion as mentioned in Sect. 5.

The final clusterings after the second iteration can be seen in Figs. 7 and 8. In these figures the dimension for user.status_count is chosen as fixed dimension for all other dimensions in order to compare them easily. Finally in the two-dimensional illustrations several clusterings around the centroids C_1, C_2, and C_3 can be seen. Note that we used some discretization of the domain values for a better overview. Figure 7 shows that the stream objects got clustered convenient regarding the dimensions user.status_count and user.followers_count. Regarding the dimensions user.followers_count and hashtag the stream objects got clustered intuitively as well as it can be seen in Fig. 8.

7 Experiments

In this section we present several experiments for our SLS algorithm and our Borda social choice clustering approach. For our experiments we used commodity hardware with a 2.53 GHz CPU and 16 GB RAM for the Java virtual machine. All algorithms are implemented in Java.

Fig. 7 Final clustering w.r.t
the dimensions
user.followers_count and
user.status_count

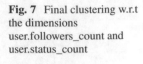
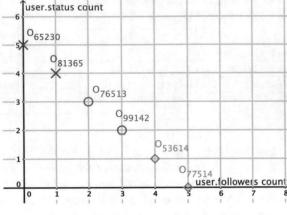

Fig. 8 Final clustering w.r.t
the dimensions
user.followers_count and
hashtags

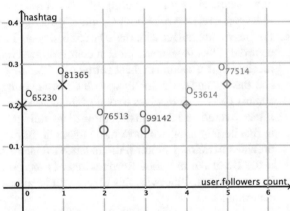

7.1 Benchmarks for Stream Lattice Skyline Algorithm

This section provides experiments on our SLS algorithm.

7.1.1 Experiments on Artifical Data

For our experiments on artificial data we generated anticorrelated data as described
in [10], because this kind of data is most challenging for Pareto queries. We wanted
to investigate how the chunk size affects the runtime of our algorithm, since this
influences the real-time behavior of SLS. In our experiments we varied the chunk
sizes from 10 to 100000 objects and tested the algorithm on an input data set of
100000 objects.

In Fig. 9 we evaluated the behavior of SLS for queries with different domains:
[2856,1,5], [2,3,5,10,10] and [2,3,7,8,4,10]. Remember, each number corresponds

Fig. 9 Runtime for different
Pareto queries

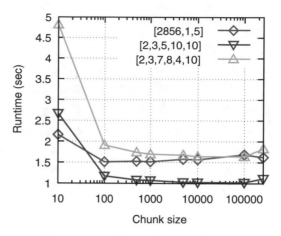

Fig. 10 Runtime for 100000
objects

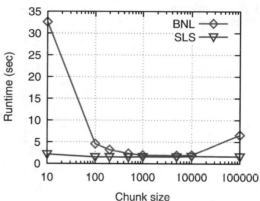

to the maximal possible values of the single domains. The algorithm is significantly slower for small chunks (up to 100 objects) than for chunks with more than 100 objects. This can be explained by the frequent repeating of the breadth-first and depth-first traversal in SLS which have to be carried out for each chunk. For the chunk size over 10000 objects the runtime of SLS increases slightly, because the adding of new objects to the BTG (Phase 2 in SLS, cp. Sect. 4.3.2) is more expensive. Therefore we claim that the optimal chunk size for the best runtime is between 100 and 10000 objects.

In Fig. 10 we used a Pareto preference query consisting of five LOWEST preferences on the domain [2,3,5,10,10]. Such a setting is typical for " real world" Pareto queries: A few categorical preferences with small maximal values (e.g., POS/POS or POS/NEG) are combined with some numerical preferences with larger domains (AROUND, AT_MOST etc).

We compared SLS to BNL mentioned in Sect. 4.3.1. The SLS algorithm outperforms BNL, but for small chunks (up to 500 objects) and very large chunks (over 10000 objects), the difference is much more significant.

Fig. 11 Runtime for 10000
objects

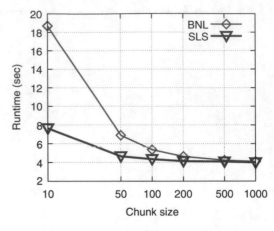

Fig. 12 Runtime for 100000
objects

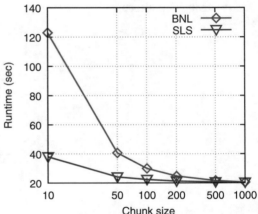

For more comprehensive experiments on synthetic data we refer to [18] and [19],
since our algorithm is an extended version of the lattice algorithms presented in
these publications.

7.1.2 Experiments on Real World Data

We also performed experiments on real-world data from Twitter. We varied the
chunk sizes from 10 to 1000 objects and the size of the input stream. Our test query
is based on the preference given in Example 1.

Figures 11 and 12 show our results, which are similar to the results of our
previous experiments. For small chunk sizes up to 200 objects our SLS algorithm is
much better than BNL. From a chunk size of 500 objects on, BNL is nearly as good
as SLS, but still worse. This can be explained by the less number of unions which
has to be carried out after each chunk evaluation, cp. Eq. (2).

Since real-time processing requires high efficient algorithms on few data objects in very short time intervals, our SLS algorithm is superior for real-time preference evaluation as found out in our experiments.

7.2 Benchmarks for Borda Social Choice Clustering

For our clustering experiments we created anti-correlated sets of multidimensional objects and varied the number of dimensions, the number of objects per set, and the number of desired clusters. Synthetic data allows us to carefully explore the behavior of our Borda social choice clustering approach.

We investigated runtime and number of iterations of our approach compared to the basic k-means algorithm using the Euclidean norm as distance measure. In order to set the focus on distances using small domains as well, the Canberra distance is a very suitable measure. It sums up the absolute fractional distances of two d-dimensional points in relation to the range of the focussed dimensions. Furthermore results of our experiments with k-means++ are also presented.

7.2.1 Runtime

In the following experiments we investigated the runtime of our Borda algorithm.

In the 3-dimensional testrow in Fig. 13 for growing number of clusters and sets of input objects (5000, 10000, 15000) the runtime is growing, too. Our approach (Borda) works in equal time compared to k-means with Euclidean (Eucl.) and Canberra (Canb.) for small numbers of clusters. For 7 and 9 clusters our approach is slower independent of the number of input objects because of a higher complexity of our approach. Benefits of a faster runtime for k-means++ (Borda++) is hardly recognizable in most cases.

Figure 14: For growing numbers of dimensions our approach reaches a better runtime compared to the Euclidean distance except for high number of clusters in all benchmarked sets of 5000, 10000, and 15000 object. In some cases our approach terminates faster than k-means using Canberra, e.g., the testrow with seven clusters, but all in all our approach mostly reaches an equal runtime in a 5-dimensional space.

A similar behavior illustrates the testrow for a 9-dimensional set of objects in Fig. 15. While both our approaches terminate in similar time compared to k-means with Canberra for 3 and 5 clusters, they are a lot faster than k-means with Euclidean distance. The trends for growing runtimes w.r.t. the number of clusters and objects can be noticed in 9-dimensional space, too.

Fig. 13 Runtime for 3 dimensions

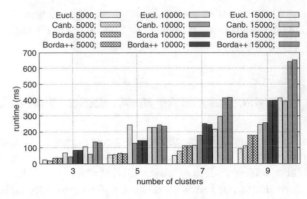

Fig. 14 Runtime for 5 dimensions

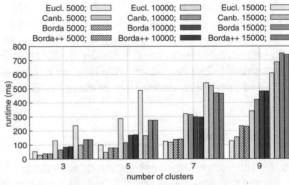

Fig. 15 Runtime for 9 dimensions

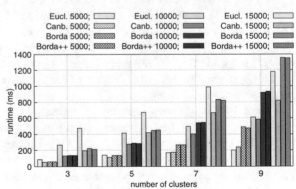

7.2.2 Iterations

For growing numbers of clusters and growing numbers of objects per set the numbers of needed iterations until termination is ascending for all dimensions as it can be seen in Figs. 16, 17 and 18.

Both our versions of k-means reach a stable clustering in clearly less iterations, especially for higher number of clusters and bigger sets of objects. While the number of iterations is growing fast for growing numbers of dimensions for k-means using

Fig. 16 Iterations for 3 dimensions

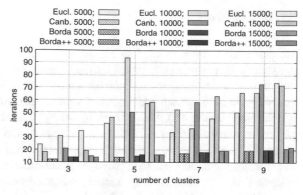

Fig. 17 Iterations for 5 dimensions

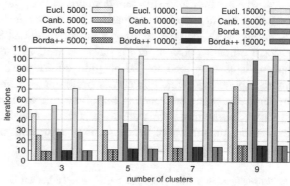

Fig. 18 Iterations for 9 dimensions

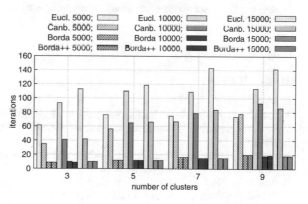

Euclidean and Canberra, our Borda approach only needs less additional iterations compared to smaller dimensions. K-means++ has only small effects on the iterations of k-means with Borda social choice for the cluster allocation.

Finally our approach works at most in equal time compared to k-means with both of the other distances, but for higher numbers of clusters it needs more time until termination because of the higher complexity of the Borda voting rule. Moreover our approaches need only a fractional part of iterations until termination for all testrows.

8 Conclusion

We presented a novel method to analyze data streams which are common in our daily lives. Our approach is based on the well-known concept of user preferences and profiles and allows personalized stream processing in order to find only relevant and valuable information. We proposed a preference stream processing framework which supports the PCQL language to query data streams as well as the SLS algorithm for real-time preference evaluation. In the case of too many Pareto-optimal objects we apply a clustering approach based on the Borda social choice rule for cluster allocation. Compared to traditional clustering approaches, Borda ensures that each dimension is treated as equal important as a Pareto preference does, whether they have different domains or not. Our experiments have shown the benefit of this approach in comparison to standard techniques.

Future work includes the development of the SLS algorithm to handle unrestricted domains, e.g., as in [19, 22], as well as the implementation of a top-k approach as described in [20, 47]. In addition we want to develop an open-source application which combines preference stream analytics with data stored in databases which still play an important role to create efficient decision support systems. Furthermore we want to perform a user-based research study in order to consider the quality of the clusterings compared to clustering-algorithms using traditional distance measures.

Acknowledgements This work has been partially funded by the German Federal Ministry for Economic Affairs and Energy according to a decision by the German Bundestag, grant no. ZF4034402LF5.

References

1. Arasu, A., Babcock, B., Babu, S., Datar, M., Ito, K., Nishizawa, I., Rosenstein, J., Widom, J.: Stream: the stanford stream data manager. In: SIGMOD '03, pp. 665–665. ACM, New York (2003)
2. Arasu, A., Babu, S., Widom, J.: CQL: A Language for Continuous Queries over Streams and Relations, pp. 1–19. Springer, Berlin (2004)
3. Arasu, A., Babu, S., Widom, J.: The CQL continuous query language: semantic foundations and query execution. VLDB J. **15**(2), 121–142 (2006)
4. Arthur, D., Vassilvitskii, S.: K-means++: the advantages of careful seeding. In: Proceedings of the Eighteenth Annual ACM-SIAM, SODA '07, pp. 1027–1035. Society for Industrial and Applied Mathematics, Philadelphia, PA (2007)
5. Babcock, B., Babu., S., Datar, M., Motwani, R., Widom, J.: Models and issues in data stream systems. In: PODS '02, pp. 1–16. New York (2002)
6. Babu, S., Widom, J.: Continuous Queries over Data Streams. SIGMOD Rec. **30**(3), 109–120 (2001)
7. Baruah, R.D., Angelov, P., Baruah, D.: Dynamically evolving clustering for data streams. In: EAIS '14 IEEE, pp. 1–6 (2014)
8. Bezerra, C.G., Costa, B.S.J., Guedes, L.A., Angelov, P.P.: A new evolving clustering algorithm for online data streams. In: EAIS '16, pp. 162–168 (2016)

9. Bonnet, P., Gehrke, J., Seshadri, P.: Towards sensor database systems. In: MDM '01, pp. 3–14. Springer, London (2001)
10. Börzsönyi, S., Kossmann, D., Stocker, K.: The skyline operator. In: ICDE '01, pp. 421–430. IEEE, Washington, DC (2001)
11. Boutilier, C., Brafman, R.I., Domshlak, C., Hoos, H.H., Poole, D.: CP-nets: A tool for representing and reasoning with conditional Ceteris Paribus preference statements. J. Artif. Intell. Res. **21**, 135–191 (2004)
12. Chen, J., DeWitt, D.J., Tian, F., Wang, Y.: NiagaraCQ: a scalable continuous query system for internet databases. In: SIGMOD '00, pp. 379–390. ACM, New York (2000)
13. Chomicki, J.: Preference formulas in relational queries. In: TODS '03: ACM Transactions on Database Systems, vol. 28, pp. 427–466. ACM Press, New York, NY (2003)
14. Chomicki, J., Ciaccia, P., Meneghetti, N.: Skyline queries, front and back. SIGMOD **42**(3), 6–18 (2013)
15. de Andrade Silva, J., Hruschka, E.R., Gama, J.: An evolutionary algorithm for clustering data streams with a variable number of clusters. Expert Syst. Appl. **67**, 228–238 (2017)
16. Döring, S., Preisinger, T., Endres, M.: Advanced preference query processing for E-commerce. In: SAC '08 Proceedings of the 2008 ACM Symposium on Applied Computing, pp. 1457–1462. ACM, New York, NY (2008)
17. Dovžan, D., Logar, V., Škrjanc, I.: Implementation of an evolving fuzzy model (eFuMo) in a monitoring system for a waste-water treatment process. IEEE Trans. Fuzzy Syst. **23**(5), 1761–1776 (2015)
18. Endres, M., Kießling, W.: High parallel skyline computation over low-cardinality domains. In: Proceedings of ADBIS '14, pp. 97–111. Springer, Berlin (2014)
19. Endres, M., Kießling, W.: Parallel skyline computation exploiting the lattice structure. J. Database Manag. **26**, 18–43 (2016)
20. Endres, M., Preisinger, T.: Behind the skyline. In: DBKDA '15. IARIA (2015)
21. Endres, M., Preisinger, T.: Beyond skylines: explicit preferences. In: DASFAA '17, pp. 327–342. Springer International Publishing, Cham (2017)
22. Endres, M., Roocks, P., Kießling, W.: Scalagon: an efficient skyline algorithm for all seasons. In: DASFAA '15 (2015)
23. Faria, E.R., Gonçalves, I.J.C.R., de Carvalho, A.C.P.L.F., Gama, J.: Novelty detection in data streams. Artif. Intell. Rev. **45**(2), 235–269 (2016)
24. Ferligoj, A., Batagelj, V.: Direct multicriteria clustering algorithms. J. Classif. **9**(1), 43–61 (1992)
25. Gama, J.: Clustering from Data Streams, pp. 226–231. Springer US, Boston, MA (2017)
26. Golfarelli, M., Rizzi, S.: Expressing OLAP preferences. In: SSDBM '09, SSDBM 2009, pp. 83–91. Springer, Berlin (2009)
27. Gu, X., Angelov, P.P.: Autonomous data-driven clustering for live data stream. In: SMC '16 IEEE, pp. 1128–1135 (2016)
28. Huang, Z., Xiang, Y., Zhang, B., Liu, X.: A clustering based approach for skyline diversity. Expert Syst. Appl. **38**(7), 7984–7993 (2011)
29. Hyde, R., Angelov, P., MacKenzie, A.: Fully online clustering of evolving data streams into arbitrarily shaped clusters. Inf. Sci. **382–383**, 96–114 (2017)
30. Ienco, D., Bifet, A., Žliobaitė, I., Pfahringer, B.: Clustering Based Active Learning for Evolving Data Streams, pp. 79–93. Springer, Berlin (2013)
31. Jain, A.K.: Data clustering: 50 years beyond K-means. Pattern Recogn. Lett. **31**(8), 651–666 (2010)
32. Jain, N., Mishra, S., Srinivasan, A., Gehrke, J., Widom, J., Balakrishnan, H., Cetintemel, U., Cherniack, M., Tibbetts, R., Zdonik, S.B.: Towards a streaming SQL standard. PVLDB **1**(2), 1379–1390 (2008)
33. Kasabov, N.K., Song, Q.: DENFIS: dynamic evolving neural-fuzzy inference system and its application for time-series prediction. IEEE Trans. Fuzzy Syst. **10**(2), 144–154 (2002). https://doi.org/10.1109/91.995117

34. Kastner, J., Endres, M., Kießling, W.: A Pareto-dominant clustering approach for Pareto-frontiers. In: DOLAP Workshops of EDBT/ICDT '17, Venice (2017)
35. Kießling, W.: Foundations of Preferences in Database Systems. In: Proceedings of VLDB '02, pp. 311–322. VLDB, Hong Kong (2002)
36. Kießling, W.: Foundations of Preferences in Database Systems. In: VLDB '02, pp. 311–322. VLDB Endowment, Hong Kong SAR (2002)
37. Kießling, W.: Preference queries with SV-semantics. In: COMAD '05, pp. 15–26. Computer Society of India, Goa (2005)
38. Kießling, W., Endres, M., Wenzel, F.: The Preference SQL system - an overview. IEEE Comput. Soc. Bull. Techn. Commitee Data Eng. **34**(2), 11–18 (2011)
39. Kontaki, M., Papadopoulos, A.N., Manolopoulos, Y.: Continuous processing of preference queries in data streams. In: SOFSEM '10, pp. 47–60. Springer, Berlin, Špindlerův Mlýn, Czech Republic (2010)
40. Krempl, G., Žliobaite, I., Brzeziński, D., Hüllermeier, E., Last, M., Lemaire, V., Noack, T., Shaker, A., Sievi, S., Spiliopoulou, M., Stefanowski, J.: Open challenges for data stream mining research. SIGKDD '14 Explor. Newsl. **16**(1), 1–10 (2014)
41. Lee, Y.W., Lee, K.Y., Kim, M.H.: Efficient processing of multiple continuous skyline queries over a data stream. Inf. Sci. **221**, 316–337 (2013)
42. Lughofer, E., Sayed-Mouchaweh, M.: Autonomous data stream clustering implementing split-and-merge concepts towards a plug-and-play approach. Inf. Sci. **304**, 54 – 79 (2015)
43. Morse, M., Patel, J.M., Jagadish, H.V.: Efficient skyline computation over low-cardinality domains. In: VLDB '07, pp. 267–278 (2007)
44. Okazaki, M., Matsuo, Y.: Semantic twitter: analyzing tweets for real-time event notification. In: BlogTalk. Lecture Notes in Computer Science, vol. 6045, pp. 63–74. Springer, Berlin (2009)
45. Pohl, D., Bouchachia, A., Hellwagner, H.: Online indexing and clustering of social media data for emergency management. Neurocomputing **172**(C), 168–179 (2016)
46. Pratama, M., Anavatti, S.G., Er, M.J., Lughofer, E.: pclass: an effective classifier for streaming examples. IEEE Trans. Fuzzy Syst. **23**(2), 369–386 (2015)
47. Preisinger, T., Endres, M.: Looking for the best, but not too many of them: multi-level and top-k skylines. Int. J. Adv. Softw. **8**(3, 4), 467–480 (2015)
48. Preisinger, T., Kießling, W.: The Hexagon algorithm for evaluating Pareto preference queries. In: MPref '07 (2007)
49. Railean, C., Moraru, A.: Discovering popular events from tweets. In: SiKDD '13. Ljubljana (2013)
50. Ribeiro, M.R., Barioni, M.C.N., de Amo, S., Roncancio, C., Labbé, C.: Reasoning with temporal preferences over data streams. In: Florida Artificial Intelligence Research Society Conference (FLAIRS '17), Marco Island
51. Roocks, P., Endres, M., Huhn, A., Kießling, W., Mandl, S.: Design and implementation of a framework for context-aware preference queries. J. Comput. Sci. Eng. **6**(4), 243–256 (2012)
52. Roocks, P., Endres, M., Mandl, S., Kießling, W.: Composition and efficient evaluation of context-aware preference queries. In: DASFAA '12: Proceedings of the 17th International Conference on Database Systems for Advanced Applications (2012)
53. Rossi, F., Venable, K.B., Walsh, T.: A Short Introduction to Preferences Between Artificial Intelligence and Social Choice. Morgan & Claypool Publishers, San Rafael (2011)
54. Rudenko, L., Endres, M.: Personalized stream analysis with PreferenceSQL. In: PPI Workshop of BTW '17, pp. 181–184, Stuttgart (2017)
55. Rudenko, L., Endres, M., Roocks, P., Kießling, W.: A preference-based stream analyzer. In: Streamvolv Workshop of ECML PKKD'16 (2016)
56. Sankaranarayanan, J., Samet, H., Teitler, B.E., Lieberman, M.D., Sperling, J.: Twitterstand: news in tweets. In: ACM '09, pp. 42–51 (2009)
57. Silva, J.A., Faria, E.R., Barros, R.C., Hruschka, E.R., de Carvalho, A., Gama, J.: Data stream clustering: a survey. ACM Comput. Surv. **46**(1) (2013). https://doi.org/10.1145/2522968. 2522981

58. Stefanidis, K., Koutrika, G., Pitoura, E.: A survey on representation, composition and application of preferences in database systems. ACM TODS **36**(3), 19:1–19:45 (2011)
59. Truong, D.T., Battiti, R.: A flexible cluster-oriented alternative clustering algorithm for choosing from the Pareto front of solutions. Mach. Learn. **98**(1), 57–91 (2015)
60. Wenzel, F., Endres, M., Mandl, S., Kießling, W.: Complex preference queries supporting spatial applications for user groups. PVLDB **5**(12), 1946–1949 (2012)

Error-Bounded Approximation of Data Stream: Methods and Theories

Qing Xie, Chaoyi Pang, Xiaofang Zhou, Xiangliang Zhang, and Ke Deng

Abstract Since the development of sensor network and Internet of Things, the volume of data is rapidly increasing and the streaming data has attracted much attention recently. To efficiently process and explore data streams, the compact data representation is playing an important role, since the data approximations other than the original data items are usually applied in many stream mining tasks, such as clustering, classification, and correlation analysis. In this chapter, we focus on the maximum error-bounded approximation of data stream, which represents the streaming data with constrained approximation error on each data point. There are two criteria for the approximation solution: self-adaption over time for varied error bound and real-time processing. We reviewed the existing data approximation techniques and summarized some essential theories such as optimization guarantee.

Reprinted by permission from Springer Nature: Springer, The VLDB Journal, Maximum error-bounded piecewise linear representation for online stream approximation, Q. Xie et al., ©Springer-Verlag Berlin Heidelberg 2014 (https://doi.org/10.1007/s00778-014-0355-0).

Q. Xie (✉)
Wuhan University of Technology, Wuhan, China
e-mail: felixxq@whut.edu.cn

C. Pang
Ningbo Institute of Technology, Zhejiang University, Ningbo, China

X. Zhou
University of Queensland, Brisbane, QLD, Australia

Soochow University, Suzhou, China
e-mail: zxf@itee.uq.edu.au

X. Zhang (✉)
CEMSE, King Abdullah University of Science and Technology (KAUST), Thuwal, Kingdom of Saudi Arabia
e-mail: xiangliang.zhang@kaust.edu.sa

K. Deng
RMIT University, Melbourne, VIC, Australia
e-mail: ke.deng@rmit.edu.au

© Springer International Publishing AG, part of Springer Nature 2019   93
M. Sayed-Mouchaweh (ed.), *Learning from Data Streams in Evolving Environments*, Studies in Big Data 41, https://doi.org/10.1007/978-3-319-89803-2_5

Two optimal linear-time algorithms are introduced to construct error-bounded piecewise linear representation for data stream. One generates the line segments by data convex analysis, and the other one is based on the transformed space, which can be extended to a general model. We theoretically analyzed and compared these two different spaces, and proved the theoretical equivalence between them, as well as the two algorithms.

1 Introduction

Since the development of information system and sensor networks, especially the increasing attention on Internet of Things (IOT) [1, 17], streaming data is attracting much attention and has now been involved in various applications, including the medical data recording the status of patients [23, 28], financial data denoting the changing trends of stock prices [29], the large amount of data during network communications [18], and those scientific data such as sun spot numbers and ocean surface temperatures. The booming web techniques and the widely popularized mobile devices bring in the impressive popularity of web data access and sharing [30, 31], which further increase the production and proliferation of streaming data. The rapid growth of streaming data, together with its high significance in the area of health, finance, entertainment, and communication in daily life, has desired novel techniques and advanced systems to manage and process the streaming data efficiently.

Characterized as continuous, random, varying, and rapid in arrival, streaming data in the data stream model differ from conventionally stored data in several ways [2]:

- The data elements in the stream arrive online.
- The system has no control over the order in which data elements arrive to be processed, either within a data stream or across data streams.
- Data streams are potentially unbounded in size.
- Once an element from a data stream has been processed, it is discarded or archived, so that it cannot be retrieved easily unless it is explicitly stored in memory, which typically is small relative to the size of the data streams.

Since the streaming data have high input rate, and are potentially unbounded in size, the query and operation on streaming data usually require approximate results. Therefore it is necessary to find a proper and compact form to represent the data stream, so that the query and operation on streaming data can be more efficiently solved from the compact representations. In practical applications, the dynamics of data such as trends, patterns, and outliers are most attractive. In this regard, many techniques such as histogram, line segment, and wavelet are employed for effective data approximation and efficient stream query processing [9, 11, 12, 33].

The intensive work on stream approximation research is the size-bounded representation [5, 8]. Its objective is to construct a prescribed number of representations that minimize the approximation error under a specified metric, where L_2 norm (i.e.,

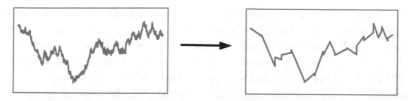

Fig. 1 An example of PLR on a time series data stream

Euclidean Distance) is mostly used. However, there are two main drawbacks on the size-bounded constraint and the use of L_2 norm for approximation: firstly, the size-bounded constraint lacks the ability to generate error-guaranteed representations for streaming data since the stream is naturally unbounded in size; secondly, the use of L_2 norm leads to the inability of controlling the approximation error on individual stream data items. To alleviate these drawbacks, researchers have made efforts in constructing the representations with guaranteed maximum allowable approximate error on each data point (L_∞ norm), which is termed the error-bounded representations. It has been engaged in many real-world applications, such as continuous queries over data streams [15, 19], sensor network management [25, 35], and monitoring physiological data for surgery operations [23, 36].

The use of line segments to represent a time series data stream, termed *Piecewise Linear Representation* (PLR) [4, 6, 16, 33], has been extensively studied for decades under different criteria [22]. The idea of PLR is to represent a complicated wave-like data stream with a number of simple line segments, so that the streaming data can be efficiently archived, and a query on the stream can be approximately answered by a query on the line segments. Compared with the stream itself, the line segments constructed from PLR provide striking visual outlines of stream trends and can be more efficiently processed and represented in the database (Fig. 1). These advantages make PLR the most popular representation technique for the data stream [13], and it has been widely applied to support date indexing [5, 14, 26], similarity search [29, 37], and correlation analysis [32, 35].

Histogram is also employed as contemporary error-bounded representation, i.e., Piecewise Constant Representation (PCR). For error-bounded PCR, Lazaridis et al. [15] provided an optimal algorithm to approximate sensor data stream. Gandhi et al. [7] proposed GAMPS that compressed multi-stream with guaranteed L_∞ error as well as a notable worst-case quality of approximation. They also extended the framework to address amnesic approximation and out-of-order approximation [8] using bucket-merging. As the PCR uses constant values (i.e., horizontal line segments) in representing a wave-like stream, it fails to reflect the trends of streaming data. Regarded as a generalized version of error-bounded PCR, the error-bounded PLR usually achieves a greater compression ratio and has more superiority for advanced applications than PCR.

In comparison to wavelet-based methods [10], PLR is more visually comprehensive and convenient for stream queries. For example, the line segments generated by error-bounded PLR on a stream can denote the stream trends and be used *directly* for answering trend queries. Such queries are crucial in monitoring patients in intensive care as medical specialists [21] believe that more accurate and earlier notifications of adverse events can be predicted from the accumulated trends and variations of the physiological streams. Even though error-bounded wavelet representations can be constructed very efficiently and effectively [23, 24], the wavelet synopses may not be ready to answer stream trends directly.

In this chapter, we introduce the state-of-the-art optimal algorithms to generate the error-bounded PLR for data stream, which aim to construct the smallest number of line segments with approximation error constrained on each data point. We will investigate the algorithm designed in time domain named as OptimalPLR [33], and also another algorithm proposed in an interesting parameter space, which is named as ParaOptimal [20]. Such innovative transformed space has been applied in the recent research for advanced applications [27, 34]. Theoretically, the optimal results can be achieved by greedy mechanism. In order to adjust a line segment to approximate as many stream points as possible, the general idea is to determine the range of all feasible line segments, which is incrementally updated during the processing of consecutive sequence points. Whenever the current point cannot be approximated within error bound, a line segment can be determined.

Both ParaOptimal and OptimalPLR are optimal solutions with linear time complexity for error-bounded PLR problem, but derived from different spaces, which provide essential and new sight on this classic problem. We further theoretically compare these two algorithms and the spaces they are based on, so as to provide deeper and more theoretical understanding for this problem. By setting up a mapping between the point and line in these two spaces, we theoretically proved the equivalence of these two spaces, and further linked the two algorithms together, which explained the optimal solution from different views.

2 Preliminary

In this section, along with some notations and new concepts for the study of error-bounded PLR, we formally provide the problem definition and the objective addressed in this chapter. We also present Theorem 1 which guarantees that the smallest number of line segments for the error-bounded PLR can be achieved by maximizing each representative line segment. The general notations used throughout this chapter are summarized in Table 1.

Let $S = \langle s_1, s_2, \ldots, s_k, \ldots \rangle$ denote a data stream where each point $s_i = (x_i, y_i)$ designates the actual value y_i at time stamp x_i. We use $S[i, j] = \langle s_i, s_{i+1}, \ldots, s_j \rangle$ to denote the stream *fragment* on time slot $[x_i, x_j]$ $(i < j)$ and $|S[i, j]| = j - i + 1$ to denote the cardinality of $S[i, j]$.

As an approximation technique, PLR approximates S with line segments. A *line segment* on time slot $[x_i, x_j]$ is representable by the linear function $y = a \cdot x + b$ for $x \in [x_i, x_j]$ with two parameters: slope a and offset b. Since a line can be

Table 1 Notations

Symbol	Description
δ	Error bound for approximation (>0)
$S = \langle s_1, s_2, \ldots, s_k, \ldots \rangle$	A data stream
$s_i = (x_i, y_i)$	Data point s_i at time stamp x_i with value y_i
$S[i, j] = \langle s_i, s_{i+1}, \ldots, s_j \rangle$	A (stream) fragment from time x_i to x_j
$\underline{s_i} = (x_i, y_i - \delta)$	Data point with deleted tolerant error
$\overline{s_i} = (x_i, y_i + \delta)$	Data point with added tolerant error
$seg[i, j]$	A δ-representative line on time slot $[x_i, x_j]$
$slp[i, j]$	The slope of $seg[i, j]$
$\underline{slp[i, j]}$ or $\overline{slp[i, j]}$	The minimum or maximum slopes of all $slp[i, j]$
$line(s_i, s_j)$	Line (segment) that passes point s_i and s_j
$line(\rho, s)$	Line (segment) with slope ρ that passes point s
$slope(s_i, s_j)$	The slope of $line(s_i, s_j)$
$\underline{cvx_k}, \overline{cvx_k}$	Reduced convex hulls for $S[1, k]$.
	$\underline{cvx} / \overline{cvx}$ bulge downward/upward

determined by its slope and a line point or two different line points alternatively, for the convenience of presentation, we also use $line(\rho, s)$ to denote the line that passes point s with slope ρ, and $line(s_i, s_j)$ to denote the line that passes the two points s_i and s_j. With this notation, we use $slope(s_i, s_j)$ as the slope of $line(s_i, s_j)$.

We define a stream fragment $S[i, j]$ ($i < j$) as *δ-representable* (at time slot $[t_i, t_j]$) if there exists a line segment identified by $y'_h = a \cdot x_h + b$, such that $|y'_h - y_h| \leq \delta$ holds for each point $s_h = (x_h, y_h)$ of $S[i, j]$ ($i \leq h \leq j$). In this situation, such line segment is defined as a *δ-representative* for fragment $S[i, j]$. (Notice that there can be more than one δ-representative.) For simplicity, we use $seg[i, j]$ to represent the δ-representative, and $slp[i, j]$ to denote the slope of $seg[i, j]$. Here, we use the maximum error metric L_∞ to guarantee the approximation quality at each stream data point. Furthermore, if fragment $S[i, j]$ is δ-representable but $S[i, j + 1]$ is not, then fragment $S[i, j]$ is *maximally* δ-representable on time slot $[t_i, t_j]^1$ and accordingly its $seg[i, j]$ is called maximal δ-representative.

With these designations, the *Error-bounded Piecewise Linear Representation* (Error-bounded PLR) problem discussed in this chapter is precisely defined as follows:

Definition 1 (Error-Bounded PLR) Given a predefined error bound $\delta > 0$ and a data stream fragment $S[1, n] = \langle s_1, s_2, \ldots, s_n \rangle$, the (optimal) error-bounded PLR is to construct a (minimal) number of δ-representative line segments $\{seg[i_1, i_2 - 1], seg[i_2, i_3 - 1], \ldots, seg[i_k, i_{k+1} - 1]\}$ to represent $S[1, n]$, where $i_1 = 1$ and $i_{k+1} - 1 = n$.

---

[1] It should be noted that $S[i - 1, j]$ can be δ-representable even if $S[i, j]$ is maximally δ-representable.

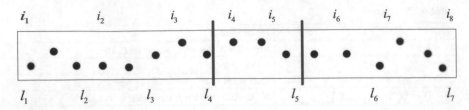

Fig. 2 The proof of Theorem 1: $\alpha = 4$

By the definition, we specify that this problem focuses on generating disconnected line segments, i.e., the consecutive segments do not share same end points.

The following theorem has been explored in literature [4, 15] which indicates that the optimal error-bounded PLR can be solved through computing maximal δ-representative line segments.

Theorem 1 *Given an error bound* $\delta > 0$ *and stream fragment* $S[1, n]$*, assume that* $seg[i_j, i_{j+1} - 1]$ *is a maximal* δ*-representative of* $S[i_j, i_{j+1} - 1]$ *on time slot* $[i_j, i_{j+1} - 1]$ *for* $1 \leq j \leq k$ *with* $i_1 = 1$ *and* $i_{k+1} - 1 = n$. *Then the error-bounded PLR on fragment* $S[1, n]$ *has at least* k *line segments.*

Proof Clearly, the claim is true for $k = 1$. For $k > 1$, assume that an optimal error-bounded PLR solution for fragment $S[1, n]$ is the set of segments $seg'[l_h, l_{h+1} - 1]$ for $1 \leq h \leq m$, where $l_1 = 1$, $l_{m+1} - 1 = n$ and $m < k$ (Fig. 2).

As claimed in the theorem, since each segment is a maximal δ-representative, both $i_1 = l_1 = 1$ and $l_2 \leq i_2$ hold. Let α be the index value such that $l_\alpha \leq i_\alpha$ and $l_{\alpha+1} > i_{\alpha+1}$ hold. Alternatively, since both $m < k$ and $l_{m+1} - 1 = n$ hold, $i_{m+1} < l_{m+1}$ holds. Hence, we have that α exists and $1 \leq \alpha \leq m$ is confirmed.

Thus, we have $l_\alpha \leq i_\alpha < i_{\alpha+1} - 1 < l_{\alpha+1} - 1$. It means that $S[i_\alpha, l_{\alpha+1} - 1]$ is δ-representable and is contradictory to the hypothesis that $seg[i_\alpha, i_{\alpha+1} - 1]$ is a maximal δ-representative line segment. Therefore, $k \leq m$ holds, and the result is proven.   □

With Theorem 1, it is inspired that the optimal error-bounded PLR results can be achieved by maximizing each δ-representative line segment, which is naturally the greedy mechanism. The algorithms introduced in this chapter all follow such mechanism.

3 OptimalPLR: An Optimal Algorithm to Generate Error-Bounded PLR

In this section, we will introduce an optimal algorithm to generate the error-bounded PLR, named OptimalPLR [33]. It can theoretically construct the minimal number of line segments, and achieve the linear time efficiency. The algorithm is based on the essential analysis on the convex outline of the data stream.

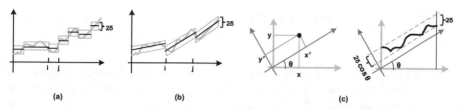

Fig. 3 Rotation examples. (**a**) Error-bounded PCR. (**b**) Error-bounded PLR. (**c**) Rotation

3.1 Extreme Slopes of Maximal δ-Representative

Generally, there can exist many δ-representative segments with various slopes for a δ-representable stream fragment. The range of the candidates' slopes depends on the tolerant error δ and the stream itself. We first discuss the complete slope range of the δ-representative segments for a δ-representable stream fragment, and then study the slope reductions and alterations when a new point is added to the stream fragment. Such analysis can provide the intuition of how to maximize the δ-representatives. Our discussion is based on the rotation of Cartesian coordinate system.

3.1.1 Slope Rotation and Extreme Slopes

We start from the methods proposed by Buragohain et al. [4] and Lazaridis et al. [15] for error-bounded PCR. As a simplified version of error-bounded PLR, error-bounded PCR uses constant values rather than general line segments for the representation. Constant values can be regarded as a line with zero slope that is denoted by a horizontal line in a Cartesian coordinate plane as in Fig. 3a. Under a predefined error bound $\delta > 0$, the optimal error-bounded PCR aims at constructing the smallest number of representation $B = \{c_{i_1}, c_{i_2}, \ldots, c_{i_h}\}$ for a given stream fragment $S[1, n] = \langle s_1, s_2, \ldots, s_n \rangle$ such that

$$\begin{cases} 0 = i_0 < i_k < i_h = n & \text{for } 0 < k < h, \\ |y_j - c_{i_k}| \leq \delta & \text{for } i_{k-1} < j \leq i_k \text{ and } 1 \leq k \leq h. \end{cases}$$

That is, using constant value c_{i_k} represents y_j for $i_{k-1} < j \leq i_k$ and $1 \leq k \leq h$.

To build B, the scheme of [4, 15] greedily checks the points of the stream fragment in order and finds the maximally δ-representable stream fragment iteratively. Here, the stream fragment is only approximated by horizontal lines. To choose the maximally δ-representable fragment started from x_i, the scheme needs to find the maximum time stamp x_j such that $\max_{i \leq i_1 < i_2 \leq j} |y_{i_1} - y_{i_2}| \leq 2\delta$. In fact, the scheme constructs the longest horizontal rectangle with 2δ width starting from x_i to cover the maximal number of steam points. The intuition is depicted in Fig. 3a in general.

Extending the idea to error-bounded PLR, we have the following observation that can be proven easily according to the error bound definition.

Observation 1 *A stream fragment $S[i, j]$ is δ-representable if and only if there exists a parallelogram of 2δ width in the vertical direction such that no points of $S[i, j]$ are placed outside of the parallelogram.*

The intuition of this observation is illustrated in Fig. 3b. We will use this observation to study the feasible slopes of a line segment approximating a δ-representable stream fragment through rotating the Cartesian coordinate plane.

To simplify the discussion, we assume that $S[1, k - 1] = \langle s_1, s_2, \ldots, s_{k-1} \rangle$ is δ-representable, and we use a line segment to represent it. For each point $s_i = (x_i, y_i)$ of $S[1, k - 1]$ where $1 \leq i < k$, its new coordinates $s_i' = (x_i', y_i')$, after having the axis rotated around the origin by an angle of $-\pi/2 < \theta < \pi/2$, must satisfy:

$$\begin{cases} x_i' = x_i \cos\theta + y_i \sin\theta, \\ y_i' = -x_i \sin\theta + y_i \cos\theta, \end{cases}$$

as illustrated in Fig. 3c, which is derived by rotation matrix. According to the observation, since fragment $S[1, k-1]$ is δ-representable, there exists $-\pi/2 < \theta < \pi/2$ such that for each pair of $\{i, j\}$, $|y_i' - y_j'| \leq 2\delta| \cos\theta|$. Since $-\pi/2 < \theta < \pi/2$, $|\cos\theta| \neq 0$ and

$$|y_i' - y_j'| = |[(-x_i \tan\theta + y_i) - (-x_j \tan\theta + y_j)] \cos\theta|$$

hold, we have $|(x_j - x_i) \tan\theta - (y_j - y_i)| \leq 2\delta$. By extending this formula, we have[2]:

$$\frac{(y_j - \delta) - (y_i + \delta)}{(x_j - x_i)} \leq \tan\theta \leq \frac{(y_j + \delta) - (y_i - \delta)}{(x_j - x_i)}. \tag{1}$$

In fact, θ is the intersection angle of x-axis and the parallelogram's non-vertical edge, which is also the inclination angle of a possible δ-representative for $S[1, k-1]$ (i.e., the centerline parallel to the parallelogram's non-vertical edge). In this sense, the $\tan\theta$ is the slope of the δ-representative line segment.

Since for each pair of $\{i, j\}$, the relationship of Eq. (1) must hold, we can derive the range of $\tan\theta$. Let $\underline{slp}[1, k - 1]$ and $\overline{slp}[1, k - 1]$ denote the minimum and maximum line slopes of the δ-representative for fragment $S[1, k - 1]$, respectively, and we have

$$\begin{cases} \underline{slp}[1, k - 1] = \max\limits_{1 \leq i < j \leq k-1} \frac{(y_j - \delta) - (y_i + \delta)}{(x_j - x_i)} \\ \qquad = \frac{(y_c - \delta) - (y_a + \delta)}{(x_c - x_a)}, \\ \overline{slp}[1, k - 1] = \min\limits_{1 \leq i < j \leq k-1} \frac{(y_j + \delta) - (y_i - \delta)}{(x_j - x_i)} \\ \qquad = \frac{(y_d + \delta) - (y_b - \delta)}{(x_d - x_b)}, \end{cases} \tag{2}$$

---

[2]Without the loss of generality, we assume that $x_i < x_j$.

Then, for a possible δ-representative line segment of $S[1, k-1]$, its slope $slp[1, k-1]$ satisfies:

$$\underline{slp}[1, k-1] \leq slp[1, k-1] \leq \overline{slp}[1, k-1]. \qquad (3)$$

In fact, Eq. (3) denotes that the stream fragment $S[1, k-1]$ is δ-representable if and only if $\underline{slp}[1, k-1] \leq \overline{slp}[1, k-1]$. In the situations of $\underline{slp}[1, k-1] < \overline{slp}[1, k-1]$, $slp[1, k-1]$ can have many feasible values within the range $[\underline{slp}[1, k-1], \overline{slp}[1, k-1]]$. Hence, it is natural to discover the feasible range of the line slope in terms of $\underline{slp}[1, k-1]$ and $\overline{slp}[1, k-1]$.

Specifically, we define the extreme points as those identify the δ-representatives with extreme slopes. Based on Eq. (2), we define:

$$\begin{cases} \{a, c\} = \underset{1 \leq i < j \leq k-1}{\arg\max} \dfrac{(y_j - \delta) - (y_i + \delta)}{(x_j - x_i)}, (a < c) \\ \{b, d\} = \underset{1 \leq i < j \leq k-1}{\arg\min} \dfrac{(y_j + \delta) - (y_i - \delta)}{(x_j - x_i)}, (b < d) \end{cases}$$

where points s_a and s_b (s_c and s_d, respectively) are the rightmost (leftmost, respectively) points that satisfy $1 \leq a < c \leq k-1$ and $1 \leq b < d \leq k-1$. Assume $s_i = (x_i, y_i - \delta)$ denotes data point with deleted tolerant error, and $\overline{s_i} = (x_i, y_i + \delta)$ denotes data point with added tolerant error, then $\overline{s_a}$, s_c, s_b, and $\overline{s_d}$ are the extreme points. As previously mentioned, we use $line(\overline{s_a}, s_c)$ to denote the line that passes points $\overline{s_a}$ and s_c, and use $line(s_b, \overline{s_d})$ to denote the line that passes points s_b and $\overline{s_d}$. In this way, $line(\overline{s_a}, s_c)$ and $line(s_b, \overline{s_d})$ are the bounding lines with extreme slopes for all δ-representatives (Fig. 4).

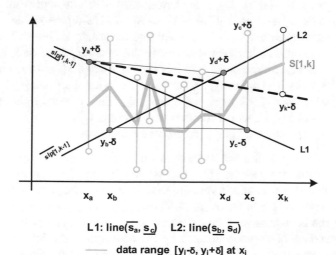

Fig. 4 Extreme slope evolution

According to Eqs. (2) and (3), we have the following lemma:

Lemma 1 *For each $1 \le i \le k - 1$, data range $[y_i - \delta, y_i + \delta]$ at time stamp x_i is intersected or met by $line(\overline{s_a}, \underline{s_c})$ and $line(\underline{s_b}, \overline{s_d})$.*

The proof is straightforward: If any range $[y_i - \delta, y_i + \delta]$ is not intersected or met by one of these two lines, it will be contradictory with the definition of $\underline{slp}[1, k-1]$ or $\overline{slp}[1, k-1]$.

As depicted in Fig. 4, Lemma 1 means that each data range $[y_i - \delta, y_i + \delta]$, denoted by the thick vertical gray line at x_i, interacts (or meets) with both $line(\overline{s_a}, \underline{s_c})$ (labeled by $L1$) and $line(\underline{s_b}, \overline{s_d})$ (labeled by $L2$).

3.1.2 Slope Evolution and Reduction

Let $S[1, k]$ be the fragment created by the addition of a new point s_k at the end of fragment $S[1, k - 1]$. As inspired by Theorem 1, we need to check if $S[1, k]$ is δ-representable greedily and resolve $slp[1, k]$, $\underline{slp}[1, k]$ and $\overline{slp}[1, k]$ to derive the potential maximal δ-representative. We will interpret how to simplify the process in three stages: Increment, Localization, and the Reduction of convex hull.

First, derived from Lemma 1, we have the following corollary [4, 6] to incrementally determine whether $S[1, k]$ is δ-representable when $S[1, k - 1]$ is.

Corollary 1 *Suppose fragment $S[1, k - 1]$ is δ-representable, fragment $S[1, k]$ is δ-representable if and only if the range $[y_k - \delta, y_k + \delta]$ at time x_k is not located below $line(\overline{s_a}, \underline{s_c})$ or above $line(\underline{s_b}, \overline{s_d})$.*

Proof For sufficiency, if the range $[y_k - \delta, y_k + \delta]$ at time x_k is not located below $line(\overline{s_a}, \underline{s_c})$ or above $line(\underline{s_b}, \overline{s_d})$, it is obvious that s_k can be approximated within error bound δ by at least one $seg[1, k - 1]$, so $S[1, k]$ is δ-representable.

For necessity, given $S[1, k]$ is δ-representable, according to Lemma 1, data range $[y_k - \delta, y_k + \delta]$ must be intersected or met by the bounding lines of $S[1, k]$. Since $S[1, k - 1]$ is always δ-representable, the bounding lines of $S[1, k]$ must be covered by those of $S[1, k - 1]$, i.e., $line(\overline{s_a}, \underline{s_c})$ and $line(\underline{s_b}, \overline{s_d})$. If the range $[y_k - \delta, y_k + \delta]$ is located below $line(\overline{s_a}, \underline{s_c})$ or above $line(\underline{s_b}, \overline{s_d})$, then it will be also out of the bounding lines of $S[1, k]$, which conflicts with Lemma 1.

Combining the above two parts, the corollary is proven.   □

Such corollary provides the method to quick check whether we should keep updating the extreme slopes for $S[1, k]$, or we can determine a fragment $S[1, k - 1]$ as maximally δ-representable. Assuming that $S[1, k]$ is verified as δ-representable, the next question is how to obtain $\underline{slp}[1, k]$ and $\overline{slp}[1, k]$. Computing them directly via Eq. (2) can be expensive in time cost as they require the complete computation of the intersection of slopes delineated by the equation. In the following, we will simplify the computation of $\underline{slp}[1, k]$ and $\overline{slp}[1, k]$ by incremental and localizing strategies in terms of $\underline{slp}[1, k - 1]$ and $\overline{slp}[1, k - 1]$.

Increment We will refine Eq. (2) and express $slp[1, k]$ in terms of $slp[1, k - 1]$. From Eq. (2), we have

$$\begin{cases} \underline{slp}[1, k] = \max_{1 \leq i < j \leq k} \frac{(y_j - \delta) - (y_i + \delta)}{(x_j - x_i)}, \\ \overline{slp}[1, k] = \min_{1 \leq i < j \leq k} \frac{(y_j + \delta) - (y_i - \delta)}{(x_j - x_i)}. \end{cases}$$

According to the definitions of $\underline{slp}[1, k - 1]$ and $\overline{slp}[1, k - 1]$, we have

$$\begin{cases} \underline{slp}[1, k] = \max_{1 \leq i < k} \left\{ \frac{(y_k - \delta) - (y_i + \delta)}{(x_k - x_i)}, \underline{slp}[1, k - 1] \right\}, \\ \overline{slp}[1, k] = \min_{1 \leq i < k} \left\{ \frac{(y_k + \delta) - (y_i - \delta)}{(x_k - x_i)}, \overline{slp}[1, k - 1] \right\}. \end{cases} \tag{4}$$

Equation (4) indicates that $\underline{slp}[1, k]$ and $\overline{slp}[1, k]$ can be derived by comparing point s_i with s_k ($i < k$) rather than each pair of points s_i and s_j ($i < j \leq k$). Therefore, the extreme slopes can be more efficiently computed from Eq. (4) than from Eq. (2). However, based on the incremental strategy, we demonstrate that this process can be further simplified, according to the Lemma 1.

Localization With the availability of $\underline{slp}[1, k - 1]$, $\overline{slp}[1, k - 1]$, and $\{\overline{s_a}, s_c, s_b, \overline{s_d}\}$ defined before, we can further reduce the range of points needed to check for the extreme slopes. We take the update of $\overline{slp}[1, k]$ as an example to explain the reduction.

As exemplified in the Fig. 5, according to Corollary 1, if $S[1, k]$ is δ-representable, $\overline{s_k}$ must be within area A_1 or A_2 (as indicated by the dotted line, and the length of A_1 is 2δ). For $i \in [1, b)$, if $\overline{s_k}$ is in A_1 area, $slope(s_i, \overline{s_k}) \geq \overline{slp}[1, k-1]$ holds; if $\overline{s_k}$ is in A_2 area, the range $[y_b - \delta, y_b + \delta]$ is above $line(s_i, \overline{s_k})$. For $i \in (c, k - 1]$, the range $[y_c - \delta, y_c + \delta]$ is always above $line(s_i, \overline{s_k})$. According to the Lemma 1 and the definition of extreme slopes, if $S[1, k]$ is δ-representable, there is no need to check the points before s_b and after s_c. The similar conclusion can be made on the case of $\underline{slp}[1, k]$ update. Thus, Eq. (4) can be rewritten into

Fig. 5 Localization of slope reduction

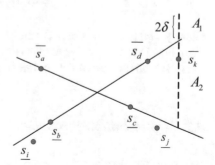

Fig. 6 Example of the convex hull points $\overline{cvx}$

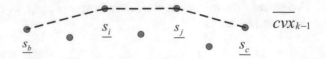

$$
\begin{cases}
\underline{slp}[1,k] = \max_{a \leq i \leq d} \left\{ \frac{(y_k - \delta) - (y_i + \delta)}{(x_k - x_i)}, \underline{slp}[1, k-1] \right\}, \\
\overline{slp}[1,k] = \min_{b \leq i \leq c} \left\{ \frac{(y_k + \delta) - (y_i - \delta)}{(x_k - x_i)}, \overline{slp}[1, k-1] \right\}.
\end{cases}
\tag{5}
$$

Convex Reduction In the following, based on the definition of the extreme lines and Lemma 1, we further indicate that the computation can be constrained to those localized points on the convex hulls.

For stream fragment $S[1, k-1]$, let $\overline{cvx}_{k-1}$ denote the set of convex points of the sequence $\{s_b, s_{b+1}, \ldots, s_c\}$, which are points with deleted tolerant error. Here, $\overline{cvx}_{k-1}$ bulges upward and is depicted by a sequence of point in the ascent time stamp order. For example, $\overline{cvx}_{k-1} = \langle s_b, s_i, s_j, s_c \rangle$ in Fig. 6. Similarly, we define $\underline{cvx}_{k-1}$ to be the points of the convex hull of $\{\overline{s}_a, \overline{s}_{a+1}, \ldots, \overline{s}_d\}$ that bulges downward. With the notations of $\underline{cvx}_{k-1}$ and $\overline{cvx}_{k-1}$, we have the following lemma.

Lemma 2 *If $\underline{slp}[1, k] = slope(\overline{s}_i, \underline{s}_k)$ and $x_a \leq x_i \leq x_d$, then $\overline{s}_i$ is in $\underline{cvx}_{k-1}$. Similarly, if $\overline{slp}[1, k] = slope(\underline{s}_j, \overline{s}_k)$ and $x_b \leq x_j \leq x_c$, then $\underline{s}_j$ is in $\overline{cvx}_{k-1}$.*

Proof We only prove the case of $\underline{slp}[1, k]$. According to the definition of convex hull, if $\overline{s}_i$ is not in the convex hull, $line(\overline{s}_i, \underline{s}_k)$ must go inside the convex hull, and there is at least one point $\overline{s}_h$ for $x_a \leq x_h \leq x_d$ below $line(\overline{s}_i, \underline{s}_k)$. If $\underline{slp}[1, k] = slope(\overline{s}_i, \underline{s}_k)$, it will be contradictory with Lemma 1 and Corollary 1, so $\overline{s}_i$ is in $\underline{cvx}_{k-1}$.   □

Lemma 2 implies that Eq. (5) can be rewritten into

$$
\begin{cases}
\underline{slp}[1,k] = \max_{\overline{s}_i \in \underline{cvx}_{k-1}} \left\{ \frac{(y_k - \delta) - (y_i + \delta)}{(x_k - x_i)}, \underline{slp}[1, k-1] \right\}, \\
\overline{slp}[1,k] = \min_{\underline{s}_j \in \overline{cvx}_{k-1}} \left\{ \frac{(y_k + \delta) - (y_j - \delta)}{(x_k - x_j)}, \overline{slp}[1, k-1] \right\}.
\end{cases}
\tag{6}
$$

Clearly, $\underline{slp}[1, k]$ and $\overline{slp}[1, k]$ can be more efficiently obtained from Eq. (6) than from Eq. (5) as the number of points in the convex hulls can be significantly smaller than the total number of data points in the relevant intervals.

3.2 Optimization Strategies

Based on the previous discussion, we will provide some useful theorems, which derive the design of the optimal algorithm for error-bounded PLR generation. The update of extreme slopes and the convex hulls are the major points we will focus on. In the following, we will study the maintenance of $\underline{cvx}_{k-1}$ and $\overline{cvx}_{k-1}$ and the derivation of $\underline{slp}[1, k]$ and $\overline{slp}[1, k]$ based on Eq. (6). The results are demonstrated in Theorem 2. For simplicity, we will only discuss the results for the update of $\underline{cvx}_k$ and $\underline{slp}[1, k]$ as the analogous results exist for $\overline{cvx}_k$ and $\overline{slp}[1, k]$.

3.2.1 Computing Extreme Slopes

In the process of computing $\underline{slp}[1, k]$, we first decide whether $\underline{slp}[1, k] > \underline{slp}[1, k - 1]$ and whether the bounding lines with extreme slopes need to update. Theorem 2 suggests that only the first and last points of the convex hulls need to be used to determine further processing. If the extreme slope $\underline{slp}[1, k]$ needs to update from $\underline{slp}[1, k - 1]$, Theorem 2 also concludes that the computation can be incrementally performed and the time cost for computing $\underline{slp}[1, k]$ is proportional to the number of removed points from $\underline{cvx}_{k-1}$.

Theorem 2 *Suppose $\underline{cvx}_{k-1} = \langle \underline{s}_{i_1}, \ldots, \underline{s}_{i_h} \rangle$, here $\underline{s}_{i_1} = \underline{s}_a$ and $\underline{s}_{i_h} = \underline{s}_d$, then the following results hold for $\underline{slp}[1, k]$ and $\underline{cvx}_k$.*

(1) If $slope(\overline{s}_{i_1}, \underline{s}_k) \le \underline{slp}[1, k - 1]$, then $\underline{slp}[1, k] = \underline{slp}[1, k - 1]$ holds.

(2) If $slope(\overline{s}_{i_1}, \underline{s}_k) > \underline{slp}[1, k - 1]$, then $\underline{slp}[1, k] = \max_e \frac{(y_k - \delta) - (y_{i_e} + \delta)}{(x_k - x_{i_e})}$ where $1 \le e \le h$, and $\overline{s}_{i_m} \notin \underline{cvx}_k$ for $1 \le m < e$.

Proof The proof of claim (1): If $slope(\overline{s}_{i_1}, \underline{s}_k) \le \underline{slp}[1, k - 1]$, then point $\underline{s}_k = (x_k, y_k - \delta)$ is either below or on the line of $line(\overline{s}_a, \underline{s}_c)$. Again, under the assumption, each point $\overline{s}_{i_e} \in \underline{cvx}_{k-1}$ is not below the line of $line(\overline{s}_a, \underline{s}_c)$. As a result,

$$\max_{\overline{s}_i \in \underline{cvx}_{k-1}} \left\{ \frac{(y_k - \delta) - (y_i + \delta)}{(x_k - x_i)} \right\} \le \underline{slp}[1, k - 1]$$

holds, which leads to $\underline{slp}[1, k] = \underline{slp}[1, k - 1]$ from Eq. (6).

The Proof of Claim (2): Clearly, if $slope(\overline{s}_{i_1}, \underline{s}_k) > \underline{slp}[1, k - 1]$ then point $\underline{s}_k = (x_k, y_k - \delta)$ is above the line of $line(\overline{s}_a, \underline{s}_c)$, and the new extreme slope can be derived from the points on the convex hull. From Eq. (6), we have $\underline{slp}[1, k] = \max_e \frac{(y_k - \delta) - (y_{i_e} + \delta)}{(x_k - x_{i_e})}$ where $1 \le e \le h$. Moreover, since we confirm the new extreme line is determined by $\overline{s}_{i_e}$ and $\underline{s}_k$, s_{i_e} will replace as new s_a, so all the points on the convex hull before $\overline{s}_{i_e}$ will be removed from the new convex hull $\underline{cvx}_k$. ◻

In fact, from the theorem and its proof, we can conclude that when we determine that the extreme slope should be updated, the new extreme slope is actually the slope of tangent line from $\underline{s}_k$ to the convex hull $\underline{cvx}_{k-1}$.

3.2.2 Updating Convex Hulls

If the extreme slopes are not updated when processing the new point, then the convex hulls keep unchanged. Or else if we have updated the extreme slopes after adding the new stream point, the convex hulls should also be updated. The maintenance of convex hull is formed by two parts: *point deletion* and *point addition*. We also take the updating of $slp[1, k]$ as an example to explain the process.

The *point deletion* part has been previously mentioned in Theorem 2(2), that is, if $\underline{slp}[1, k] = slope(\overline{s_l}, \underline{s_k})$ for $\overline{s_l} \subset \underline{cvx}_{k-1}$, then the earlier points of $\underline{cvx}_{k-1}$ before $\overline{s_i}$ are NOT in $\underline{cvx}_k$.

For the *point addition* part, we have $\overline{cvx}_k \subseteq \overline{cvx}'_{k-1} \cup \{\underline{s_k}\}$, where $\overline{cvx}'_{k-1}$ denotes the updated $\overline{cvx}_{k-1}$ after removing those earlier points in *point deletion*. If $\underline{s_k}$ determines the new extreme slope, it should also be added into $\overline{cvx}_k$, since it will replace as new $\underline{s_c}$ as defined before. After adding $\underline{s_k}$ to the tail of $\overline{cvx}'_{k-1}$, some previous convex points may have to be deleted in order to keep the convex characteristic. Such process can be derived from the triangle check technique in [3], which will be explained later.

3.3   Error-Bounded PLR Algorithm

Now we present the linear-time algorithm OptimalPLR for the error-bounded PLR problem. Let us assume that the given stream fragment has n points, and the error bound for each data point is δ. The OptimalPLR algorithm outputs a set of maximal δ-representative line segments with a minimized space cost upper bounded by n, and achieves the minimized number of segments. For the convenience of algorithm presentation, we use ρ or $\overline{\rho}$ to represent the extreme slopes during processing. In general, the strategy used in this algorithm is to progressively propagate the extreme slopes $\underline{slp}[1, k]$ and $\overline{slp}[1, k]$ upon new stream points in the process of generating maximal δ-representative segments.

3.3.1   Description of OptimalPLR

Given a stream fragment $S[1, n]$, the OptimalPLR algorithm generates maximal δ-representative line segments starting from x_1 successively. In the process of generating the maximal δ-representative segment from x_i (we assume $i = 1$ without the loss of generality), the OptimalPLR algorithm needs to maintain $\underline{cvx}_{k-1}, \overline{cvx}_{k-1}$, $\underline{\rho}$, and $\overline{\rho}$ to compute $\underline{slp}[1, k]$ and $\overline{slp}[1, k]$ as indicated in Sect. 3.2. In this part, we first describe the OptimalPLR algorithm and then discuss its time and space complexities.

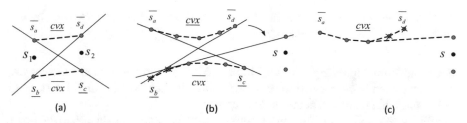

Fig. 7 Optimal algorithm: (**a**) Initialization; (**b**) Updating of extreme slopes; (**c**) Updating of convex hull

---

Algorithm 1: OptimalPLR

Input: S: stream fragment starts from x_1; δ: the specified error bound
Output: A maximal δ-representative line segment starts from x_1
1 % **Initialization**
2 $s_1 = (x_1, y_1)$; $s_2 = (x_2, y_2)$;
3 $\overline{s_a} = (x_1, y_1 + \delta)$; $\underline{s_c} = (x_2, y_2 - \delta)$;
4 $\underline{s_b} = (x_1, y_1 - \delta)$; $\overline{s_d} = (x_2, y_2 + \delta)$;
5 $\underline{\rho} = slope(\overline{s_a}, \underline{s_c})$; $\overline{\rho} = slope(\underline{s_b}, \overline{s_d})$;
6 $\underline{cvx} = \langle \overline{s_a}, \overline{s_d} \rangle$; $\overline{cvx} = \langle \underline{s_b}, \underline{s_c} \rangle$;
7 % **Processing**
8 **while** s is NOT outside the two lines, $line(\overline{s_a}, \underline{s_c})$ and $line(\underline{s_b}, \overline{s_d})$, more than δ **do**
9   % Maintain $\underline{\rho}, \overline{\rho}, \underline{cvx}$ and $\overline{cvx}$
10   **if** $y + \delta < \overline{\rho}(x - x_b) - y_b$ **then**
11   find the point q of $\overline{cvx}$ that minimizes $slope(q, \overline{s})$;
12   let $\underline{s_b}$ be q and $\overline{s_d}$ be $\overline{s}$;
13   delete all the points of $\overline{cvx}$ prior to point q;
14         $\overline{\rho} = slope(\underline{s_b}, \overline{s})$;
15   insert $\overline{s}$ to the tail of $\underline{cvx}$, and update cvx by triangle check([3]);
16   **end**
17   **if** $y - \delta > \underline{\rho}(x - x_a) - y_a$ **then**
18   find the point q of $\underline{cvx}$ that maximizes $slope(q, \underline{s})$;
19   let $\overline{s_a}$ be q and $\underline{s_c}$ be $\underline{s}$;
20   delete all the points of $\underline{cvx}$ prior to point q;
21         $\underline{\rho} = slope(\overline{s_a}, \underline{s})$;
22   insert $\underline{s}$ to the tail of $\overline{cvx}$, and update $\overline{cvx}$ by triangle check;
23   **end**
24 **end**
25 % **Producing a line segment**
26 Let $s_o = (x_o, y_o)$ be the intersection of $line(\overline{s_a}, \underline{s_c})$ and $line(\underline{s_b}, \overline{s_d})$;
27 $\rho = (\underline{\rho} + \overline{\rho})/2$;
28 **return** a line segment: pass point $s_o = (x_o, y_o)$ with slope ρ

The general steps of OptimalPLR are interpreted in Fig. 7, and we depict the procedure for constructing a maximal δ-representative line segment in Algorithm 1. Referring to the figure and the algorithm, we describe the OptimalPLR in detail.

Initialization Primely we initialize the extreme lines and the convex hulls by the first two stream points s_1 and s_2 by setting $\overline{cvx} = \langle s_b, s_c \rangle$ and $\underline{cvx} = \langle \underline{s_a}, \overline{s_d} \rangle$. ($a = b = 1, c = d = 2$)

Extreme Slope Updating As stated in Sect. 3.2, the extreme lines and the convex hulls may need to be updated when a new point s is read in. The condition of Line (8) implies that point s is in the current segment.

Taking the update of $\overline{\rho}$ as an example, since the approximate value for point s is within the range of $[\underline{s}, \overline{s}]$, if $\overline{s}$ is NOT under the extreme line $line(s_b, \overline{s_d})$, the maximum slope will not be updated and the extreme line $line(s_b, \overline{s_d})$ will be used as the new extreme line. Otherwise, the maximum slope should be reduced. As exemplified in Fig. 7b, the strategy is to find the point q of $\overline{cvx}$ that minimizes $slope(q, \overline{s})$, which can be found by the tangent line of the convex hull from $\overline{s}$. The new extreme line is then defined as $line(q, \overline{s})$ and the new $\overline{cvx}$ is updated from the old one by removing the points before q.

Convex Hull Updating After updating the extreme line, $\overline{s}$ should be merged into the upper convex hull $\underline{cvx}$. Figure 7c depicts the merging strategy. After inserting $\overline{s}$ into the tail of $\underline{cvx}$, the triangle check [3] needs to be carried out to maintain the convex characteristic. It starts by examining the three most recent consecutive points and then moving backwards. If the middle point is above or on the line formed by the other two points, then the middle point is removed. This process is continued for the remaining three most recent consecutive points until the middle point is no longer being removed (refer to [3] for the details of convex hull algorithm).

The correctness of OptimalPLR algorithm is evidenced from both Theorem 1 and the derivation of Eq. (6) as the deduction process preserves the maximum range of slopes for δ-representative segments.

3.3.2 Complexity Analysis

In the following, we discuss the time and space complexity of OptimalPLR.

Time Complexity To show that the time complexity of the OptimalPLR algorithm is $O(n)$ for stream fragment $S[1, n]$, it is sufficient to show the time complexity of Algorithm 1 is $O(k)$ for maximally δ-representable fragment $S[1, k]$.

Clearly, the iteration times of while loop is bounded by k for fragment $S[1, k]$. In each loop, the extreme slopes (slp and $\overline{slp}$) and convex hulls (cvx and $\overline{cvx}$) need to be updated for the newly inserted stream point. As indicated in Line (10–15), the costs of updating extreme slopes are dominated by the costs of updating convex hulls. In the process of updating convex hulls, each convex hull (cvx or $\overline{cvx}$) needs to be maintained by deleting some earlier stream points from it (e.g., Line (13)) and/or inserting a recent stream point into it (e.g., Line (15)). Once a point is deleted from cvx (or $\overline{cvx}$), the point will not be inserted back into cvx (or $\overline{cvx}$). Therefore, the total costs for maintaining cvx (or $\overline{cvx}$) in the process of constructing the segment are bounded by $2k$. Thus, the time cost of Algorithm 1 on fragment $S[1, k]$ is bounded by $(2k + 2k) + ck = (4 + c)k$, where c is a constant number that summarizes other costs in the loop. We conclude that the time complexity of OptimalPLR is $O(n)$.

Space Complexity Since each early obtained segment is not used for latter computation and can be output directly, the space cost of the OptimalPLR algorithm on stream fragment $S[1, n]$ is proportional to the space cost on generating a maximal δ-representative segment of $S[1, n]$.

During the process of generating a segment, we only need to maintain cvx plus a constant number of points such as ρ, at each while loop of Algorithm 1. Therefore, the space complexity of the OptimalPLR algorithm is $O(n_{cx})$, where n_{cx} is the size of maximum cvx. Let n_{sg} denote the maximum number of stream points in the derived maximal δ-representative segments of $S[1, n]$. Then, we have $n_{cx} \le n_{sg} \le n$ holds. Therefore, the space cost of the OptimalPLR algorithm is also bounded by $O(n)$.

In summary, we have the following major result for the OptimalPLR algorithm.

Theorem 3 *Given a stream fragment $S[1, n]$ and the error bound δ, the OptimalPLR algorithm is an optimal algorithm for error-bounded PLR with $O(n)$ time and $O(n_{cx})$ space complexities where $n_{cx} \le n$.*

3.3.3 Discussions of OptimalPLR

In this section, we will discuss some important issues of OptimalPLR algorithm, including the application for high-dimensional data and the adaption for varied error bound.

Extension for High-Dimensional Data

High-dimensional data are popular in streaming data nowadays, so here we discuss whether OptimalPLR can work on high-dimensional data and how it can be extended effectively.

Since the aim is to generate error-bounded PLR for data stream, it is essential how we define the error bound. Different from the case that the stream point value is a single number, for high-dimensional data, we have to consider whether the error bound is defined on each dimension, or on the data point in high-dimensional space.

Case 1: Error Bound Defined on Each Dimension

If the error bound is same for each dimension, it will be easy to extend. We can simply treat the data values of each dimension as a data stream and carry out the original OptimalPLR algorithm to generate line segments. To combine the segments of different dimensions together, when we find the first maximal δ-representative for any dimension, we stop and determine a maximally representable segment, and then start a new segmentation. Such strategy is also applied in [6].

Case 2: Error Bound Defined on High-Dimension Point

If the error bound is defined in high-dimensional space, it will be more complicated. We will only discuss the data stream in 3D space, in which each stream point at time stamp x_k is identified by $\{y_k, z_k\}$. In this case, the error bound δ will be defined as follows: For each stream point s_k, the Euclidean distance between original point and the approximate point satisfies $\sqrt{(y_k' - y_k)^2 + (z_k' - z_k)^2} \leq \delta$. From geometrical view, the error bound describes a circle with radius δ around the point s_k, so the OptimalPLR cannot be applied for this situation directly. However, we can still find an approximated way to embed OptimalPLR.

First we have the following relationship:

$$\sqrt{2|y_k' - y_k||z_k' - z_k|} \leq \sqrt{(y_k' - y_k)^2 + (z_k' - z_k)^2} \leq \delta.$$

If we define $|y_k' - y_k| \leq \delta / \sqrt{2}$ and $|z_k' - z_k| \leq \delta / \sqrt{2}$, then it can be guaranteed that the above relationship always holds. In this way, instead of dealing with the error bound in high-dimensional space, we can define alternative error bound in each dimension, and the solution can be designed as Case 1.

It should be noticed that such strategy cannot guarantee optimal results (minimized number of segments), since we are not maintaining all feasible line segments. However, it can be executed efficiently and promise the bounded approximation error in high-dimensional space.

Adaption to Varied Error Bound

In practical applications, the real-time approximation criteria may vary during the stream processing, so the error bound for different data points may change according to the current data status. For example, when the sampling rate increases, the system may need to reduce the approximation error, so the error bound should be reduced accordingly. In those applications with periodic variation, e.g., the traffic monitoring system, the approximation error margin for traffic flow in rush hour may be much smaller than that in slack hours.

Fortunately, OptimalPLR algorithm can be easily extended to adapt to varied error bound. During the algorithm, assume the approximate error bound for each data point s_i is δ_i. In the algorithm procedure, the parameter δ can be replaced by δ_i when processing s_i. In this way, the algorithm can execute as usual and at the same time, adapt to different error bounds.

4 ParaOptimal: An Optimal Algorithm in Transformed Space

In this section, we will introduce an interesting optimal algorithm to generate the error-bounded PLR in transformed space, named ParaOptimal [20]. It designs the algorithm in a special slope-offset parameter space, also with linear time efficiency. Furthermore, we will introduce the generalization of such algorithm model.

4.1 Description of ParaOptimal

Similarly, the data stream is a sequence of data points which are numerical numbers, and we denote it as $S = \langle s_1, s_2, \ldots, s_n \rangle$. Each data point s_i consists of the value y_i and the time stamp x_i. The approximate value of s_i can be determined by $y_i' = a \cdot x_i + b$, where the a and b, respectively, stand for the slope and the offset of the line segment approximating s_i. The error bound δ is defined to restrict the approximate error on single point, i.e., $\forall i \in [1, n]$, $|y_i' - y_i| \leq \delta$.

The ParaOptimal is a greedy and incremental algorithm in which the line segments are generated and adjusted to approximate as many points as possible. The basic idea of ParaOptimal is to determine the line function by maintaining the feasible region of all line candidates in the parameter space. More specifically, for a point s_i, the approximate value should be within the error bound, that is,

$$ y_i - \delta \leq a \cdot x_i + b \leq y_i + \delta. $$

It means any pair of parameters $\{a, b\}$ that meets the above inequalities can identify a candidate line function, which composes the solution set. If more data points are approximated, the solution set of line functions approximating all data points will be the intersection of all solution sets for individual points.

Considering the parameter pair in the parameter space (Fig. 8), each pair can be viewed as a point, and the error bound provides linear constraints on the parameters. The ParaOptimal algorithm runs in this way: Each new point provides two linear constraints; the algorithm incrementally updates the feasible region of candidate line parameters by intersecting the feasible region with the solution region of new linear constraints; a new line segment is determined if the feasible region is empty, and then, the algorithm initializes the feasible region for a new segmentation. Figure 8 exemplifies the feasible region update in parameter space.

4.1.1 Theoretical Preparation

Primely, we provide some basic properties and lemmas for the further description.

Fig. 8 Feasible region
update in parameter space

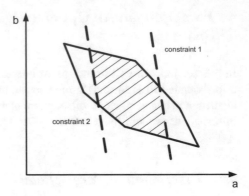

Lemma 3 *For the candidate line parameters, any edge of the feasible region belongs to a boundary of a linear constraint.*

Proof Since the feasible region is the intersection of all the solution regions for all linear constraints, it is obvious that the boundaries of the feasible region are composed of the boundaries of linear constraints, so any edge is part of a boundary of a linear constraint.   □

Corollary 2 *The feasible region of the candidate line parameters is a convex polygon.*

Proof Given an edge of the feasible region, it is part of a boundary of certain linear constraint. Since the feasible region is the intersection of all solution regions of linear constraints, it must be the subset of the solution region of corresponding linear constraint, so all the feasible region lies on one side of the given edge. According to the convex definition, the feasible region is a convex polygon.   □

Lemma 4 *For the boundary of linear constraints provided by error bound, the slopes are less than zero, and decreasing monotonously as the order of data points.*

Proof Given an upper error bound of point s_i, the linear constraint is identified by $a \cdot x_i + b \leq y_i + \delta$, so the slope of the linear constraint boundary is $-x_i$. Since $x_i > 0$ always holds, the slope is less than zero.

For any two points s_i and s_j, if $x_i < x_j$, then the slope of the constraint boundary corresponding to s_i is greater than that of s_j.   □

Corollary 3 *The left most point of the feasible region is the highest point; the right most point of the feasible region is the lowest point.*

Proof Given a point on the feasible region boundary, since all the edge slopes are less than zero, it must be lower than the boundary point to its left. The rest can be done in the same manner, and we will have the point is lower than the leftmost point of the feasible region. So, the leftmost point is the highest point. Similarly, we can have the rightmost point of the feasible region is the lowest point.   □

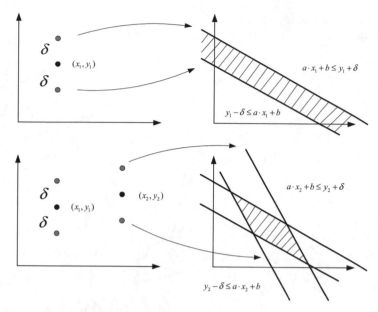

Fig. 9 Initialization of feasible region

Now, we describe the ParaOptimal algorithm step by step in the following parts.

4.1.2 Initialization

In the parameter space, the error bound of a data point will provide two linear constraints for the candidate line parameters, which compose a pair of parallels. When the segmentation starts, to restrict the feasible region with bounded size, the algorithm considers the first two points for initialization and generates the feasible region. Figure 9 demonstrates this process.

4.1.3 Feasible Region Update

When the algorithm processes a new data point, it updates the current feasible region according to the new linear constraints and checks whether it is possible to process more points, or else a line segment can be determined. For simplification, we only discuss the linear constraint by the upper error bound.

Assume we process a new data point s_i, and the linear constraint provided by the upper error bound is $a \cdot x_i + b \leq y_i + \delta$. Since x_i is always greater than zero, the solution area for the above linear constraint is the half space to the left of the boundary, which is identified by linear function $a \cdot x_i + b = y_i + \delta$. As exemplified in Fig. 10 in the clockwise order, we divide the edge points of feasible region into

Fig. 10 Points definition of
feasible region

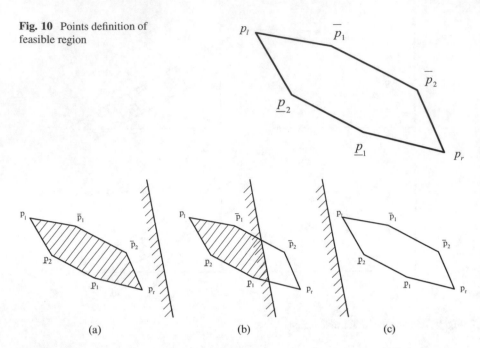

Fig. 11 Three cases for feasible region update. (**a**) Case 1. (**b**) Case 2. (**c**) Case 3

$\langle \overline{p}_1, \overline{p}_2, \ldots, \overline{p}_t \rangle$ and $\langle \underline{p}_1, \underline{p}_2, \ldots, \underline{p}_s \rangle$ by the leftmost point p_l and rightmost point p_r. Each point is a 2-tuple $\{a, b\}$ in parameter space.

The feasible region update can be categorized into three cases (Fig. 11).

Case 1: $p_r.a \cdot x_i + p_r.b < y_i + \delta$. In this case, the whole feasible region satisfies the linear constraint, so the updated feasible region will remain the same and no extra processing is needed.

Case 2: $p_l.a \cdot x_i + p_l.b \leq y_i + \delta \leq p_r.a \cdot x_i + p_r.b$. In this case, the solution area of the linear constraint will intersect with the current feasible region, and the updated feasible region should be determined. Since we discuss the linear constraint from the upper error bound, we can first determine the new p_r for the new feasible region. We start from the first point of the sequence $\langle \underline{p}_1, \underline{p}_2, \ldots, \underline{p}_s \rangle$, and check backwards until the current point $\underline{p}_i$ lies on the right side of the boundary of linear constraint, that is, $\underline{p}_i.a \cdot x_i + \underline{p}_i.b \leq y_i + \delta$. Then, we calculate the intersection point generated by the boundary line of linear constraint, and the line identified by point $\underline{p}_i$ and point $\underline{p}_{i-1}$. (For $\underline{p}_1$, $\underline{p}_{i-1}$ is p_r.) We set the intersection point as new p_r. Then, the sequence of $\langle \overline{p}_i \rangle$ should be updated similarly as the previous procedure did, and then, the feasible region can be updated according to the new boundary points.

Case 3: $p_l.a \cdot x_i + p_l.b > y_i + \delta$. In this case, the whole feasible region will fall outside the solution area of linear constraint, so the updated region will be empty. The line segment can be determined, and new segmentation can be initialized.

Algorithm 2: ParaOptimal

Input: S: stream fragment starts from x_1; δ: the specified error bound
Output: A maximal δ-representative line segment starts from x_1

1 % **Initialization**
2 $s_1 = (x_1, y_1)$; $s_2 = (x_2, y_2)$;
3 $p_l = (\frac{y_2 - y_1 - 2\delta}{x_2 - x_1}, \frac{x_2 y_1 - x_1 y_2 + x_2\delta + x_1\delta}{x_2 - x_1})$; $p_r = (\frac{y_2 - y_1 + 2\delta}{x_2 - x_1}, \frac{x_2 y_1 - x_1 y_2 - x_2\delta - x_1\delta}{x_2 - x_1})$;
4 $\overline{p}_1 = (\frac{y_2 - y_1}{x_2 - x_1}, \frac{x_2 y_1 - x_1 y_2 + x_2\delta - x_1\delta}{x_2 - x_1})$; $\underline{p}_1 = (\frac{y_2 - y_1}{x_2 - x_1}, \frac{x_2 y_1 - x_1 y_2 - x_2\delta + x_1\delta}{x_2 - x_1})$;
5 Maintain the edge points of feasible region in clockwise order as $p_l, \langle \overline{p}_1 \rangle, p_r, \langle \underline{p}_1 \rangle$;
6 % **Processing a new point** $s = (x, y)$
7 **while** $p_l.a \cdot x + p_l.b \leq y + \delta$ **do**
8 $\quad$ % Update the edge points of feasible region
9 $\quad$ **if** $p_l.a \cdot x + p_l.b \leq y + \delta \leq p_r.a \cdot x + p_r.b$ **then**
10 $\quad\quad$ set the new p_r as the intersection point of line $b = -x \cdot a + y + \delta$ with the sequence $\langle \underline{p}_1, \underline{p}_2, \ldots, \underline{p}_s \rangle$;
11 $\quad\quad$ update the sequence $\langle \overline{p}_1, \overline{p}_2, \ldots, \overline{p}_t \rangle$ after intersecting with line $b = -x \cdot a + y + \delta$;
12 $\quad\quad$ set the new p_l as the intersection point of line $b = -x \cdot a + y - \delta$ with the sequence $\langle \overline{p}_1, \overline{p}_2, \ldots, \overline{p}_t \rangle$;
13 $\quad\quad$ update the sequence $\langle \underline{p}_1, \underline{p}_2, \ldots, \underline{p}_s \rangle$ after intersecting with line $b = -x \cdot a + y - \delta$;
14 $\quad$ **end**
15 **end**
16 % **Producing a line segment**
17 Let (a_o, b_o) be one point in the feasible region;
18 **return** *a line segment:* $y = a_o \cdot x + b_o$

After the feasible region update, if the new feasible region is empty, we can randomly choose a point in the old feasible region and apply the line segment identified by the chosen parameter to approximate the streaming points until the current point. The current point will be applied to initialize a new segmentation. Otherwise, the feasible region is not empty, and the algorithm will continue reading points and updating feasible region. All the procedures will work until the data stream ends.

Formally, the procedure for constructing a maximal δ-representative line segment by ParaOptimal is described in Algorithm 2.

4.2 Generalization of ParaOptimal

The core idea of ParaOptimal is to deal with the problem in the transformed space, which can inspire a more general model to approximate the data stream.

Generally speaking, an application may need to approximate the data with an k-degree polynomial, i.e., the approximate value of s_i is determined by $y_i' = a_0 + a_1 x_i + a_2 x_i^2 + \cdots + a_k x_i^k$. Given the error bound δ, we have

$$y_i - \delta \le a_0 + a_1 x_i + a_2 x_i^2 + \cdots + a_k x_i^k \le y_i + \delta.$$

These inequalities identify two hyperplanes in the parameter space of $\{a_i\}$ ($i = 1, 2, \ldots, k$), which are the linear constraints of the solution set. Similarly, if more data points are considered, the intersection of all sub-spaces identified by the hyperplanes can compose the final solution set. In this way, the ParaOptimal can be generalized to process the polynomial approximation of any order. Specifically, when $k = 0$, the problem is the PCR (Piecewise Constant Representation); when $k = 1$, the problem is exactly the PLR. It should be noticed that when $k \ge 2$, the algorithm will be executed in high-dimensional space, and the complexity will increase to $O(n^k)$ [34].

In addition, ParaOptimal algorithm can also support varied error bound, i.e., for point s_i, its corresponding error bound is defined as δ_i. During the processing, only the offset of the linear constraints will be adjusted accordingly, which will not affect the algorithm mechanism.

5 Theoretical Analysis of the Equivalence

The OptimalPLR and ParaOptimal both achieve the optimal results with linear complexity, but process the data points in different spaces, i.e. time-value stream space and slope-offset parameter space. As a result, we are motivated to compare these two algorithms theoretically. In this part, we will formally analyze the relationship and difference between the stream space and the parameter space, and discuss the equivalence between the two spaces and these two algorithms, which is able to provide a deeper understanding for the optimal algorithm.

5.1 Mapping of Two Spaces

First, we put our effort to discover the relationship between the stream point space and line parameter space, namely S-space and P-space. We have the following theorem:

Theorem 4 *The line function identified by two points in S-space is the intersection point of two lines in P-space, which are corresponding to the linear constraints brought by related points in S-space and vise versa.*

Proof As denoted in Fig. 12, consider the stream points s_i and s_j, and for the line function to approximate the data points, the function parameters must satisfy the following inequality set for upper error bound:

$$\begin{cases} a \cdot x_i + b \le y_i + \delta \\ a \cdot x_j + b \le y_j + \delta \end{cases}$$

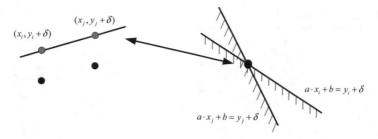

Fig. 12 Mapping between two spaces

Therefore, the boundary lines for corresponding linear constraints in P-space are $a{\cdot}x_i+b = y_i+\delta$ and $a{\cdot}x_j+b = y_j+\delta$, respectively. In this way, we can calculate the intersection point of these two lines by combining and solving the above equations.

Notice that in S-space, the line identified by the upper error bound points of s_i and s_j is the line crossing $(x_i, y_i + \delta)$ and $(x_j, y_j + \delta)$. So, the line function parameters are just mapped to the intersection point in the P-space as described above.

With the same argument, we claim that the line function identified by two points in P-space is the intersection point of corresponding lines in S-space. □

5.2 Equivalence Discussion

Based on Theorem 4, we can formally establish the equivalence between OptimalPLR and ParaOptimal algorithms.

For OptimalPLR, we model the convex hull which constrains the line candidates in the following way. As depicted in Fig. 13, the segment sequence composing the upper convex hull is denoted as $\langle \overline{q}_1, \overline{q}_2, \ldots, \overline{q}_s \rangle$, and the sequence composing the lower convex hull is $\langle \underline{q}_1, \underline{q}_2, \ldots, \underline{q}_t \rangle$. The line segments with maximum slope and minimum slope are q_h and q_l, respectively.

Define $p \sim q$ as the mapping relationship between point p and segment q, and the following theorem illustrates the equivalence between OptimalPLR and ParaOptimal in different spaces.

Theorem 5 *The points on the feasible region in P-space are one-to-one mapped with those segments composing the convex hull in S-space. More specifically, $\overline{p}_i \sim \overline{q}_i$, $\underline{p}_i \sim \underline{q}_i$, $p_r \sim q_h$ and $p_l \sim q_l$.*

We apply mathematical induction method to prove the theorem.

Proof

Step 1: For the first two points of initialization, the theorem holds obviously (Fig. 14).

Step 2: Assume for the first k points, the theorem holds.

Fig. 13 Segment definition
of convex hull in S-space

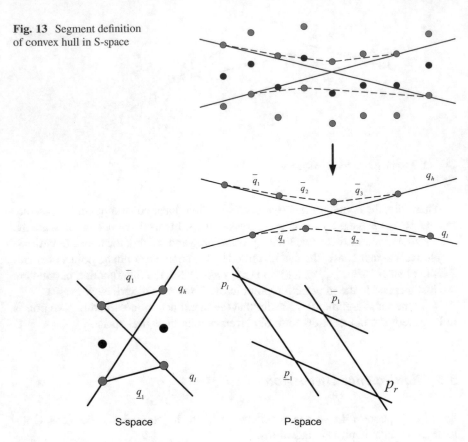

Fig. 14 Mapping for the first two points

Step 3: Now, we discuss when processing the $k + 1$ point and still only consider the upper error bound of s_{k+1}. The case of lower error bound can be proved with similar argument.

- **Case 1:** $q_h.a \cdot x_{i+1} + q_h.b < y_{i+1} + \delta$. In this case, the segment indicating the maximum slope q_h will not change. Since $q_h \sim p_r$, Case 1 describes the same situation as the first case in Fig. 11.

- **Case 2:** $q_l.a \cdot x_{i+1} + q_l.b \leq y_{i+1} + \delta \leq q_h.a \cdot xi + 1 + q_h.b$.

  In this case, q_h should be adjusted. According to OptimalPLR, the new q_h is the tangent line from the point $(x_{i+1}, y_{i+1} + \delta)$ toward the lower convex hull. As described in Fig. 15a, the point of tangency is the intersection point of q_2 and q_3, one of which is out of the upper error bound, and the other of which is within the upper error bound. According to Theorem 4, the intersection point is the segment connecting p_2 and p_3. Therefore, the new q_h is mapped with new p_r. For the lower convex hull, q_1 and q_2 should be deleted, and also for the feasible region in the P-space, p_1 and p_2 are deleted. So, the rest of q_i and p_i is still mapped.

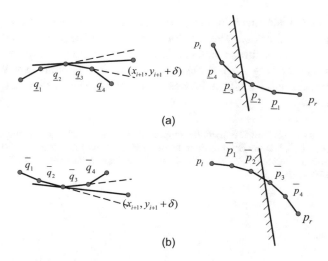

Fig. 15 Mapping for $k + 1$ points

Furthermore, the upper convex hull should also change according to the new q_h. The OptimalPLR proposes to find the tangent line from point $(x_{i+1}, y_{i+1} + \delta)$ toward the upper convex hull. Similar as lower convex hull, we can find the tangency point, and $\overline{q}_3$ and $\overline{q}_4$ are deleted. Accordingly, in P-space, $\overline{p}_3$ and $\overline{p}_4$ are deleted, and the rest of $\overline{q}_i$ and $\overline{p}_i$ is still mapped.

- **Case 3:** $q_l.a \cdot x_{i+1} + q_l.b > y_{i+1} + \delta$. In this case, the convex hull to indicate the possible line segment candidates is empty, which is exactly same as the third case in Fig. 11.

Based on the above analysis, we prove that ParaOptimal and OptimalPLR are theoretically equivalent.   □

The theoretical meaning above is that it guarantees that the two methods are basically the same, which reveals the inherent reason deriving the optimal approximation with linear complexity and also provides the understanding of the algorithm from different view. However, the two algorithms may have different performance in practice, due to the different ways of recording intermediate data during the processing, which results from the space they are based on. Such difference may affect the processing efficiency in both memory and time cost under certain situation.

6 Summary

In this chapter, the significant piecewise linear representation (PLR) for online stream approximation problem is investigated with given error bound in L_∞ norm. Based on the theoretical analysis, highly practical algorithms in time domain and parameter domain are introduced to achieve the optimal results with linear time and space complexity. The relationship between the time-value space and the parameter space is further investigated, which provides complete and better theoretical understanding for this classic problem. In the future, we will further study the application of PLR, and focus on the predictive stream analysis based on PLR patterns.

Acknowledgements This work is partially supported by Natural Science Foundation of China (Grant No. 61602353), Natural Science Foundation of Hubei Province (Grant No. 2017CFB505), and the Fundamental Research Funds for the Central Universities (Grant No. WUT:2017IVA053 and WUT:2017IVB028).

References

1. Atzori, L., Lera, A., Morabito, G.: The internet of things: a survey. Comput. Netw. **54**, 2787–2805 (2010)
2. Babcock, B., Babu, S., Datar, M., Motwani, R., Widom, J.: Models and issues in data stream systems. In: Proceedings of the 21st ACM SIGMOD-SIGACT-SIGART Symposium on Principles of Database Systems, pp. 1–16 (2002)
3. Berg, M.D., Cheong, O., van Kreveld, M., Overmars, M.: Computational Geometry Algorithms and Applications. Springer, Berlin (2008)
4. Buragohain, C., Shrivastava, N., Suri, S.: Space efficient streaming algorithms for the maximum error histogram. In: Proceedings of the 23rd International Conference on Data Engineering, pp. 1026–1035 (2007)
5. Chen, Q., Chen, L., Lian, X., Liu, Y., Yu, J.X.: Indexable pla for efficient similarity search. In: Proceedings of the 33rd International Conference on Very Large Data Bases, pp. 435–446 (2007)
6. Elmeleegy, H., Elmagarmid, A.K., Cecchet, E., Aref, W.G., Zwaenepoel, W.: Online piecewise linear approximation of numerical streams with precision guarantees. Proc. VLDB Endow. **2**, 145–156 (2009)
7. Gandhi, S., Nath, S., Suri, S., Liu, J.: Gamps: compressing multi sensor data by grouping and amplitude scaling. In: Proceedings of the ACM SIGMOD International Conference on Management of Data, pp. 771–784 (2009)
8. Gandhi, S., Foschini, L., Suri, S.: Space-efficient online approximation of time series data: streams, amnesia, and out-of-order. In: Proceedings of IEEE 26th International Conference on Data Engineering, pp. 924–935 (2010)
9. Gilbert, A.C., Kotidis, Y., Muthukrishnan, S., Strauss, M.J.: Surfing wavelets on streams: one-pass summaries for approximate aggregate queries. In: Proceedings of the International Conference on Very Large Data Bases, pp. 79–88 (2001)
10. Guha, S., Harb, B.: Approximation algorithms for wavelet transform coding of data streams. IEEE Trans. Inf. Theory **54**, 811–830 (2008)
11. Guha, S., Shim, K.: A note on linear time algorithms for maximum error histograms. IEEE Trans. Knowl. Data Eng. **19**, 993–997 (2007)
12. Jagadish, H.V., Jin, H., Ooi, B.C., Tan, K.L.: Global optimization of histograms. In: Proceedings of the ACM SIGMOD International Conference on Management of Data, pp. 223–234 (2001)
13. Keogh, E., Chu, S., Hart, D., Pazzani, M.: An online algorithm for segmenting time series. In: Proceedings of the 1st IEEE International Conference on Data Mining, pp. 289–296 (2001)

14. Keogh, E., Chakrabarti, K., Mehrotra, S., Pazzani, M.: Locally adaptive dimensionality reduction for indexing large time series databases. In: Proceedings of the ACM SIGMOD International Conference on Management of Data, pp. 151–162 (2001)
15. Lazaridis, I., Mehrota, S.: Capturing sensor-generated time series with quality guarantees. In: Proceedings of the 19th International Conference on Data Engineering, pp. 429–440 (2003)
16. Li, G., Li, J., Gao, H.: ε-Approximation to data streams in sensor networks. In: Proceedings of IEEE INFOCOM, pp. 1663–1671 (2013)
17. Li, S., Xu, L.D., Zhao, S.: The internet of things: a survey. Inf. Syst. Front. **17**, 243–259 (2015)
18. Nguyen, B., Abiteboul, S., Cobena, G., Preda, M.: Monitoring xml data on the web. In: Proceedings of the ACM SIGMOD International Conference on Management of Data, pp. 437–448 (2001)
19. Olston, C., Jiang, J., Widom, J.: Adaptive filters for continuous queries over distributed data streams. In: Proceedings of the ACM SIGMOD International Conference on Management of Data, pp. 563–574 (2003)
20. O'Rourke, J.: An on-line algorithm for fitting straight lines between data ranges. Commun. ACM **24**(9), 574–578 (1981)
21. Paix, A.D., Williamson, J.A., Runciman, W.B.: Crisis management during anaesthesia: difficult intubation. Qual. Saf. Health Care **14**(3), e5 (2005)
22. Palpanas, T., Vlachos, M., Keogh, E.: Online amnesic approximation of streaming time series. In: Proceedings of the 20th International Conference on Data Engineering, pp. 339–349 (2004)
23. Pang, C., Zhang, Q., Hansen, D., Maeder, A.: Unrestricted wavelet synopses under maximum error bound. In: Proceedings of the 12th International Conference on Extending Database Technology: Advances in Database Technology, pp. 732–743 (2009)
24. Pang, C., Zhang, Q., Zhou, X., Hansen, D., Wang, S., Maeder, A.: Computing unrestricted synopses under maximum error bound. Algorithmica **65**, 1–42 (2013)
25. Sathe, S., Papaioannou, T.G., Jeung, H., Aberer, K.: A survey of model-based sensor data acquisition and management. In: Managing and Mining Sensor Data, pp. 9–50. Springer, Berlin (2013)
26. Shatkay, H., Zdonik, S.B.: Approximate queries and representations for large data sequences. In: Proceedings of the 12th International Conference on Data Engineering, pp. 536–545 (1996)
27. Soroush, E., Wu, K., Pei, J.: Fast and quality-guaranteed data streaming in resource-constrained sensor networks. In: Proceedings of the 9th ACM International Symposium on Mobile Ad Hoc Networking and Computing, pp. 391–400 (2008)
28. Vullings, H.J.L.M., Verhaegen, M.H.G., Verbruggen, H.B.: Ecg segmentation using time-warping. In: Advances in Intelligent Data Analysis Reasoning About Data, vol. 2, pp. 275–285 (1997)
29. Wu, H., Salzberg, B., Zhang, D.: Online event-driven subsequence matching over financial data streams. In: Proceedings of the ACM SIGMOD International Conference on Management of Data, pp. 23–34 (2004)
30. Xie, Q., Huang, Z., Shen, H., Zhou, X., Pang, C.: Efficient and continuous near-duplicate video detection. In: Proceedings of the 12th International Asia-Pacific Web Conference, pp. 260–266 (2010)
31. Xie, Q., Huang, Z., Shen, H.T., Zhou, X., Pang, C.: Quick identification of near-duplicate video sequences with cut signature. World Wide Web J. **15**, 355–382 (2012)
32. Xie, Q., Shang, S., Yuan, B., Pang, C., Zhang, X.: Local correlation detection with linearity enhancement in streaming data. In: Proceedings of the ACM International Conference on Information and Knowledge Management, pp. 309–318 (2013)
33. Xie, Q., Pang, C., Zhou, X., Zhang, X., Deng, K.: Maximum error-bounded piecewise linear representation for online stream approximation. VLDB J. **23**, 915–937 (2014)
34. Xu, Z., Zhang, R., Kotagiri, R., Parampalli, U.: An adaptive algorithm for online time series segmentation with error bound guarantee. In: Proceedings of the 15th International Conference on Extending Database Technology, pp. 192–203 (2012)
35. Yu, L., Li, J., Gao, H., Fang, X.: Enabling ϵ-approximate querying in sensor networks. Proc. VLDB Endow. **2**(1), 169–180 (2009)

36. Zhang, Q., Pang, C., Hansen, D.: On multidimensional wavelet synopses for maximum error bounds. In: Proceedings of 14th International Conference on Database Systems for Advanced Applications, pp. 646–661 (2009)
37. Zhou, M., Wong, M.H.: A segment-wise time warping method for time scaling searching. Inf. Sci. **173**, 227–254 (2005)

Ensemble Dynamics in Non-stationary Data Stream Classification

Hossein Ghomeshi, Mohamed Medhat Gaber, and Yevgeniya Kovalchuk

Abstract Data stream classification is the process of learning supervised models from continuous labelled examples in the form of an infinite stream that, in most cases, can be read only once by the data mining algorithm. One of the most challenging problems in this process is how to learn such models in non-stationary environments, where the data/class distribution evolves over time. This phenomenon is called *concept drift*. Ensemble learning techniques have been proven effective adapting to concept drifts. Ensemble learning is the process of learning a number of classifiers, and combining them to predict incoming data using a combination rule. These techniques should incrementally process and learn from existing data in a limited memory and time to predict incoming instances and also to cope with different types of concept drifts including incremental, gradual, abrupt or recurring. A sheer number of applications can benefit from data stream classification from non-stationary data, including weather forecasting, stock market analysis, spam filtering systems, credit card fraud detection, traffic monitoring, sensor data analysis in Internet of Things (IoT) networks, to mention a few. Since each application has its own characteristics and conditions, it is difficult to introduce a single approach that would be suitable for all problem domains. This chapter studies ensembles' dynamic behaviour of existing ensemble methods (e.g. addition, removal and update of classifiers) in non-stationary data stream classification. It proposes a new, compact, yet informative formalisation of state-of-the-art methods. The chapter also presents results of our experiments comparing a diverse selection of best performing algorithms when applied to several benchmark data sets with different types of concept drifts from different problem domains.

H. Ghomeshi (✉) · M. M. Gaber · Y. Kovalchuk
School of Computing and Digital Technology, Birmingham City University, Birmingham, UK
e-mail: Hossein.Ghomeshi@mail.bcu.ac.uk; Mohamed.Gaber@bcu.ac.uk;
Yevgeniya.Kovalchuk@bcu.ac.uk

© Springer International Publishing AG, part of Springer Nature 2019
M. Sayed-Mouchaweh (ed.), *Learning from Data Streams in Evolving Environments*, Studies in Big Data 41, https://doi.org/10.1007/978-3-319-89803-2_6

1 Introduction

Over the past few years, data stream classification has been playing an important role in the area of knowledge discovery and big data analytics. The goal of classification, in the context of data streams, is to predict the class label of incoming instances from continuous data records that, generally, can be read only once in a limited time and memory. This is done by extracting useful knowledge from the past data inside the stream by using machine learning techniques.

As our digital world is growing rapidly, there are more data available in the form of data streams (e.g. World Wide Web, Internet of Things, etc.). That fact justifies the importance of paying attention to the mentioned domain of research since knowledge discovery is more complex in data streams. Suppose a sensor network that produces data related to credit card transactions of a bank from different types of devices (ATM, POS, online shopping, etc.) in the form of a data stream. A credit card fraud detection system can detect the fraudulent transactions using a data stream classification technique. The same task usually takes a long time or a high cost of resources (manual works) in traditional systems. Other applications of data stream classification include stock market analysis and prediction, weather forecasting, spam detection and filtering, traffic and forest monitoring, electricity management systems, web search pattern detection, sensor data analysis in an Internet of Things (IoT) network, among many other applications.

In such tasks, the characteristics of different types of data streams should be taken into careful consideration in order to have a successful data stream classification. General characteristics of data streams as seen by Babcock et al. [1] include the unlimited size of data streams, on-line arrival of data elements, order of data elements that is not governable, and finally, the restrictions about processing the elements only one time (it is possible to process an element more than once, but with a high cost of storing elements).

From the data distribution point of view, there are two types of data streams: *stationary* (stable) data streams, where the probability distribution of instances is fixed, and *non-stationary* (evolving) data streams, where the probability distribution of incoming data evolves or target concepts (labelling mechanism) change over time. This later phenomena is called *concept drift*. Existence of concept drifts in data streams makes classification tasks more complex and difficult to handle. This chapter is focused on non-stationary data stream classification.

As stated by Gama et al. [10], concept drifts may manifest in different forms over time. These forms can be divided into four general types: abrupt (sudden), gradual, incremental, and recurrent (recurring). Different types of concepts are depicted in Fig. 1. In abrupt or sudden concept drifts, the data distribution at time t is replaced suddenly with a new distribution at time $t+1$. Incremental concept drifts occur when the data distribution changes and stays in a new distribution after going through some new, unstable, median data distributions. In gradual concept drifts, the amount of new probability distribution of incoming data increases, while the amount of data that belong to the former probability distribution decreases over time. Recurring concepts happen when the same old probability distribution of data reappears after some time of a different distribution.

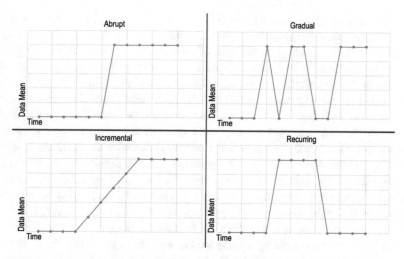

Fig. 1 Different types of concept drifts. Adapted from [10]

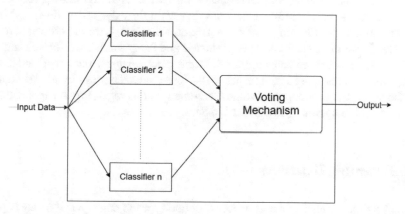

Fig. 2 An ensemble learning system, adapted from [18]

In order to cope with the concept drift problem in a data stream, it is important to build a classification system that adapts to different concept drifts as quickly as possible. *Ensemble learning* techniques are among the most effective approaches to do data stream classification [11], especially when dealing with non-stationary environments and concept drifts [18].

Ensemble learning is a machine learning approach in which multiple classifiers are created and combined with each other using a voting mechanism. In other words, as can be seen in Fig. 2, a voting mechanism is used to combine different classifiers' outputs in order to establish a single class label as the output of the ensemble. This is done in order to cover different types of features in a data stream. The combination is usually done by majority voting or weighted majority voting.

Adaptation to concept drifts can be achieved via different methods, with the most common ways being adding new classifiers into the ensemble, removing old classifiers from it, updating the weights of classifiers (assigning higher weights for more accurate classifiers at each iteration) and resetting the ensemble to an initial state. All of these methods are related to the dynamic behaviour of the ensemble.

This chapter aims to study ensembles' dynamic behaviour of existing ensemble methods in non-stationary data stream classification. In the authors' point of view, the key to building a successful ensemble is to understand different ensembles' dynamic behaviour and their reaction towards different concept drifts and environments. Furthermore, this chapter presents a new, compact, yet informative formalisation of the state-of-the-art methods. The authors argue that understanding the dynamic behaviour of different ensembles, along with the introduced formalisation, can facilitate the development of new as well as the current ensembles.

The rest of this chapter is organised as follows. In Sect. 2, the current ensemble approaches for non-stationary environments are introduced and their dynamic behaviour is discussed. A novel taxonomy for classification in such environments based on dynamic behaviour is proposed in Sect. 2. A formalisation, along with some of the current ensemble techniques (based on their dynamism diversity) using the proposed formalisation, is included in Sect. 3. In Sect. 4, the experimental results of several ensemble techniques are presented and analysed using different data sets. In Sect. 5, the observed behaviour of different mechanisms is discussed and several suggestions are proposed with respect to various characteristics. Finally, a summary of the experiments, along with some recommendations regarding the application of each ensemble approach, is provided in Sect. 6.

2   Ensemble Dynamics

In non-stationary environments, where different types of concept drifts may happen, it is expected that an ensemble adapts to a new concept drift swiftly. Since the adaptation in such environments is being done by adding a new classifier to the ensemble, removing old classifiers and changing the weights of current classifiers, understanding the dynamic behaviour of an ensemble toward different types of concept drifts can help us to choose the best approach for a specific application domain and develop new ensemble learning techniques for the required purpose.

This section discusses the aspects of an ensemble technique that form the ensemble dynamics of an approach. These aspects are addition, removal and updating of classifiers in an ensemble. The following subsections describe each of these aspects in detail, along with comparing different algorithms based on the dynamism related criteria. Over 20 different ensemble methods for non-stationary environments are studied and compared for this purpose. It has been tried to include the most recent and diverse ensemble techniques in this study.

2.1 Addition

Adding new classifiers that have been trained with recent instances in a stream is one of the most important actions that needs to be applied to the ensemble when data is evolving. The aim of this operation is adapting to drifting data, as well as improving classification accuracy of the ensemble based on the fact that, in most cases, incoming data is more likely to be similar to upcoming instances. The lack of this action might result in severe decrease in accuracy of the ensemble especially when concept drift is happening. One decision that needs to be taken when making a strategy for the ensemble is to decide when to add new classifiers to the ensemble, or in other words, what time frame needs to be taken for the addition operation. Some algorithms use a *fixed* time, while others use a *dynamic* time for it.

2.1.1 Fixed Time of Addition

Algorithms that use a fixed time to train and add new classifiers usually use a similar strategy; they do the addition operation after receiving a new block of data or after receiving a predefined p instances. A considerate number of existing algorithms are using this strategy to add new classifiers. The main challenge to build or use such algorithms is to pick a decent size of *blocks* or p in order to have the best possible output. Picking a large size might decelerate the adaptation while using a small size might make the ensemble sensitive to noise.

2.1.2 Dynamic Time of Addition

The algorithms that use dynamic time of training and adding are more diverse than the ones that use fixed time. Some of them use a method based on concept drift detection to determine when to train and add new classifiers. These types of algorithms start to train a new classifier when the concept drift detector signals and identifies a concept drift. Such mechanism is called *detection based dynamic* approach for addition operation. Some algorithms start to build a new classifier once the ensemble misclassifies an example. This strategy is called *misclassified based dynamic* in this chapter. Other mechanisms include adding a new classifier based on an acceptance factor [22]. This approach adds a new classifier when the threshold of the acceptance factor has been passed and a new classifier is needed. Another approach trains and adds a new classifier once an old classifier has been removed and a free space is available [24].

All of the studied algorithms and their addition mechanisms are shown in Table 1.

Table 1 Overview of dynamic behaviour of studied algorithms

Algorithm	Addition	Removal	Update	Train	Reference
SEA	Fixed	Full	No update	No	[25]
AWE	Fixed	Performance	Fixed	No	[26]
CDC	Other	Performance	Fixed	No	[24]
Aboost	Fixed	Full	Dynamic	No	[7]
CBEA	Fixed	Full	No update	No	[23]
AddExp	Misclassify	No removal	Dynamic	No	[16]
ACE	Detection	No removal	Dynamic	No	[20]
DWM	Misclassify	Performance	Fixed	Yes	[17]
TRE	Other	No removal	Fixed	No	[22]
Adwin Bag	Detection	Detection	No update	No	[2]
BWE	Detection	Detection	Fixed	No	[8]
Learn++	Fixed	No removal	Fixed	No	[9]
Heft-Stream	Fixed	Full	Fixed	No	[19]
WAE	Fixed	Full	Fixed	No	[27]
RCD	Detection	No removal	Dynamic	No	[12]
DACC	Fixed	Full	Fixed	Yes	[15]
ADACC	Fixed	Full	Fixed	Yes	[15]
AUE	Fixed	Full	Fixed	Yes	[6]
OAUE	Fixed	Full	Fixed	Yes	[5]
Fast-AE	Fixed	Full	Fixed	No	[21]

Fixed: *Fixed time of adding/updating the classifiers,* Detection: *Detection based (dynamic) times,* Misclassify: *Misclassified based (dynamic) times of adding classifiers,* Full: *Removing old classifier when the ensemble size is full,* Performance: *removing when the performance of a classifier drops from the predefined threshold*

2.2 Removal

Removing classifiers is a strategy to forget previously gained knowledge from a data stream that is unhelpful in the current situation, in order to adjust the ensemble to an updated state. In the majority of cases, removing classifiers from an ensemble happens when a predefined ensemble size is reached. However, in some algorithms, classifiers are being removed when their accuracy drops below a predefined threshold. In yet other algorithms, the size of ensemble is set unlimited, hence no classifier will be eliminated from the ensemble unless a pruning method is utilised. In this chapter, the removing strategy of algorithms is categorised into the following four types as can be seen in Table 1:

• *Full*: is performed when the set ensemble size is reached, and there is a new classifier that needs to be added to the ensemble. Such algorithms eliminate classifiers based on the classifiers' age in the ensemble or their performance on the recent data. All of the algorithms that use this mechanism for removing classifiers are the ones that use 'fixed' strategy for adding new classifiers.

- *Performance based*: is performed when the performance of a classifier in the last predefined k example drops below a specified threshold in the stream. In this mechanism, when a classifier becomes 'unhelpful' in a new concept, it is considered as an obsolete classifier and is removed from the ensemble.
- *Drift detection based*: when a concept drift detection method identifies a 'concept drift', a classifier is chosen to be eliminated. According to this approach, when the ensemble is full, for every new concept drift that would be detected by a drift detection method, only one classifier will be eliminated based on its accuracy over the recent instances. This happens in order to make room for a new classifier that needs to be added to the ensemble. All of such algorithms use a detection dynamic mechanism for adding new classifiers.
- *No removal*: a considerable amount of algorithms do not remove any old classifiers from the ensemble, and only the weights of classifiers are changed in order to avoid 'unhelpful' classifiers. The main reason behind this strategy is that when a classifier becomes weak in an environment, it can again be a useful classifier once a drift has happened, especially when that drift is a recurring concept drift. The algorithms that use such mechanisms need to have a pruning method in place, in order to avoid memory overload (since no classifier is being removed from the ensemble).

2.3 Update

Updating an ensemble can be referred to two main operations: the first one is updating the weight or ranking each classifier in the ensemble, and the second is whether or not to train old classifiers with incoming data. Most of the current algorithms use the 'updating weight' mechanisms in order to improve accuracy, however, only a few algorithms use the 'training old classifiers' mechanism, as a high load of memory is needed to train all of the classifiers with incoming data. The existing algorithms and their updating strategies are depicted in Table 1.

Updating the power of each classifier is an efficient way of improving the accuracy of an ensemble, especially when a concept drift happens and there are diverse classifiers in the ensemble. This is usually done by evaluating the positive effectiveness of each classifier in an environment and changing the weight, or the rank, of the classifier, so that the classifier with a higher accuracy towards the current condition has a bigger impact to the ensemble's output than a weaker classifier. Note that the algorithms that use a simple majority voting method for selecting the output of the ensemble are unable to employ this procedure, as there is no weight or rank set for each classifier. Similar to the addition stage, the mechanisms for updating weights of classifiers are categorised to *fixed* times and *dynamic* times. The methods that use dynamic times for updating classifiers are usually used when a drift is detected, except for AddExp algorithm [16], where updating is done when a classifier misclassified an example.

2.4   Ensemble Dynamics Taxonomy

To summarise the above operations, we propose a taxonomy for defining ensemble's dynamics in non-stationary data stream classification (Fig. 3). According to the proposed taxonomy, the dynamic behaviour of ensemble techniques is categorised into three main sections of addition, removal and update as mentioned in Sect. 2. The addition mechanisms are partitioned into fixed and dynamic methods and dynamic ones are then divided into detection based, performance based and others (such as using acceptance factor, etc.). The removal techniques are partitioned into full (which remove a classifier whenever the ensemble is full), performance based, detection based and no removal (methods that do not remove classifiers). Finally, the update section is divided into two subsections of updating the classifiers' weights (or ranks) and training old classifiers. The first updating subsection is partitioned into fixed times, dynamic times and no update, while the second one (training) is simply divided into the algorithms that do train the old classifiers (yes) and the ones that do not do so (no).

In order to compare and analyse the existing algorithms with regard to their dynamic behaviour (as presented in this chapter), six representative algorithms are selected based on their diversity across the elements of the proposed taxonomy. The selected algorithms are: Adaptive Boosting (Aboost) [7], Dynamic Weighted Majority (DWM) [17], Track Recurring Ensemble (TRE)[22], Adwin Bagging (AdwinBag) [2], Recurring Concept Drift (RCD) [12] and Online Accuracy Update Ensemble (OAUE) [5]. These algorithms and their dynamic characteristics are shown in Fig. 4. As can be observed from Fig. 4, none of the chosen algorithms follow the same path across all four phases of addition, removal, updating and training. Furthermore, there are no two algorithms with more than two common dynamic characteristics in this selection.

3   Formalisation

Formalising algorithms is a suitable way to comprehend and modulate the existing approaches in order to develop novel methods. In this chapter, a formalised version of the selected algorithms (as specified in Sect. 2) is presented with the intention to simplify the process of examining and building new approaches.

The following functions are used in the presented algorithms. Note that the sequence of the functions is the matter of importance in this formalisation, and the specific implementation of each function might be different for every algorithm.

- *Classify()*: The ensemble classifies data according to its combinational rule (e.g., weighted majority vote or majority vote).
- *Eval()*: Evaluating the whole ensemble or classifiers using an evaluation method.
- *Update()*: Updating the weights (or ranks) of all or one classifier with regard to its own evaluation and updating mechanism.

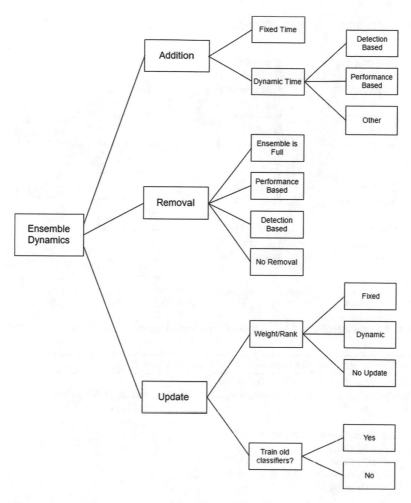

Fig. 3 The proposed taxonomy for ensemble's dynamics in non-stationary data stream classification

- *Build()*: Building a new classifier using the recently received data.
- *Add()*: Adding the newly built classifier to the ensemble.
- *Remove()*: Removing one or some classifiers based on the ensemble's specific removal mechanism.
- *Train()*: Training all or some old classifiers using the new data or data block.
- *DriftDetection()*: Detecting drifts using a concept drift detection method.

Adaptive boosting (Aboost) algorithm [7] presented in Algorithm 1 takes blocks of data, classifies the instances and then evaluates the ensemble's performance. If a concept drift is detected, it updates all the classifiers and assigns the default weight of 'one' to them. Otherwise, it assigns a weight to each instance in the block

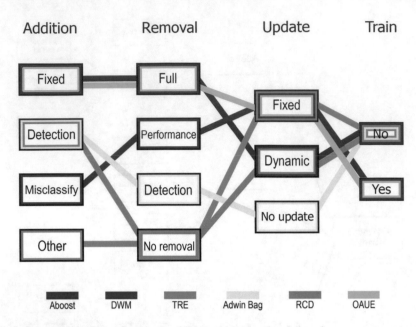

Fig. 4 Selected algorithms diversity in addition, removal and updating phases

---

Algorithm 1: ABOOST Adaptive Boosting Algorithm

Input: Continuous data blocks, $DB = \{db_1, db_2, .., db_n\}$
Output: C: A set of classifiers $c = \{c_1, c_2, .., c_m\}$ and their corresponding weights
$\qquad w = \{w_1, w_2, .., w_m\}$

1 $i := 1$
2 **while** *data stream is not empty* **do**
3 $\quad$ Classify(db_i)
4 $\quad$ Eval(Ensemble)
5 $\quad$ **if** *DriftDetection()=1 (drift is detected)* **then**
6 $\quad\quad$ Update(c)

7 $\quad$ **else**
8 $\quad\quad$ Eval(db_i)
9 $\quad\quad$ Update(db_i)

10 $\quad$ Build(c_{i+1})
11 $\quad$ Add(c_{i+1})
12 $\quad$ Remove() //remove oldest classifier
13 $\quad$ $i = i + 1$

---

according to whether or not that instance has been classified correctly (lines 8–9). If an instance is misclassified, a higher weight would be assigned. If the instance is classified correctly, the default weight of 'one' would be assigned. Finally, the oldest classifier in the ensemble will be removed and a new classifier will be built and added to the ensemble (based on the weighted instances in the block).

Algorithm 2: DWM Dynamic Weighted Majority algorithm

Input: A Data Stream, $DS = \{d_1, d_2, .., d_n\}$
l_i: Real label of the ith example
1 Θ: Threshold for removing classifiers
2 p: specified period for addition, removal and update of classifiers.
Output: A set of classifiers $c = \{c_1, c_2, ..., c_m\}$ and their corresponding weights
            $w = \{w_1, w_2, .., w_m\}$
3
4 $i := 1$
5 **while** *data stream is not empty* **do**
6 $\quad$ **for** $j = 1$ *to* $j = m$ **do**
7 $\quad\quad$ Classify(d_i)
8 $\quad\quad$ **if** *Output(b_j)$\neq l_i$ and i mod p = 0* **then**
9 $\quad\quad\quad$ Update()
10 $\quad\quad$ **if** *i mod p = 0* **then**
11 $\quad\quad\quad$ **while** $w_j < \theta$ **do**
12 $\quad\quad\quad\quad$ Remove(c_j)
13 $\quad\quad$ Train(c_j)
14 $\quad$ **if** *Classify(d_i) $\neq l_i$* **then**
15 $\quad\quad$ Build()
16 $\quad\quad$ Add()
17 $\quad$ $i := i + 1$

In Dynamic Weighted Majority (DWM) algorithm [17] shown in Algorithm 2, the data comes in an online form and after a predefined period p, if a classifier misclassifies an instance, the weight of that classifier will be reduced by a constant value regardless of the ensemble's output (lines 8–9). After this period, all the weights will be normalised and the classifiers with lower weights than a threshold (θ) will be removed from the ensemble. Finally, when the ensemble misclassifies an instance, a new classifier will be built and added to the ensemble. Note that all classifiers are trained incrementally with incoming samples.

In Tracking Recurrent Ensemble algorithm (TRE) [22], a new classifier will be added only when the ensemble error reaches a predefined permitted error (τ). Each classifier's weight would be updated once its performance drops below an acceptance factor (θ). This approach does not remove old classifiers unless a pruning method is used. The formalised version of this algorithm is depicted in Algorithm 3.

Adwin Bagging (AdwinBag) [2] is an approach that uses a concept drift detection method to specify when a new classifier is needed. When the new classifier is built and there is no more room in the ensemble, the worst performing classifier will be removed in order to make room for the new one. The formalised version of this algorithm is shown in Algorithm 4.

The Recurring Concept Drift framework (RCD) [12] presented in Algorithm 5 uses a buffer to store the context related to each data distribution in the stream. When the concept drift detector signals a warning, a new classifier is created and trained

---

Algorithm 3: TRE Tracking Recurrent Ensemble

Input: Continuous data blocks, $DB = \{db_1, db_2, .., db_n\}$
τ: Permitted error θ: Acceptance factor
Output: A set of classifiers $c = \{c_1, c_2, .., c_m\}$ and their corresponding weights
$\qquad\qquad w = \{w_1, w_2, .., w_m\}$

1 $i := 1$
2 **while** *data stream is not empty* **do**
3 $\quad$ Classify(db_i)
4 $\quad$ **for** $j = 1$ *to* $j = m$ **do**
5 $\quad\quad$ Eval(c_j)
6 $\quad\quad$ **if** *Eval(c_j) $< \theta$* **then**
7 $\quad\quad\quad$ Update(c_j)

8 $\quad$ Eval(Ensemble)
9 $\quad$ **if** *Ensemble error $> \tau$* **then**
10 $\quad\quad$ Build()
11 $\quad\quad$ Add()

12 $\quad$ $i := i + 1$

Algorithm 4: ADWINBAG Adwin Bagging algorithm

Input: A Data Stream, $DS = \{d_1, d_2, .., d_n\}$
M: Ensemble size
Output: A set of classifiers $c = \{c_1, c_2, .., c_m\}$

1 $i := 1$
2 **while** *data stream is not empty* **do**
3 $\quad$ Classify(d_i)
4 $\quad$ **if** *DriftDetection()=1* **then**
5 $\quad\quad$ Build()
6 $\quad\quad$ Add()

7 $\quad$ **for** $j = 1$ *to* $j = m$ **do**
8 $\quad\quad$ Eval(c_j)

9 $\quad$ **if** *Ensemble size = M* **then**
10 $\quad\quad$ Remove() //remove worst performing classifier
11 $\quad$ $i := i + 1$

---

alongside with a new buffer. If the concept drift detector signals a drift, which means the concept drift is certain, the framework checks whether or not the new concept drift is similar to previous concepts in the buffer (in case it is a recurring concept drift). If the new concept is similar to an old concept based on a statistical test, the framework uses the classifier created with that concept to classify incoming data and starts to train that classifier. If data distribution (concept) is new to the framework, it stores the new buffer and classifier in the system and uses the new classifier to classify incoming data. Otherwise, if the signal was a false alarm, the system ignores the stored data and continues to classify using the last classifier. In this approach, only one classifier is active at a time and does the classification task.

Algorithm 5: RCD Recurring Concept Drift framework

Input: A Data Stream, $DS = \{d_1, d_2,.., d_n\}$
Output: A set of classifiers $c = \{c_1, c_2,.., c_m\}$, Buffer list $b = \{b_1, b_2,.., b_m\}$
1 c_a= Active classifier, b_a= Active buffer
2 c_n= New classifier, b_n= New buffer
3 $i := 1$
4 **while** *data stream is not empty* **do**
5   Classify(d_i)
6   DriftDetection()
7   **switch** *Drift Detection* **do**
8   **case** *DriftDetection()= Warning and c_n=null* **do**
9   Build(c_n)
10   Build(b_n)
11   **case** *DriftDetection()= Warning and $c_n \neq null$* **do**
12   Train(c_n)
13   **case** *DriftDetection()= Drift* **do**
14            $c_a \leftarrow c_n$
15            $b_a \leftarrow b_n$
16   **otherwise do**
17            $c_n = b_n = null$
18      $i := i + 1$

Online Accuracy Update Ensemble (OAUE) Algorithm [5] is designed to incrementally train all of the old classifiers and weight them based on their error in constant time and memory. Since this approach needs a high load of memory due to training all classifiers with incoming data, a threshold for memory is assigned— so that, whenever the threshold is met, a pruning method is used to decrease the size of classifiers. The formalised version of this approach is depicted in Algorithm 6.

4 Experimental Study

To evaluate and analyse the selected algorithms (as specified in Sect. 2), and to observe the behaviour of different mechanisms with respect to various concept drifts, a set of experiments are conducted using several data sets. For this purpose, two artificial and two real world data streams are employed and the algorithms are compared using different criteria including classification accuracy, training time, memory usage, average adaptation time to concept drifts and average accuracy drop upon concept drifts. Each evaluation run in these experiments involves passing one of the chosen data sets described below through a specific algorithm in a form of data stream with a specified number of instances per interval.

Algorithm 6: OAUE Online Accuracy Updated Ensemble algorithm

Input: A continuous blocks of data, $DB = \{db_1, db_2, .., db_n\}$
M: Ensemble size, θ: Memory threshold
Output: A set of classifiers $c = \{c_1, c_2, .., c_m\}$ and their corresponding weights
$\quad\quad w = \{w_1, w_2, .., w_m\}$

1 $i := 1$
2 **while** *data stream is not empty* **do**
3 $\quad$ Classify(db_i)
4 $\quad$ Eval(c)
5 $\quad$ Build(c_i)
6 $\quad$ **if** $i < M$ **then**
7 $\quad\quad$ Add(c_i)

8 $\quad$ **else**
9 $\quad\quad$ Remove() //remove least accurate classifier
10 $\quad\quad$ Add(c_i)

11 $\quad$ **for** $j = 1$ *to* $j = m$ **do**
12 $\quad\quad$ Update(c_j)
13 $\quad\quad$ Train(c_j)

14 $\quad$ **if** *Memory usage* $> \theta$ **then**
15 $\quad\quad$ Prune(c) //decrease size of classifiers

16 $\quad$ $i := i + 1$

All of the experiments are implemented by Massive Online Analysis (MOA) framework [3]. MOA is an open source framework for data stream mining in evolving environments implemented at the University of Waikato. Aboost, DWM, OAUE and AdwinBag algorithms are already included in MOA framework and TRE and RCD algorithms are added using *classifiers and drift detection methods* extension.[1] The experiments were performed on a machine equipped with an Intel Core i7-4702MQ CPU @ 2.20 GHz and 8.00 GB of Installed memory (RAM).

4.1 Data Sets

4.1.1 Hyperplane Generator

Hyperplane generator is a synthetic data stream with drifting concepts based on a rotating hyperplane. A hyperplane in d-dimensional space is the set of points that satisfy $\sum_{i=1}^{d} w_i x_i = w_0$ where x_i is the ith coordinator of point x. Instances

---

[1]http://sites.google.com/site/moaextensions.

with $\displaystyle\sum_{i=1}^{d} w_i x_i \geq w_0$ are labelled as positive and $\displaystyle\sum_{i=1}^{d} w_i x_i < w_0$ are labelled as negative. Rotating Hyperplane Generator was introduced by Hulten et al. [14] and is a good way to simulate concept drift by changing the location of the hyperplane and additionally to change the smoothness of drifting data by specifying the magnitude of the changes.

For this experiment, the number of classes and attributes are set to four and fourteen respectively, and the magnitude of change is set to 0.01.

4.1.2 SEA Data Stream Generator

SEA generator is a data set inspired by four SEA concepts as described in [25]. The data set is a set of random points in a three-dimensional feature space. All three features have the value between 0 and 10, but only the first two are relevant to classification. These points are then divided into four blocks with different concepts. This is done to specify different concept drifts by assigning different conditions and goals for each class.

For this experiment along with the normal concept drifts that are being generated in the data stream, three abrupt concept drifts are added manually in three predefined times in order to be able to analyse the behaviour of different algorithms in the exact same situation specifically towards abrupt concept drifts.

4.1.3 Forest Cover-Type Data Set

Forest Cover-type data set [4] from the UCI Machine Learning Repository[2] contains the forest cover type of 30×30 meter cells obtained from the US Forest Service (USFS) Region 2 Resource Information System (RIS) data. It contains 581,012 instances and 54 attributes. The goal with this data set is to predict the forest cover type from cartographic variables.

4.1.4 Electricity Data Set

Electricity is a widely used data set by Harries and Wales [13] collected from the Australian New South Wales Electricity Market. In this market, prices are not fixed and are affected by demand and supply. The Electricity data set contains 45,312 instances. Each instance contains 8 attributes and the target class specifies the change of the price (whether going up or down) according to a moving average of the last 24 h.

---

[2]http://archive.ics.uci.edu/ml.

4.2   Results and Analysis

To evaluate performance of the selected algorithms, three generic criteria are used, including 'Accuracy', 'Execution time' and 'Memory usage'. Accuracy is the percentage of correctly classified instances in the given interval. Execution time and memory usage represent how much time and memory overall it takes for an algorithm to complete an evaluation run. For the second experiment with SEA data stream, two more criteria are utilised in order to study algorithms' behaviour in the presence of abrupt concept drift. These criteria are accuracy drop upon a concept drift and recovery time from a concept drift (adaptation time). Accuracy drop is calculated as a ratio (in %) between the last interval's accuracy rate before the new drift is introduced and the next interval's accuracy rate. Recovery time is the average number of instances it takes for each algorithm to achieve its average accuracy again (after an abrupt concept drift).

Figure 5 shows the percentage of classification accuracy of the selected algorithms over Hyperplane data set. Algorithms' behaviour in identical scenarios are demonstrated in Fig. 6. The elapsed time and memory usage are shown in Figs. 7 and 8.

Accuracy rates of the selected algorithms over SEA generator data stream are shown in Fig. 9. In order to create abrupt concept drifts at specified times, one million instances are generated from SEA data generator [25] with three different parameters that happen every 250 thousand instances. Figure 10 compares algorithms' accuracy rates in one chart.

Figure 11 demonstrates behaviour of the tested algorithms towards one of the added abrupt concept drifts. In Fig. 12, the selected algorithms are compared according to their accuracy drop, recovery time, average accuracy, average memory usage and the overall time of an experiment.

The average accuracy of the selected algorithms over Forest Cover-type data set is depicted in Fig. 13. Comparison of the algorithms over this data set is demonstrated in Fig. 14 and the memory usage and execution time of the algorithms is shown in Figs. 15 and 16, respectively. Average accuracy, performance comparison, memory usage and overall execution time of all algorithms over Electricity data set is demonstrated in Figs. 17, 18, 19, and 20 respectively.

Finally, the overall results of the above mentioned experiments are summarised in Table 2 according to the three main criteria: classification accuracy, execution time, and memory usage.

As it can be observed from Table 2 along with the above mentioned figures, the RCD algorithm [12] has the lowest memory usage and execution time in all experiments, but it has the poorest classification accuracy for the majority of the data sets (Hyperplane, SEA and Forest-cover type). This algorithm has a long recovery time from concept drifts (average of 25,900 instances), and its accuracy drops drastically upon abrupt concept drifts (average drop of 7.2% (Fig. 12)). The OAUE algorithm [5] has the best classification accuracy for the majority of the

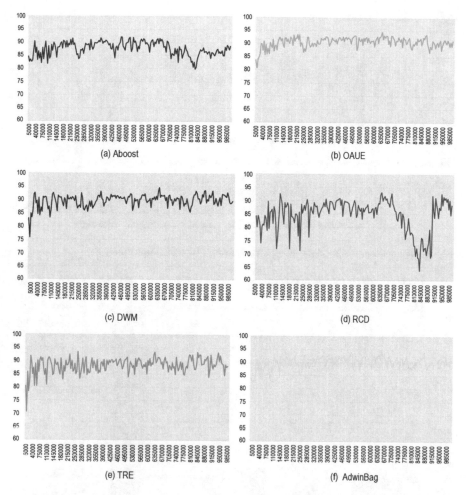

Fig. 5 Classification accuracy of different algorithms for Hyperplane data stream generator (one million instances). X-axis: Instance number; Y-axis: Accuracy in % (calculated every 5000 instances). (**a**): Adaptive Boosting algorithm, (**b**): Online Accuracy Updated Ensemble, (**c**): Dynamic Weighted Majority algorithm, (**d**): Recurring Concept Drift framework, (**e**): Tracking Recurrent Ensemble, (**f**): Adwin Bagging algorithm

data sets (Hyperplane, SEA and Forest-cover type) with an average execution time, but a relatively high memory usage, especially for Electricity data set (Fig. 19). Furthermore, it has the lowest recovery time from concept drifts (average of 4940 instances) and a medium performance drop upon abrupt concept drifts (average drop of 5.2%). Aboost algorithm [7] has a high classification performance (with the best observed accuracy over Electricity data set), along with an average execution time. However, it has the highest level of memory load for Hyperplane, SEA and Forest Cover-Type data sets. This algorithm has an average time of recovery from concept

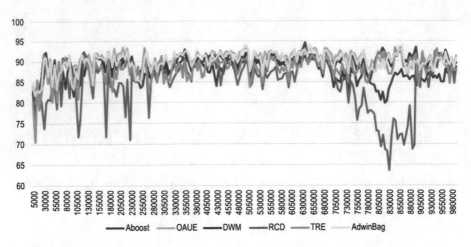

Fig. 6 Comparison of algorithms' accuracy rates over Hyperplane generator

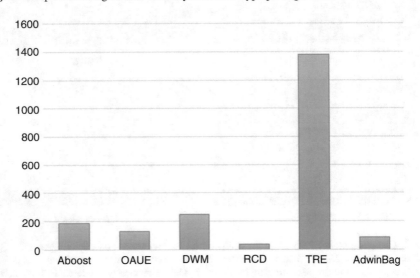

Fig. 7 Overall execution time of the selected algorithms over Hyperplane data generator (in seconds)

drifts (average of 23,800 instances) and the poorest classification performance upon abrupt concept drifts (7.9% drop). In DWM algorithm [17], memory usage and execution time are average and classification performance is acceptable in the majority of the cases (except for Electricity data set, where the average accuracy is 70.7%). The accuracy decreases slightly in the presence of abrupt concept drifts (average drop of 2.1%), and the average time of adaptation is mediocre (average of 20,900 instances). TRE algorithm [22] has the longest execution time for all data sets, however, the accuracy and memory usage are medium in most cases (except for accuracy in Electricity data set, which is 71.7%). This algorithm has

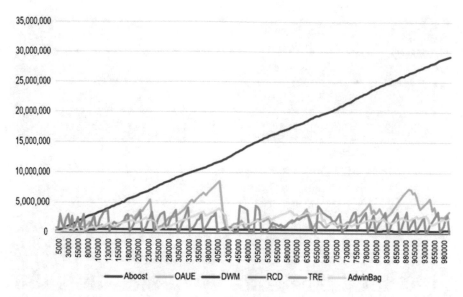

Fig. 8 Memory usage of the selected algorithms (calculated every 5000 instances). X-axis: Instance number; Y-axis: Memory in bytes

an average adaptation time (average of 16,000 instances) and the lowest accuracy drop upon abrupt concept drifts among the other selected algorithms (1.1%). Finally, in Adwin Bagging algorithm [2], the classification accuracy is high in all cases and memory usage and execution time are relatively low for the majority of the data sets. However, it has the highest value of adaptation time in concept drifts (average of 34,100 instances), and the accuracy drops drastically once an abrupt concept drift happens (5.9%).

5 Discussion

It is observed from the first experiment (over the Hyperplane Generator data set) that in RCD and Aboost algorithms, where the update phase happens in dynamic times (upon drift detection), the fluctuation of accuracy is relatively high (Fig. 5). This might be due to the fact that such algorithms are sensitive to concept drifts and also prone to false alarms, where noise can be detected as a concept drift. As can be seen from Fig. 6 for example, accuracy rates of both RCD and Aboost algorithms during the instance numbers 215,000–250,000 drop drastically, while for other algorithms, accuracy remains the same or increases. This can be explained by inability of the algorithms to distinguish between the true signal and noise. In the instance number 775,000, the accuracy of all algorithms drops smoothly, while in Aboost and especially RCD this drop is more severe. Furthermore, in Fig. 5, it is clear that accuracies of OAUE and DWM algorithms (b,c) have the lowest rate of

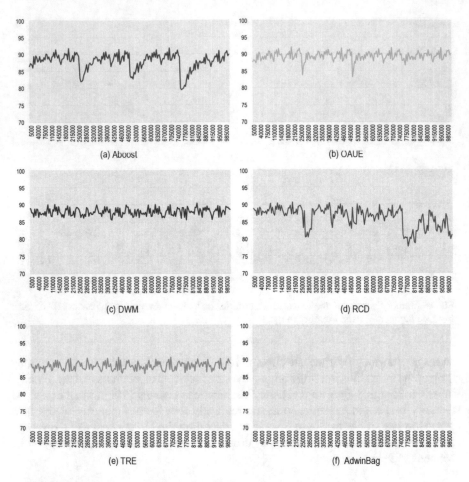

Fig. 9 Classification accuracy of the algorithms over SEA data stream generator (one million instances). X-axis: Instance number; Y-axis: Accuracy rate (calculated every 5000 instances in %). (**a**): Adaptive Boosting algorithm, (**b**): Online Accuracy Updated Ensemble, (**c**): Dynamic Weighted Majority algorithm, (**d**): Recurring Concept Drift framework, (**e**): Tracking Recurrent Ensemble, (**f**): Adwin Bagging algorithm

fluctuation among the others. This is possibly the result of training old classifiers as new examples are passing through the ensemble.

As can be noticed from the second experiment (over SEA generator), where three abrupt concept drifts are added in the points 250,000, 500,000 and 750,000 in Figs. 9, 10, 11, and 12, the accuracy of TRE algorithm is the most consistent when the concept drifts happen. The reason for such behaviour might be due to the fact that in TRE no classifier is removed from the ensemble and the algorithm regularly checks to see if a new concept drift is similar to an old one. Note that while RCD algorithm has the same mechanism as TRE for recurring concept drifts, a drift detection method is used in RCD, which makes it sensitive to concept drifts.

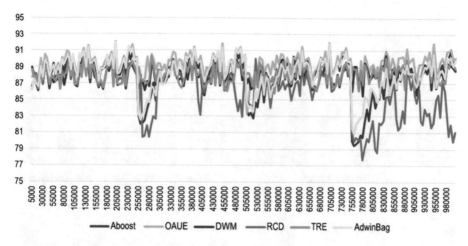

Fig. 10 Comparison of algorithms' accuracy rates over SEA generator

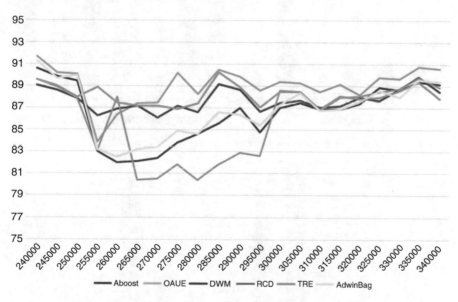

Fig. 11 A closer look at one of the added abrupt concept drifts and the behaviour of different algorithms towards this drift. X-axis: Instance number; Y-axis: Classification accuracy rate (calculated every 5000 instances in %)

In addition, only one classifier at a time is active in RCD algorithm. According to Fig. 10, accuracies of OAUE and DWM algorithms drop upon concept drifts, however, they recover from (adapt to) those concepts swiftly (Fig. 12). This is due to training old classifiers with incoming data in these algorithms. Similar to the previous experiment (Hyperplane data set), the algorithms that use concept drift detection methods (AdwinBag, RCD and Aboost) adapt to concept drifts slowly and their accuracy drops drastically upon concept drifts (Fig. 12).

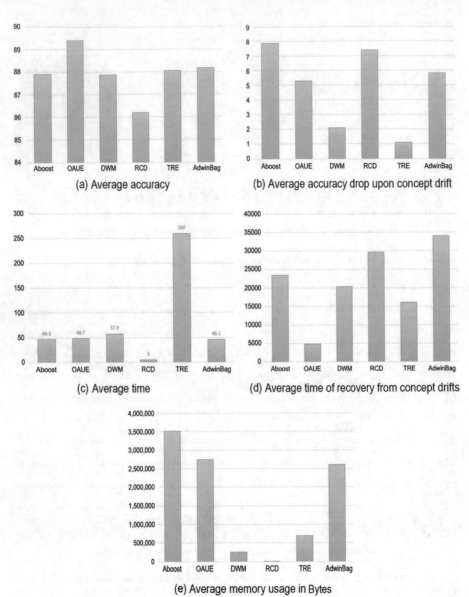

Fig. 12 Comparison of the algorithms in SEA generator data stream in the presence of different concept drifts in one million instances. (**a**): Average accuracy in %, (**b**): Average drop of accuracy upon concept drifts, (**c**): Average time of recovery (adaptation) from concept drift (number of instances to pass in order to achieve the average performance again), (**d**): Overall time, (**e**): Average memory usage

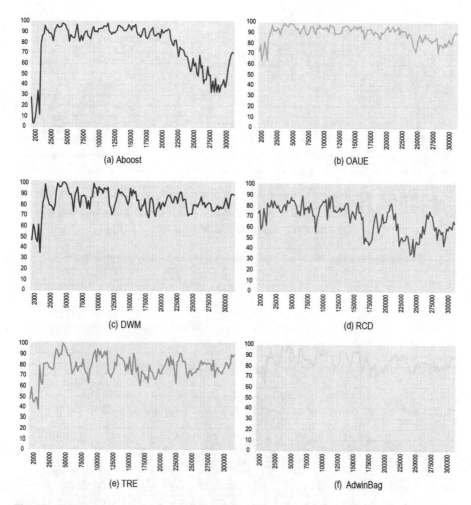

Fig. 13 Classification accuracy of the tested algorithms over the Forest Cover-type data set (581,012 instances). X-axis: Instance number, Y-axis: Accuracy rate (calculated every 2000 instances in %). (**a**): Adaptive Boosting algorithm, (**b**): Online Accuracy Updated Ensemble, (**c**): Dynamic Weighted Majority algorithm, (**d**): Recurring Concept Drift framework, (**e**): Tracking Recurrent Ensemble, (**f**): Adwin Bagging algorithm

In Fig. 11, which shows the first added concept drift more closely, it is interesting to see that four algorithms with different mechanisms (Aboost, OAUE, RCD and AdwinBag) have exactly the same reaction to the concept drift in the first 5000 instances after the concept drift happened (time 250,000–255,000). However after this time, each algorithm has its own reaction to the concept drift. This shows that these algorithms either do not have an immediate reaction to concept drifts or they detect and approve concept drifts with a delay. The consistency of TRE and DWM algorithms in Fig. 11 is significant as their accuracy rates do not drop from 86%,

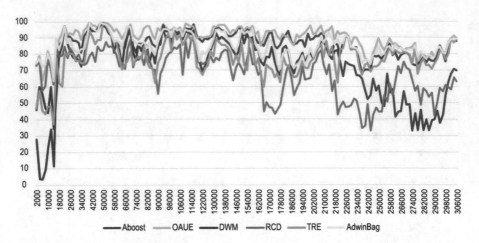

Fig. 14 Comparison of algorithms' accuracy rates over the Forest Cover-type data set

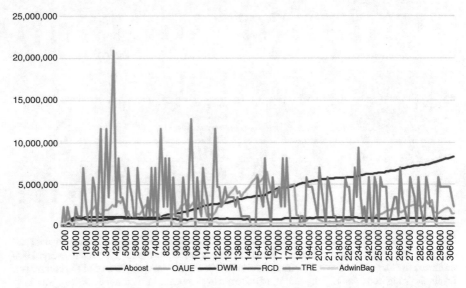

Fig. 15 Memory usage of the selected algorithms over Forest Cover-type data set (calculated every 2000 instances). X-axis: Instance number, Y-axis: memory used (in bytes)

which shows a promising reaction to such drifts. This is due to the fact that in these algorithms more new classifiers will be built and added to the ensemble when data is evolving. DWM adds a new classifier when an example is misclassified and TRE does the same when the threshold of an acceptance factor is passed. Upon concept drifts, both conditions happen often, which leads to adding new classifiers more frequently.

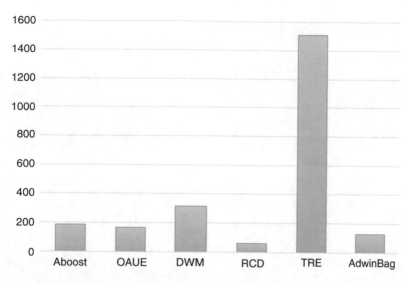

Fig. 16 Overall execution time of the selected algorithms over Forest Cover-type data set (in seconds)

The accuracy of all algorithms over Forest Cover-type data set fluctuates a lot according to Figs. 13 and 14 (when compared with other experiments). This shows that Forest Cover-type data set has more severe drifting data than the other data sets. However, the mentioned behaviour toward concept drifts remains the same and only the drop of accuracy is more drastic than in previous experiments (Fig. 13), particularly at points 164,000, 218,000 and 326,000. The initial accuracy of algorithms DWM, TRE and especially Aboost that have the average accuracy rates of about 50, 49 and 18% in the first 18,000 instances, shows that these algorithms need some time in order to achieve an initial consistency. Furthermore, it can be noticed from Fig. 14 that RCD and Aboost algorithms have inconsistent performance in highly evolving data sets.

In the last experiment done over Electricity data set, the fluctuation of accuracy as depicted in Fig. 17 is less than in previous experiments, which proves the fact that the number of concept drifts in this data set is less or concept drifts are more smooth (gradual) in this data set. This result is more prominent in Aboost algorithm which has consistent accuracy over Electricity data set, unlike for the rest of the algorithms (Fig. 18). Accuracy rates of the algorithms that use fixed times of addition (OAUE and Aboost) are higher and have more stability than the other algorithms. Furthermore, in DWM algorithm with average accuracy of 70.7% the performance is not satisfactory. This is possibly due to the addition operation in DWM algorithm happening when an instance is misclassified by the whole ensemble and the removal operation is based on the performance of each classifier at specific times.

As can be seen from Figs. 7, 12c, 16, and 20, TRE algorithm has the longest execution time by far. This is due to the fact that in TRE algorithms there is no

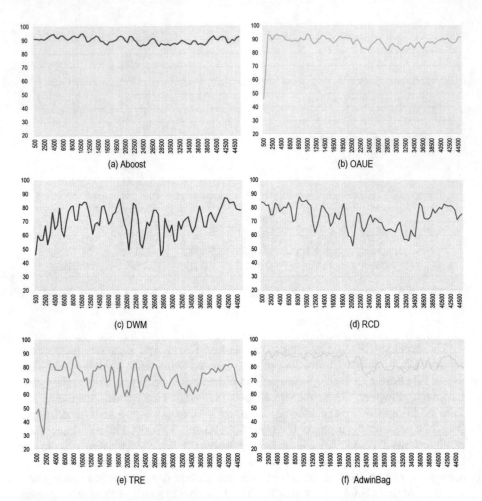

Fig. 17 Classification accuracy of different algorithms over Electricity data set (45,312 instances). X-axis: Instance number; Y-axis: Accuracy rate (calculated every 500 instances in %). (**a**): Adaptive Boosting algorithm, (**b**): Online Accuracy Updated Ensemble, (**c**): Dynamic Weighted Majority algorithm, (**d**): Recurring Concept Drift framework, (**e**): Tracking Recurrent Ensemble, (**f**): Adwin Bagging algorithm

mechanism for removing old classifiers and new incoming instances are being compared with previous ones in order to find recurring concept drifts. Furthermore, DWM and OAUE algorithms mostly have the longest time of execution after TRE, since in these algorithms all the classifiers are trained using new data. As opposed to TRE, RCD algorithm has the shortest execution time because in this algorithm, only one classifier is active at a time and a new classifier is being built at the same time. Finally, AdwinBag algorithm has relatively low execution times in all experiments (Table 2), as this approach does not update the weight or rank classifiers.

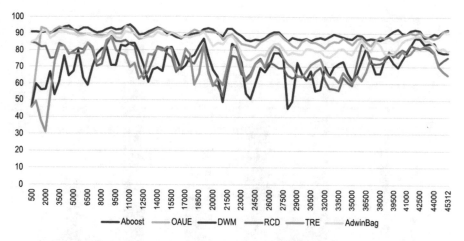

Fig. 18 Comparison of algorithms' accuracy rates over Electricity data set

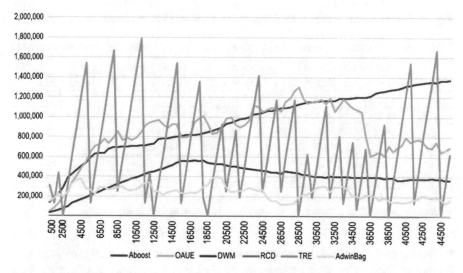

Fig. 19 Memory usage of the selected algorithms over Electricity data set (calculated every 500 instances). X-axis: Instance number; Y-axis: Memory in bytes

Memory usage of different experiments are shown in Figs. 8, 12, 15, and 19. It is clear that RCD algorithm is the least memory greedy method, with an average memory usage of around one kilobyte for each classification task. This is obviously due to its addition mechanism and output determination. A new classifier in RCD is built only upon new concept drifts, and for each example, only one classifier specifies the output. AdwinBag algorithm is the most efficient algorithm after RCD in terms of memory usage. This is because a limited amount of classifiers is involved in each iteration and in addition, previously built classifiers are not being trained or updated in the procedure. Aboost algorithm has a low memory usage at the

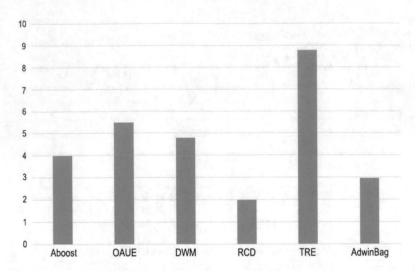

Fig. 20 Overall execution time of the selected algorithms over Electricity data set (in seconds)

Table 2 Overview of all experiments with respect to accuracy, execution time and memory usage

	Average accuracy (%)				Execution time (second)				Memory usage (KB)			
Algorithm	Plane	SEA	Forest	Elec	Plane	SEA	Forest	Elec	Plane	SEA	Forest	Elec
Aboost	87.56	87.91	75.48	**89.77**	188	46	184	4	14,807	3515	5631	550
OAUE	**90.69**	**89.42**	**90.11**	87.5	128	49	164	3.98	2888	2754	1947	844
DWM	89.66	87.87	80.26	70.73	247	58	317	5.48	609	249	932	330
RCD	84.7	86.2	62.66	73.45	**34**	**5**	**62**	**1.98**	**4.4**	**1.1**	**24.7**	**2.6**
TRE	88.33	88.05	77.3	71.68	1378	260	1509	8.77	1924	694	3423	666
AdwinBag	90.06	88.17	84.81	84.34	86	46	124	2.97	2059	2612	428	238

Bold values represent best performance in accuracy, time and memory criteria over different data streams (higher average accuracy, lower execution time and lower memory usage)

beginning of the process, but as new instances come, it grows incrementally. This feature of Aboost algorithm makes it heavy for long lasting tasks and light for short time classification tasks. TRE and OAUE algorithms use a high value of memory. This is because in TRE algorithm, no classifier is being removed, and in OAUE algorithm, all classifiers are incrementally trained. This leads to both algorithms needing a pruning method to shrink the number of classifiers in TRE, and to shrink the size of each classifier in OAUE.

Overall, for applications where the overall accuracy is an important factor and classification time is not restricted, OAUE and Aboost algorithms demonstrate better results. For applications where memory and time are limited, AdwinBag and RCD algorithms are recommended. In applications where consistency of accuracy is important, algorithms such as TRE, DWM and OAUE should be used. Finally, for applications where good performance upon concept drift is required, TRE and DWM algorithms are recommended as they demonstrate the most consistent results.

6 Summary

In this chapter, dynamic changes of different ensemble-based approaches for data stream classification in non-stationary environments have been studied. A novel taxonomy has been proposed based on dynamic behaviour of these approaches, in order to establish different types of reactions to concept drifts. To simplify the process of understanding the current approaches' dynamics and to encourage the development of novel algorithms, a formalisation method for classification algorithms in streaming analytics has been presented and characteristics of some of the current algorithms have been represented using this method. Finally, six algorithms out of the studied twenty algorithms are selected based on their diverse dynamic behaviour and four experiments have been designed for the purpose of this chapter. These experiments have been conducted to analyse the consequences of employing different types of dynamic behaviour towards different applications and concept drifts.

Based on the experimental results, the most significant observations are as follows:

- For the tasks where accuracy is the most important factor and the target data stream is being evolved frequently and severely, it is suggested to use algorithms with frequent updating and training phases, such as Aboost, OAUE and DWM.
- For applications where only a small amount of memory is available (such as IoT and sensor networks) or the time of output needs to be short, it is suggested to use RCD, Adwin Bagging and other algorithms with less updating or adding procedures.
- For the applications where frequent recurring concept drifts happen and memory usage is not crucial, algorithms such as TRE algorithm, where no classifier is deleted, are recommended to use.
- For the tasks where the memory capacity is limited and the job needs to be done in a short time with a satisfactory level of accuracy, AdwinBag and DWM algorithms are suggested.
- For least evolving data streams, algorithms such as Aboost and OAUE demonstrate the best performance, especially in terms of accuracy.
- For applications where recovery time (time of adaptation) is a critical factor, OAUE and DWM algorithms that train all classifiers incrementally seem to be the best option.
- For applications where consistency of accuracy rate is important, algorithms such as DWM, TRE and OAUE that update their classifiers frequently are the best choices.
- For applications where the accuracy over concept drifts is the most important factor, algorithms that add more classifiers or have more 'adding' procedures in evolving environments (e.g., misclassified-based and performance-based mechanisms of adding), such as TRE and DWM are proved to be the best options.

References

1. Babcock, B., Babu, S., Datar, M., Motwani, R., Widom, J.: Models and issues in data stream systems. In: Proceedings of the Twenty-First ACM SIGMOD-SIGACT-SIGART Symposium on Principles of Database Systems, pp. 1–16. ACM, New York (2002)
2. Bifet, A., Holmes, G., Pfahringer, B., Kirkby, R., Gavaldà, R.: New ensemble methods for evolving data streams. In: Proceedings of the 15th ACM SIGKDD International Conference on Knowledge Discovery and Data Mining, pp. 139–148. ACM, New York (2009)
3. Bifet, A., Holmes, G., Kirkby, R., Pfahringer, B.: Moa: massive online analysis. J. Mach. Learn. Res. **11**, 1601–1604 (2010)
4. Blackard, J.A., Dean, D.J.: Comparative accuracies of artificial neural networks and discriminant analysis in predicting forest cover types from cartographic variables. Comput. Electron. Agric. **24**(3), 131–151 (1999)
5. Brzezinski, D., Stefanowski, J.: Combining block-based and online methods in learning ensembles from concept drifting data streams. Inf. Sci. **265**, 50–67 (2014)
6. Brzezinski, D., Stefanowski, J.: Reacting to different types of concept drift: the accuracy updated ensemble algorithm. IEEE Trans. Neural Netw. Learn. Syst. **25**(1), 81–94 (2014)
7. Chu, F., Zaniolo, C.: Fast and light boosting for adaptive mining of data streams. In: Pacific-Asia Conference on Knowledge Discovery and Data Mining, pp. 282–292. Springer, Berlin (2004)
8. Deckert, M.: Batch weighted ensemble for mining data streams with concept drift. In: International Symposium on Methodologies for Intelligent Systems, pp. 290–299. Springer, Berlin (2011)
9. Elwell, R., Polikar, R.: Incremental learning of concept drift in nonstationary environments. IEEE Trans. Neural Netw. **22**(10), 1517–1531 (2011)
10. Gama, J., Žliobaitė, I., Bifet, A., Pechenizkiy, M., Bouchachia, A.: A survey on concept drift adaptation. ACM Comput. Surv. (CSUR), **46**(4), 44 (2014)
11. Gomes, H.M., Barddal, J.P., Enembreck, F., Bifet, A.: A survey on ensemble learning for data stream classification. ACM Comput. Surv. (CSUR) **50**(2), 23 (2017)
12. Gonçalves, P.M., Jr., De Barros, R.S.M.: RCD: a recurring concept drift framework. Pattern Recogn. Lett. **34**(9), 1018–1025 (2013)
13. Harries, M., Wales, N.S.: Splice-2 comparative evaluation: Electricity pricing (1999)
14. Hulten, G., Spencer, L., Domingos, P.: Mining time-changing data streams. In: Proceedings of the Seventh ACM SIGKDD International Conference on Knowledge Discovery and Data Mining, pp. 97–106. ACM, New York (2001)
15. Jaber, G.: An approach for online learning in the presence of concept change. PhD thesis, Citeseer (2013)
16. Kolter, J.Z., Maloof, M.A.: Using additive expert ensembles to cope with concept drift. In: Proceedings of the 22nd International Conference on Machine Learning, pp. 449–456. ACM, New York (2005)
17. Kolter, J.Z., Maloof, M.A.: Dynamic weighted majority: an ensemble method for drifting concepts. J. Mach. Learn. Res. **8**, 2755–2790 (2007)
18. Krawczyk, B., Minku, L.L., Gama, J., Stefanowski, J., Woźniak, M.: Ensemble learning for data stream analysis: a survey. Inf. Fusion **37**, 132–156 (2017)
19. Nguyen, H.-L., Woon, Y.-K., Ng, W.-K., Wan, L.: Heterogeneous ensemble for feature drifts in data streams. In: Advances in Knowledge Discovery and Data Mining, pp. 1–12 (2012)
20. Nishida, K., Yamauchi, K.: Adaptive classifiers-ensemble system for tracking concept drift. In: Machine Learning and Cybernetics, 2007 International Conference on, vol. 6, pp. 3607–3612. IEEE, New York (2007)
21. Ortíz Díaz, A., del Campo-Ávila, J., Ramos-Jiménez, G., Frías Blanco, I., Caballero Mota, Y., Mustelier Hechavarría, A., Morales-Bueno, R.: Fast adapting ensemble: a new algorithm for mining data streams with concept drift. Sci. World J. **2015**, 1–15 (2015)

22. Ramamurthy, S., Bhatnagar, R.: Tracking recurrent concept drift in streaming data using ensemble classifiers. In: Machine Learning and Applications, 2007. ICMLA 2007. Sixth International Conference on, pp. 404–409. IEEE, New York (2007)
23. Rushing, J., Graves, S., Criswell, E., Lin, A.: A coverage based ensemble algorithm (CBEA) for streaming data. In: Tools with Artificial Intelligence, 2004. ICTAI 2004. 16th IEEE International Conference on, pp. 106–112. IEEE, New York (2004)
24. Stanley, K.O.: Learning concept drift with a committee of decision trees. Informe técnico: UT-AI-TR-03-302, Department of Computer Sciences, University of Texas at Austin, USA (2003)
25. Street, W.N., Kim, Y.: A streaming ensemble algorithm (SEA) for large-scale classification. In: Proceedings of the Seventh ACM SIGKDD International Conference on Knowledge Discovery and Data Mining, pp. 377–382. ACM, New York (2001)
26. Wang, H., Fan, W., Yu, P.S., Han, J.: Mining concept-drifting data streams using ensemble classifiers. In: Proceedings of the Ninth ACM SIGKDD International Conference on Knowledge Discovery and Data Mining, pp. 226–235. ACM, New York (2003)
27. Woźniak, M.: Application of combined classifiers to data stream classification. In: Computer Information Systems and Industrial Management, pp. 13–23. Springer, Berlin (2013)

Processing Evolving Social Networks for Change Detection Based on Centrality Measures

Fabíola S. F. Pereira, Shazia Tabassum, João Gama, Sandra de Amo, and Gina M. B. Oliveira

Abstract Social networks have an evolving characteristic due to the continuous interaction between users, with nodes associating and disassociating with each other as time flies. The analysis of such networks is especially challenging, because it needs to be performed with an online approach, under the one-pass constraint of data streams. Such evolving behavior leads to changes in the network topology that can be investigated under different perspectives. In this work we focus on the analysis of nodes position evolution—a node-centric perspective. Our goal is to spot change-points in an evolving network at which a node deviates from its normal behavior. Therefore, we propose a change detection model for processing evolving network streams which employs three different aggregating mechanisms for tracking the evolution of centrality metrics of a node. Our model is space and time efficient with memory less mechanisms and in other mechanisms at most we require the network of current time step T only. Additionally, we also compare the influence on different centralities' fluctuations by the dynamics of real-world preferences. Consequently, we apply our model in the user preference change detection task, reaching competitive levels of accuracy on Twitter network.

1 Introduction

Social networks evolve at a fast rate of edge arrival. Due to this characteristic processing and analyzing such networks is challenging. Usually, the most feasible approach to process them is to consider network streams and perform analysis under online approaches and one-pass constraints [8]. When considering such evolving

F. S. F. Pereira (✉) · S. de Amo · G. M. B. Oliveira
Federal University of Uberlandia, Uberlandia, Brazil
e-mail: fabiola.pereira@ufu.br; deamo@ufu.br; gina@ufu.br

S. Tabassum · J. Gama
University of Porto, LIAAD INESC TEC, Porto, Portugal
e-mail: shazia.tabassum@inesctec.pt; jgama@fep.up.pt

© Springer International Publishing AG, part of Springer Nature 2019
M. Sayed-Mouchaweh (ed.), *Learning from Data Streams in Evolving Environments*, Studies in Big Data 41, https://doi.org/10.1007/978-3-319-89803-2_7

aspect, analysis can be performed from the perspective of changes on the structure of the underlying network.

According to [11], change detection refers to techniques and mechanisms for explicit drift detection characterized by the identification of change points or small time intervals during which changes occur. In evolving networks context, these changes can be detected observing the whole network, for instance communities [7] and motifs [4] evolution; or the changes can be analyzed in a node-centric way, where nodes centrality and roles are observed during network evolution [17]. In this work, our focus is on node-centric network evolving analysis.

Semantically, social networks structures model users' relationships and are able to reveal their behaviors and preferences. In an evolving environment, all the time users are facing others' opinions and being socially influenced, making their preferences fairly dynamic. This scenario dispatches several research efforts to investigate the interplay between user preferences and social networks [1, 13].

In this paper we propose a model for processing evolving network streams to calculate node-centric centrality scores and further to detect change points based on these centrality values. Essentially, our model consists in processing edges' stream, calculating nodes centrality, and employing aggregating mechanisms such as moving window average, weighted moving window average, and Page–Hinckley test [16] for tracking the evolution of the centrality measures for a given user u. We show that this evolution is related to u's preference changes.

Massive networks are difficult to be stored in memory as a total aggregated network and processing them as aggregate makes no sense with evolution. Our model uses a memory less Page–Hinckley test and two window based mechanisms with only window size of memory required for centrality measures of corresponding nodes. For updating centralities such as betweenness and closeness we only need to store the network of current time step T as far as for updating degree centrality we don't even need to store any network.

Rest of the paper is organized as follows: initially, in Sect. 2, we explain what are user preferences and how they evolve in a temporal environment. In Sect. 3 we present our method for processing evolving network streams to calculate node-centric centrality scores. Section 4 outlines main algorithms designed to implement the proposed aggregating mechanisms. In Sect. 5 we present the evolving networks and preference extraction strategy used to run the experiments, which are then described in Sect. 6. Finally, related work is highlighted in Sects. 7 and 8 concludes the paper.

2   User Preference Dynamics

We present background definitions that ground the applicability of our proposed model in this article for processing evolving network stream. The following concepts have been proposed in [18].

2.1 User Preferences

According to [14], a preference is a comparative opinion that establishes an order relation between two objects. For example, when a user says "I prefer to read news about politics than sports" we can identify a preference of politics over sports.

A preference relation on a finite set of objects on a running domain $A = \{a_1, \ldots, a_n\}$ is a strict partial order over A, represented by $\succ$. We denote by $a_1 \succ_t a_2$ the fact that a_1 is preferred to a_2 at time t. The transitive closure (TC) of all preference relations of a user u at t composes u preferences at t, denoted by TC_t^u. As example, let $A = \{sports, politics, religion\}$ be the set of objects in our running domain. We have $TC_4^u = \{sports \succ_4 politics, politics \succ_4 religion, sports \succ_4 religion\}$ and $TC_6^u = \{politics \succ_6 sports\}$.

2.2 Preference Changes in Evolving Environments

We have interest in analyzing preferences in evolving environments. User Preference Dynamics (UPD) refer to the observation of how a user evolves his/her preferences over time. Two preferences $a_1 \succ_{t'} a_2$ and $a_2 \succ_t a_3$, for $t' < t$, can be unified to infer a third preference $a_1 \succ_t a_3$ at t once considering transitivity of both, preference relation $\succ$ and timestamp order. We have defined this as user profile union Ω_t^u. Remarking on above example, $\Omega_6^u = \{sports \succ_4 politics, politics \succ_4 religion, sports \succ_4 religion, politics \succ_6 sports, politics \succ_6 politics, sports \succ_6 s ports\}$.

A key property of preferences in evolving environments is the irreflexivity. We say that Ω_t^u is inconsistent when there is a preference $a_1 \succ_t a_1 \in \Omega_t^u$. It would mean that "$u$ prefers a_1 better than a_1!", which does not hold for strict partial orders. When this scenario occurs, we say that there is preference change event. Therefore, we have defined a preference change based on the consistency of a user profile union.

Definition 1 (Preference Change δ_t^u) If Ω_t^u is inconsistent, a preference change event has been detected at time t for user u.

$$\delta_t^u = \begin{cases} 1 & \text{if } \Omega_t^u \text{ is inconsistent} \\ 0 & \text{otherwise} \end{cases} \tag{1}$$

Remarking on our running example, we have a preference change at $t = 6$, as $politics \succ_6 politics \in \Omega_6^u$.

3 Preference Change Detection

In this section we present our model for processing evolving network streams to calculate node-centric centrality scores and further employing aggregating mechanisms such as moving window average, weighted moving window average, and Page–Hinckley test for tracking the vacillation of preferences by users based on their temporal streams of centrality scores. Additionally we use a change point scoring function and change point detection threshold to quantify the change points obtained. Some assumptions for the above model such as handling of insertions and deletions are briefed in Sect. 3.8. Lastly in this section we detail the evaluation methodology for gauging our results.

3.1 Processing Streaming Network

For change detection in traditional uni-variate time series data, multiple change points were detected from a vector of elements distributed temporally. We have a similar temporal data, which is networked and evolving that makes it complex. In this work we consider not just one vector of uni-variate time series data but we have such $|V|$ number of multiple streams of uni-variate time series considering the centrality measures of each node in the network. This problem is different to multivariate time series as the centrality score stream of each node is treated independent to each other and the change points are detected per node independently. The only dependence is considered while computing centralities. To elucidate the aforementioned process, we delineate the model below.

Definition 2 (Evolving Network Stream) We consider an edge stream S which is a continuous and unbounded flow of objects $E_1, E_2, E_3 \ldots ..$, where each edge E_i is defined by (v, u, t) which represents a connection between vertices/nodes u and v at time t. The vertices $\{u, v, \ldots .\} \in V$ and get added to or deleted from V at anytime t.

For every incoming edge (v, u, t) from the above defined stream S, the centrality scores $C^m(v)$ and $C^m(u)$ for nodes u and v are updated in the order of t (In case of degree centrality, the degree of nodes u and v is updated for every incoming edge $\{u, v\}$ at t where as in the case of betweenness and closeness, the centrality of nodes are updated only after every T). Another variable T is a discrete time-step/time-interval with granularity defined by the user. In our experiments we considered the granularity of T as 1 day. After every T the centrality scores are reset and start accumulating again. Therefore, we keep track of centrality score of nodes per day. Consequently we store a set of nodes (with changing cardinality) and a streaming vector of its associated centralities per time step T. As a result, we have an independent non-stationary stream of centrality scores $\{C_{T_1}^m, C_{T_2}^m, C_{T_3}^m \ldots \ldots .\}$ for every node v in S after every time step T. To get a normalized version of scores, after every time-step T the centrality of a node is divided by the number of nodes in graph at T. Therefore we have normalized centrality scores $\{C_{T_1}^m, C_{T_2}^m, C_{T_3}^m \ldots \ldots\}$ in the vector stream. For notational simplicity in the below equations we use C_T for $C_T^m(v)$ as all notations for the techniques below are considered for a stream of centrality scores per node per centrality metric. Further we employed the smoothing mechanisms below (Sects. 3.3–3.5) to the above streams of centrality scores per node.

As the centrality score stream of every node is independent of each other, parallel implementation of the above aggregating mechanisms per node centrality feature stream is practicable. Though here we employed them sequentially for every node in the graph, as computing mean for |V| number of nodes is not expensive.

3.2   Computing Centralities

We have considered three centralities for our experiments, degree, betweenness and closeness which are explained below. The notion of employing three types of measures is to compare their efficiency for predicting preference changes of a node while considering the trade-offs between efficiency, time, and space complexity. The process of calculating centralities is explained below.

3.2.1   Degree Centrality

Degree Centrality of a node is the measure of number of edges adjacent to it. Degree Centrality can be computed on a fly for streaming data. As explained above the centrality score of a node is updated for every incoming edge. It is then stored in a queue as it should follow first in first out principle for window based approaches, then the edges are discarded. Degree centrality is space efficient with $O(V_T)$ (where V_T is the number of nodes from the time-step T) as we do not need to store the network. For updating centrality at the arrival of edge the cost is negligible with $O(V_T)$ as it only needs to increment a counter for degree centrality. For window based approaches the length of queue storing degree centralities is always equal to the window size W_S and the space used is constant, whereas for PH test we only need to store the current degree centrality score.

3.2.2   Betweenness Centrality

Number of shortest paths passing through a node is the Betweenness Centrality measure of that node. For computing this measure we follow the strategy described in [5] which is implemented by Gephi API.[1] Different from degree centrality,

---

[1] github.com/gephi.

betweenness is not computed in a stream fashion. This measure is not updated incrementally for every incoming edge, but the edges/network are stored for each T. After every T the betweenness centrality is batch calculated and current centrality score C_T generated. The edges are then discarded and the process restarts. Betweenness centrality requires $O(V_T + E_T)$ space and run in $O(V_T E_T)$ time on unweighted networks, where V_T and E_T is the number of nodes and edges from time-step T. Note that in this approach, centrality score is not computed incrementally, but after being generated we maintain the streaming strategy by adding in a queue the centrality score (after every T) for window based approaches (constant used space of size W_S) and maintaining only the current Betweenness score for PH test.

3.2.3 Closeness Centrality

Closeness centrality is the inverse of the average shortest path length between a node and all the other nodes in the graph. The smaller the average shortest path length, the higher the centrality for the node. Computing closeness centrality follows the same strategy above described for betweenness as it is also a shortest-path based centrality. We need to store incoming edges/network at each T, batch process closeness [5] and then discard edges and restart the process. The space complexity is $O(V_T + E_T)$ and requires $O(V_T E_T)$ time. In PH test just current closeness score is maintained and for window based approaches the centrality score is stored, consuming W_S space.

3.3 Moving Window Average (MWA)

A window of size W_S consists of data points from the latest temporal time steps $\{T, T-1, T-2, \ldots, T-(W_S-1)\}$. The window keeps on sliding to always maintain the latest W_S time steps and the data points from $T - W_S$ are forgotten. Alongside, the mean of data points within the window is calculated by using simple Eq. (2) where C_{T-i} is the stream of centrality scores at time-step $T-i$ using measure m per node. In this approach all the data points in the window are assigned equal weights.

$$\mu_T = \frac{1}{W_S} \sum_{i=0}^{W_S-1} C_{T-i} \tag{2}$$

As the window slides the mean of data points in the window is updated, by using the above Eq. (2) for small window sizes and Eq. (3) for large window sizes.

$$\mu_T = \mu_{T-1} W_S - C_{T-W_S} + C_T \tag{3}$$

3.4 Weighted Moving Window Average (WMWA)

Weighted moving window average follows the same window sliding strategy as in MWA and computes average over the data points in the window. The improvement over MWA is that the accumulated data points per time step T in the window are weighted linearly as given in Eq. (4). The oldest data points in the window attain a least weight and the latest data point acquires the highest weight linear to the least one. Weights are updated, when the window slides. Assignment of weights per data point depends on the size of window.

$$\mu_T = \sum_{i=0}^{W_S-1} \frac{C_{T-i}(W_S - i)}{W_S - i} \tag{4}$$

3.5 Page–Hinckley Test (PH)

Page–Hinckley [16] is one of the memory less sequential analysis techniques typically used for change detection [10, 11, 15, 21]. We use it as a non-parametric test, as the distribution is non-stationary and not known. This test considers a cumulative variable m_T, defined as the cumulated difference between the latest centrality score at T and the previous mean till the current moment, as given in Eq. (5) below:

$$m_T = \sum_{i=1}^{T} |C_T - \mu_{T-1}| - \alpha \tag{5}$$

where $\mu_T = 1/|T| \sum_{i=1}^{T} C_i$, $\mu_0 = 0$ and α = magnitude of changes that are allowed. For calculating μ_T we also need to store the number of time-steps passed.

Relative α Equation (5) given above uses fixed α value, which is not pertinent with our multiple vector streams of centralities per node, where the centrality scores of few active nodes are way higher than some least active nodes. Therefore, using same value of α over differing node centralities would not be fair enough. Hence, we use a relative α, which is relative with the differing centrality scores per node. **Relative α** is a point percentage of previous aggregated mean of that node, as given in Eq. (6). **Example:** Consider a node from stream S with a current centrality score $C_T = 2$ and $\mu_{T-1} = 3$ and fixed $\alpha = 0.1$. Using Eq. (5) we get m_T as $3 - 2 - 0.1 = 0.9$. Using relative alpha ($0.1 * 3 = 0.3$) in Eq. (6) we get m_T as $3 - 2 - 0.3 = 0.7$. Consider another node of high activity from the same stream S with $C_T = 60$ and $\mu_{T-1} = 50$, with fixed α $m_T = |50 - 60| - 0.1$ which doesn't make a proper sense, while using relative α, i.e ($0.1*60$) we get $m_T = |50 - 60| - 6$.

$$m_T = \sum_{i=1}^{T} |C_T - \mu_{T-1}| - \alpha \mu_{T-1} \tag{6}$$

Further to calculate change point score we need a variable M_T which is the minimum value of m_T and is always maintained and updated for every new time step T as given in Eq. (7)

$$M_T = \min(m_T; i = 1 \dots T) \tag{7}$$

3.6 Change Point Scoring Function

To detect the change points and their magnitude after every time-step T in MWA and WMWA, we use a change point scoring function given in Eq. (8)

$$\Gamma_T = \frac{|C_T - \mu_{T-1}|}{\max(C_T, \mu_{T-1})} \tag{8}$$

where C_T is the current centrality score and μ_{T-1} is the mean of previous centrality scores in the window. The change point scoring function gives the percentage point increase or decrease of the current centrality score with the previous mean. It takes values $0 \le \Gamma_T \le 1$.

For a PH test, after every time-step T the change points are scored using Eq. (9).

$$\Gamma_T = m_T - M_T \tag{9}$$

3.7 Change Point Detection

We can decide the magnitude of change allowed by the above change point scoring function. For this we use a threshold θ on Γ to signal an alarm of change in the preference of the user/node. It takes values either 0 or 1. "1" indicates a preference change and "0" indicates no change.

$$\epsilon_T = \begin{cases} 1, & \text{if } \Gamma_T \ge \theta. \\ 0, & \text{otherwise.} \end{cases} \tag{10}$$

Relative θ As a relative α given in Sect. 3.6 (Eq. (9)), we also apply a relative θ for detecting change points in PH Test only, as the change point scores from windowed approaches are already normalized in Eq. (8). Therefore to normalize threshold over multiple streams of centrality scores in PH test we use a relative threshold θ by multiplying the threshold θ with M_T of that node at time T as in Eq. (11).

$$\epsilon_T = \begin{cases} 1, & \text{if } \Gamma_T \geq (\theta \times M_T). \\ 0, & \text{otherwise.} \end{cases} \qquad (11)$$

3.8 Assumptions

While carrying out the above-mentioned mechanisms we considered the following assumptions.

- For window based approaches change detection starts only after the window of size W_S is filled.
- If there exists no edges for a node in a time step T, then the mean is calculated assuming a "0" centrality score.
- If a node is newly introduced (with edges) in the stream in the time interval T then the previous mean at $T - 1$ is considered "0" during change point scoring.
- In window based approaches if a node does not appear in the stream for a W_S time steps, the node is deleted to save space.

3.9 Evaluation

The output of our change detection model is a kind of binary classification problem. For every centrality stream, after every T we label C_T as a preference change point or not. We compare this with our ground truth of preference changes in the data set using a strategy described in Sect. 5.2. Therefore we use Recall, Precision, and F-measure. The experimental results are presented in Sect. 6. The purpose of our model is not just to evaluate the efficiency of our model but we intend to compare the different centrality measures and the impact of their changes on the dynamics of real-world preferences. The motive of using aggregate mechanisms is to investigate the deviations of current centrality values with the past values over preference changes.

4 Algorithms

We now present the algorithms for processing evolving network streams to calculate node-centric centrality scores and detect change points. The goal of all algorithms is to process the evolving network handled as edge stream and store in a binary

Algorithm 1 MWA

Input: Target node v, an edge stream $E_1...E_r$, window size W_S, threshold θ, centrality metric m
Output: A binary vector ε^v containing v's events for each time-step T
1: $V \leftarrow \emptyset, E \leftarrow \emptyset, N = (V, E)$
2: $\mu \leftarrow 0, C_T \leftarrow 0$
3: $W \leftarrow \emptyset$ // W is a queue structure representing the window
4: $T \leftarrow t$ // T is the initial time-step
5: **for** each incoming edge stream object $E_i = (u, z, t)$ **do**
6:      **if** $t \geq T + 1$ **then**   // next time-step
7:          **if** $m \in \{\text{betweenness, closeness}\}$ **then**
8:              $C_T \leftarrow \text{computeCentrality}(N, v)$
9:          **if** $W.size \leq W_S$ **then**   // window is not full
10:             $\mu \leftarrow \mu + C_T / W_S$
11:         **else**   // slides window
12:             $C_{T-W_S} \leftarrow W.head$
13:             $\mu \leftarrow \mu - (C_{T-W_S} / W_S) + (C_T / W_S)$
14:             $Dequeue(W)$
15:             $\Gamma_T \leftarrow |C_T - \mu| / max(C_T, \mu)$ // change point scoring function
16:             $\varepsilon_T^v \leftarrow \Gamma_T \geq \theta \; ? \; 1 : 0$ // change point detection
17:             $Enqueue(W, C_T)$
18:             $T \leftarrow T + 1$
19:             $E \leftarrow \emptyset, V \leftarrow \emptyset, C_T \leftarrow 0$
20:     **if** $m \in \{\text{betweenness, closeness}\}$ **then**
21:         $E \leftarrow E \cup \{E_i\}, V \leftarrow V \cup \{u, z\}$
22:     **else if** $m \in \{\text{degree}\}, z = v$ or $u = v$ **then**   // update degree incrementally
23:         $C_T \leftarrow C_T + 1$
24: **return** ε^v

vector ε^v events detected for node v at each time-step T. The centrality metric $m \in$ {degree, closeness, betweenness} slightly impacts on network processing approach. Algorithms complexity is ruled by computing centralities algorithms, previously described in Sect. 3.2.

Algorithm 1 refers to Moving Window Average (MWA) strategy. The main loop from line 5 indicates the network evolving. In lines 20–23 we distinguish centralities, as degree is incrementally calculated whereas closeness and betweenness need to store edges for each T to be calculated. Line 6 indicates a transition to next time-step. Lines 7–8 compute betweenness or closeness centrality according to input m using algorithms described in Sects. 3.2.2 and 3.2.3, respectively. While window W is not full, the moving window average μ is accumulated. When W is full (from line 11), $T - W_S$ centrality scores are forgotten and μ updated with current centrality score (lines 12–14). Finally, line 15 implements the change point scoring function (Eq. (8)) and line 16 detects a change point based on θ (Eq. (10)).

Algorithm 2 describes the Weighted Moving Window Average (WMWA), which follows the same sliding window strategy as in MWA. The difference is that the accumulated moving window average μ is composed by linearly weighted centrality scores (lines 13–17).

Algorithm 2 WMWA

Input: Target node v, an edge stream $E_1...E_r$, window size W_S, threshold θ, centrality metric m
Output: A binary vector ε^v containing v's events for each time-step T
   – *copy from line 1 – 8 of MWA algorithm* –
9: **if** $W.size > W_S$ **then**   // slides window
10:    $sum \leftarrow 0, denominator \leftarrow 0$
11:    **for** $i \leftarrow 0...W_S - 1$ **do**
12:       $sum \leftarrow sum + W[i] * (i + 1)$
13:       $denominator \leftarrow denominator + (i + 1)$
14:    $\mu \leftarrow sum/denominator$
15:    $Dequeue(W)$
16:    $\Gamma_T \leftarrow |C_T - \mu|/max(C_T, \mu)$ // change point scoring function
17:    $\varepsilon^v_T \leftarrow \Gamma_T \geq \theta ? 1 : 0$ // change point detection
   – *continue from line 17 of MWA algorithm* –

Algorithm 3 Page–Hinckley Test (PH)

Input: Target node v, an edge stream $E_1...E_r$, threshold θ, centrality metric m
Output: A vector ε^v containing v's events for each time step T
1: $V \leftarrow \emptyset, E \leftarrow \emptyset, N = (V, E)$
2: $T \leftarrow t$ // T is the initial time step
3: $m_T \leftarrow 0, M_T \leftarrow MAXVALUE, \mu \leftarrow 0, C_T \leftarrow 0, instancesSeen \leftarrow 0$
4: **for** each incoming edge stream object $E_i = (u, z, t)$ **do**
5:    **if** $t \geq T + 1$ **then**   // next time step
6:       **if** $m \in \{betweenness, closeness\}$ **then**
7:          $C_T \leftarrow computeCentrality(N, v)$
8:       $instancesSeen \leftarrow instancesSeen + 1$
9:       $percentualValue \leftarrow \mu * \alpha$
10:       $m_T \leftarrow m_T + |C_T - \mu| - percentualValue$
11:       $\mu \leftarrow \mu + C_T/instancesSeen$
12:       **if** $m_T < M_T$ **then**
13:          $M_T \leftarrow m_T$
14:       $\Gamma_T \leftarrow m_T - M_T$ // change point scoring function
15:       $\varepsilon^v_T \leftarrow \Gamma_T \geq \theta * M_T ? 1 : 0$ // change point detection
16:       $T \leftarrow T + 1$
17:       $E \leftarrow \emptyset, V \leftarrow \emptyset, C_T \leftarrow 0$
18:    **if** $m \in \{betweenness, closeness\}$ **then**
19:       $E \leftarrow E \cup \{E_i\}, V \leftarrow V \cup \{u, z\}$
20:    **else if** $m \in \{degree\}, z = v$ or $u = v$ **then**   // update degree incrementally
21:       $C_T \leftarrow C_T + 1$
22: **return** ε^v

The Page–Hinckley strategy is implemented by Algorithm 3. As in previous algorithms, network stream processing is represented by the main loop in line 4, considering the differences among centralities computation (lines 18–21 and lines 6–7). The cumulative variable m_T represents the cumulated difference between the latest centrality score C_T and the previous mean μ. Remark that in this strategy we do not handle window. Finally, at line 14 change point scoring function is applied and line 15 implements PH test (Eq. (11)).

5 Methodology

5.1 Dataset and Evolving Networks

We used Twitter data to run experiments. Through Twitter Streaming APIs,[2] during the course of 95 days, we collected tweets related to Brazilian news. All tweets, retweets, and quoted-status[3] containing some mention to the Brazilian newspaper, whose Twitter user is @*folha* were considered. In all, we collected 1,771,435 tweets and 292,310 distinct users in a time span of tweets posting times from Aug 7, 2016 to Nov 9, 2016. From the collected data, we built two different evolving networks.

5.1.1 Homogeneous Network

The first one, homogeneous network H, is based on retweets. Nodes are Twitter users and two nodes u_1 and u_2 have a direct edge (u_2, u_1, t) if u_1 retweeted u_2 at time t (note that edge direction represents the information flow). Figure 1a summarizes the network building strategy. This strategy is similar to the one used in [17].

An important characteristic of our H network is that it has a low average path length. This is a consequence of the fact that in Twitter a retweet always comes from the original post, not mattering from where the user read that post—from the user who originally posted it or from an intermediate user who already retweeted it. As our dataset has a high diversity of Twitter users, such as celebrities, common users, and commercial users, in this network we can identify nodes with different centrality roles. There are nodes that maintain high out-degree during the whole evolving period (mostly is retweeted, thus a content producer), nodes with high in-degree (mostly retweets, thus content consumer), and nodes with balanced behavior. Figure 2 describes nodes and edges evolution behavior. On average, H contains 10,189 nodes and 14,662 edges per day.

5.1.2 Bipartite Network

The second one is a bipartite network B where nodes are Twitter users or topics. Topics represent the main themes that users are *tweeting* about and have been extracted using LDA (Latent Dirichlet Allocation) model. In [17] more details about this topic extraction process can be found. An edge (u, p, t, w) means that user u tweeted/retweeted about topic p at time t, w times. As our time granularity is 1 day,

---

[2]https://dev.twitter.com/streaming/.
[3]Quoted-status are retweets with comments.

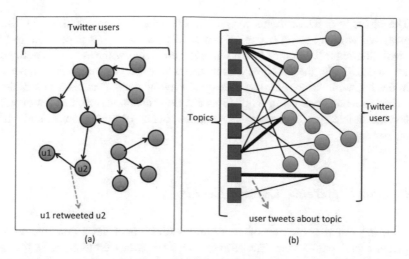

Fig. 1 Strategy to build (**a**) homogeneous network and (**b**) bipartite network from Twitter data

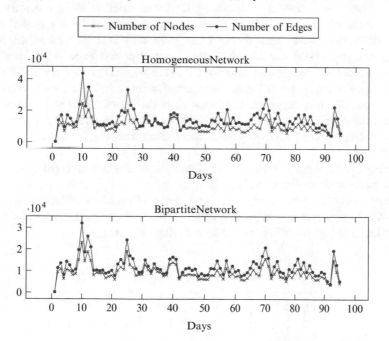

Fig. 2 Evolving behavior description of homogeneous (up) and bipartite (down) networks

a user can post many times at t and this volume of daily-post is represented through edges weights w. Thus B is a weighted bipartite network. Figure 1b summarizes this network building strategy.

The bipartite network represents essentially what is being "talked" in Twitter (topics) and who is talking (users). From a user perspective it is possible to track his evolution regarding topics that he mostly interacted with. In all, we defined 10 topics nodes. Thus, these nodes, during the whole network evolution, have high degree value, while user nodes, on average, have low degree values (as users do not talk about several topics in a day). Figure 2 describes nodes and edges evolution behavior. On average, B contains 9176 nodes, being 10 nodes topics, and 11,892 edges per day.

5.2 User Preference Change Events

Considering our dataset, rich of user's interactions over news content, we have extracted users' preferences. The strategy used to extract preferences is the same proposed in [17]. If user u tweets (or retweets) about o at time t, then u has more interest in o over the remaining topics in domain in that moment. We also considered a weight $w_t^u(o)$ based on the number of tweets posted at the same time about some topic o (our time granularity is 1 day; therefore, a user can post many tweets at t). In this case, the top posted topic is preferred over others, the second top posted topic is preferred over remaining ones and so on. $A = \{politics, international, corruption, sports, religion, entertainment, education, e - conomy, security, others\}$ is the set of topics in the preference domain on which we extract user preferences. These are the topics extracted by LDA model previous described (Sect. 5.1.2) and each tweet is labeled with one topic $o \in A$.

In Fig. 3 we illustrate the preference change evolution of user u_1 (id = 14594813) based on concepts explained in Sect. 2.

Once extracted users' preferences and detected preference change events, we used these change events as ground truth to evaluate our methods for node event detection based on centrality metrics in evolving networks.

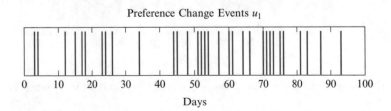

Fig. 3 Ground truth of preference change events for user u_1

6 Experiments

6.1 Experimental Environment

Nodes We have selected ten specific nodes to perform our analysis. These nodes were randomly selected among users in the dataset. For the sake of simplicity we refer to them as $u_1, u_2, \ldots, u_{10}$ and the corresponding dataset ids are: u_1 (14594813), u_2 (334345564), u_3 (3145222787), u_4 (343820098), u_5 (122757872), u_6 (28958495), u_7 (260856271), u_8 (636368737), u_9 (279635698), u_{10} (58488491). All the presented results correspond to the average value among these 10 nodes.

Threshold θ The threshold corresponds to the magnitude of changes allowed. We varied $\theta = \{0.01, 0.1, 0.2, 0.4\}$ in order to explore how this magnitude impacts on our findings. In PH method we set the factor $\alpha = 0.1$ which is also related to the magnitude of changes we deal with.

Window Size W_S The methods MWA and WMWA are based on sliding window mechanism. We varied $W_S = \{2, 5\}$ and analyzed how this window size can influence in our findings.

Evolving Networks We run experiments considering the homogeneous network H and the bipartite network B.

Centrality Metrics The node centralities we used are degree for B and in-degree for H, betweenness and closeness. During the analysis our target is to compare their efficiency for predicting preference changes of a node.

Methods Finally, we compare the three proposed methods in this paper MWA, WMWA, and PH. Table 1 summarizes the experimental environment.

Table 1 Experimental environment

Feature	Variation	Default
Node	$u_1, u_2, \ldots, u_{10}$	u_1
θ	0.01, 0.1, 0.2, 0.4	0.1
α (PH)	0.1	0.1
W_S (MWA, WMWA)	2, 5	2
Centrality metric	Degree/in-degree, betweenness, closeness	In-degree
Evolving network	Bipartite, homogeneous	Homogeneous
Methods	MWA, WMWA, PH	MWA

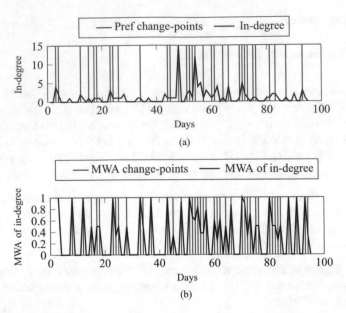

Fig. 4 Comparative analysis between preference change-points and MWA detected change-points considering default setup for node u_1. The accuracy is F-measure = 0.61. (**a**) u_1 in-degree values against u_1 preference ground-truth change-points. (**b**) MWA values for u_1 in-degree against MWA detected change-points

6.2 Detecting u_1 Change-Points

As illustrative example, we describe the change-point detection process for user u_1 considering our default scenario. As time flies, current u_1 in-degree centrality and MWA of past u_1 in-degree centralities are calculated and compared to each other. Then, the change-point scoring function is computed raising alarms when change-points are detected. In the end, the detected change-points and the preference ground-truth are compared to obtain the accuracy of the method.

In Fig. 4a we show the relation between u_1 in-degree centrality evolution and preference ground truth change-points in homogeneous network (default setup). Intuitively, the expectation is that centrality peaks/valleys overlap preference change-points. We can observe that there are some overlaps despite the number of preference change events is higher than peaks/valleys. In Fig. 4b we depicted the evolution of MWA of u_1 in-degree centrality in the same scenario previously described. The performance of the method is directly related to the balance between the magnitude of changes that we are looking for, i.e. threshold θ value, and the past average values that should be considered, i.e. window size W_S. Considering our default setup, we reach precision 0.46, recall 0.9, and F-measure 0.61 for node u_1.

6.3   Performance of Proposed Methods

Here we come to the task of evaluating our proposed methods considering different scenarios. Figures 5 and 6 present the results for homogeneous and bipartite networks, respectively. In a general way, bipartite network got lower accuracy than homogeneous network, specially when considering high threshold and window sizes. As in bipartite network topic nodes occupy extremely central positions, the notion of centrality cannot be efficient for the task of detecting events.

When observing centrality measures, betweenness clearly got the worst performance. This can be explained by the fact that bridge nodes usually do not vary significantly their positions. Moreover, we can conclude that the notion of bridge itself is not a good centrality metric to correlate with preference change. Even so, in most of the scenarios betweenness got performance superior to a naive random baseline (F-measure = 0.5). Comparing degree and closeness, we observe that degree is slightly more accurate than closeness. Considering that the notion of preference change has been defined based on the number of tweets, and consequently number of edges incoming in a node, it was expected that degree centrality fit well in the context.

We also observe that for homogeneous graph, degree centrality performs better and bipartite graphs closeness is superior. This difference in behaviors is based on the graph data, as homogeneous graph is a multi-graph incorporating frequency of edges i.e edge weights are considered while using this graph. In bipartite graph the weights of edges are not considered. Hence we see that when frequency of an edge is considered, it is favorable to employ degree centrality, while when we do not know the frequency of edges or for un-weighted graphs it is beneficial to use closeness. Finally, comparing the performance of our proposed methods, Page–Hinckley got the most significant results, with homogeneous behavior in different scenarios. We can conclude that the idea of accumulating values and base the evolution on the minimum obtained value so far is the most suitable. On average, PH degree performances reach F-measure = 0.75. There is no consensus between MWA and WMWA performances and the choice for one method is not conclusive.

6.4   Impact of Parameters

The goal here is to analyze the behavior of parameters trade off. Varying W_S means that we are considering the recent past for low values (short-term events) or a big historic for high values (long-term events). θ adjusts the intensity of the events, varying from smooth to drastic events. From a general viewpoint, we observed that performances keep the proportions according to parameters setup for MWA and WMWA, but not for PH, which got similar results independent of the parameter setup. Another observation is that recall is always higher than precision, except for the highest $\theta = 0.4$. This behavior indicates that performances decrease as the magnitude of allowed changes increase.

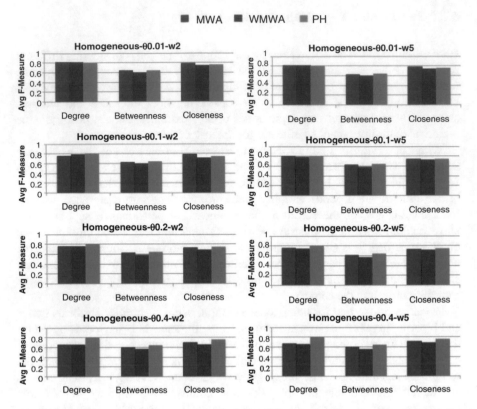

Fig. 5 Performance of our proposed methods in different scenarios for different centrality measures in homogeneous network

7 Related Work

We discuss related work in the directions of event and change detection in networks, and graph stream processing. Our proposal is innovative when considering *change detection* from a *node-centric perspective* in a *stream processing* environment.

Event Detection in Networks The most studied events in evolving networks are anomalies and bursts [7]. Anomaly detection refers to the discovery of rare occurrences in datasets and has been largely explored when considering dynamic networks [3, 19]. The pioneer work in anomaly detection for dynamic graphs is [12]. It addresses the problem considering a time sequence of graphs (graph sequences). First, authors extract activity vectors from the principal eigenvector of dependency matrix. Next, via singular value decomposition, it is possible to find a typical activity pattern (in $t - 1$) and the current activity vector (t). In the end, the angular variable between the vectors defines the anomaly metric. The network processing is through snapshots, not in a streaming fashion. Moreover, this Eigen Behavior based Event

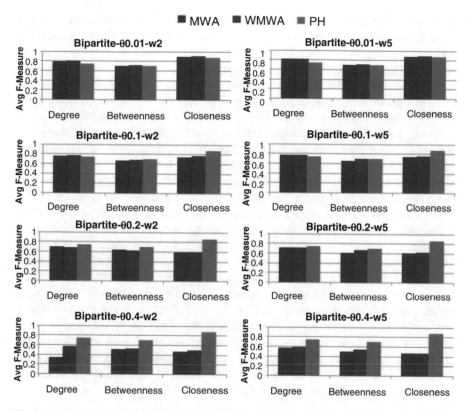

Fig. 6 Performance of our proposed methods in different scenarios for different centrality measures in bipartite network

Detection (EBED) method is orthogonal to ours as it detects events in a global perspective of the network, while ours is node-centric.

Burst events are generally related to topic evolving detection and tracking [7]. These works are looking for events like hot buzz words, what are users' sentiments about a product release or how is a specific topic evolving. In [6] the goal is to track interest profiles in real time by detecting bursts in Twitter's social media stream in real time using linear regression. These approaches are orthogonal to ours because are focused on the content of the network (texts, topics) not on the topology evolution analysis. The work [2] incorporates network structure in event discovery over purely content-based methods. Each text message is associated with at least a pair of actors in the social network. The events detected are also related with topics evolving. Finally, in [20] authors consider the problem of mining activity networks in order to identify interesting events, such as a big concert in a city, or a trending keyword in a user community in a social network. The algorithms are founded in geo-spatial event detection information. Any stream processing strategy is addressed.

Change Detection Our model is similar to [22] where three window based aggregation techniques are employed, namely simple aggregation, linear aggregation, and exponential aggregation based on assigning weights to time-steps in the window. These methods were used to aggregate the centrality scores per node inside the window. The aggregations were called persistence p. The comparison of aggregations of time-step p_t to $p_t - 1$ of a node v was called Emergence. Further they used a regression model over the above outcome to predict the future centrality scores (based on previous window based aggregated scores) and evaluate the results with the actual ground truth centrality scores. While in our work we used three techniques from which two are window based and the other is Page–Hinckley test for change detection. For the window based techniques, we used two weighting mechanisms inside the window one is, equal weights to all time-steps other is linearly increasing weights to the latest time step. The two window based techniques not just aggregate the centrality scores per node inside the window but calculate the mean of scores. As the Emergence function in [22] we use a change point scoring function which calculates the percentage change of new score at t to $t - 1$ per node. Our scenario is similar to a binary classification problem, therefore we used precision, recall, and F-measure to evaluate by comparing the change points detected in ground truth. The motive of our work is not just to evaluate the correlation between previous scores and the latest score but how well the centrality scores depict the real-world preference changes. Additionally we present the comparison between three techniques (briefed above) in a streaming perspective for change detection. Further more we analyze which of the three centrality metrics portrays the real preference change points.

Graph Streams Processing Processing graphs as streams is an incoming problem. The work [4] is one of the most complete works when considering data mining in evolving graph streams. The focus, however, is on mining closed graphs, not on event detection. In [8] a framework for processing graphs as streams is proposed for the link prediction task. This framework considers the cumulative grown of the graph, not addressing the space saving feature [9].

8 Conclusion

We have proposed a model for processing evolving network streams and detect change points based on centrality measures. We have explored two window-based aggregating mechanisms—moving window average (MWA) and weighted moving window average (WMWA), and a third memory less mechanism—Page–Hinckley (PH). Moreover, we have implemented algorithms considering degree, betweenness, and closeness centrality measures. We have applied our proposed model in the user preference change detection problem and evaluated the performance of our algorithms on homogeneous and bipartite Twitter networks. As a result, degree centrality in homogeneous network using PH approach has performed with the highest F-measure values.

Acknowledgements This work was supported by the research project "TEC4Growth - Pervasive Intelligence, Enhancers and Proofs of Concept with Industrial Impact/NORTE-01-0145-FEDER-000020," financed by the North Portugal Regional Operational Programme (NORTE 2020). This work was also supported by the Brazilian Research Agencies CAPES, CNPq, and Fapemig.

References

1. Abbasi, M.A., Tang, J., Liu, H.: Scalable learning of users' preferences using networked data. In: Proceedings of the 25th ACM Conference on Hypertext and Social Media, HT '14, pp. 4–12. ACM, New York, NY (2014)
2. Aggarwal, C.C., Subbian, K.: Event detection in social streams. In: 12th SIAM International Conference on Data Mining, pp. 624–635 (2012)
3. Akoglu, L., Tong, H., Koutra, D.: Graph based anomaly detection and description: a survey. Data Min. Knowl. Disc. **29**(3), 626–688 (2015)
4. Bifet, A., Holmes, G., Pfahringer, B., Gavaldà, R.: Mining frequent closed graphs on evolving data streams. In: 17th ACM SIGKDD International Conference on Knowledge Discovery and Data Mining, KDD '11, pp. 591–599 (2011)
5. Brandes, U.: A faster algorithm for betweenness centrality. J. Math. Sociol. **25**, 163–177 (2001)
6. Buntain, C., Lin, J.: Burst detection in social media streams for tracking interest profiles in real time. In: 39th International ACM SIGIR Conference (2016)
7. Cordeiro, M., Gama, J.: Online Social Networks Event Detection: A Survey, pp. 1–41. Springer International Publishing, Cham (2016)
8. Fairbanks, J., Ediger, D., McColl, R., Bader, D.A., Gilbert, E.: A statistical framework for streaming graph analysis. In: IEEE/ACM International Conference on Advances in Social Networks Analysis and Mining, ASONAM '13, pp. 341–347 (2013)
9. Gama, J.: Knowledge Discovery from Data Streams. Chapman & Hall/CRC, Boca Raton (2010)
10. Gama, J., Sebastião, R., Rodrigues, P.P.: On evaluating stream learning algorithms. Mach. Learn. **90**(3), 317–346 (2013)
11. Gama, J., Žliobaitė, I., Bifet, A., Pechenizkiy, M., Bouchachia, A.: A survey on concept drift adaptation. ACM Comput. Surv. (CSUR) **46**(4), 44 (2014)
12. IDÉ, T., KASHIMA, H.: Eigenspace-based anomaly detection in computer systems. In: Proceedings of the Tenth ACM SIGKDD International Conference on Knowledge Discovery and Data Mining, KDD '04, pp. 440–449 (2004)
13. Li, J., Ritter, A., Jurafsky, D.: Inferring user preferences by probabilistic logical reasoning over social networks. CoRR (2014). http://arxiv.org/abs/1411.2679
14. Liu, B.: Sentiment Analysis and Opinion Mining. Morgan & Claypool Publishers, San Rafael (2012)
15. Mouss, H., Mouss, D., Mouss, N., Sefouhi, L.: Test of page-hinckley, an approach for fault detection in an agro-alimentary production system. In: Control Conference, 2004, 5th Asian, vol. 2, pp. 815–818. IEEE, New York (2004)
16. Page, E.S.: Continuous inspection schemes. Biometrika **41**(1/2), 100–115 (1954)
17. Pereira, F.S.F., de Amo, S., Gama, J.: Detecting events in evolving social networks through node centrality analysis. Workshop on Large-scale Learning from Data Streams in Evolving Environments Co-located with ECML/PKDD (2016)
18. Pereira, F.S.F., de Amo, S., Gama, J.: On Using Temporal Networks to Analyze User Preferences Dynamics, pp. 408–423. Springer International Publishing, Cham (2016)
19. Ranshous, S., Shen, S., Koutra, D., Harenberg, S., Faloutsos, C., Samatova, N.F.: Anomaly detection in dynamic networks: a survey. Wiley Interdiscip. Rev. Comput. Stat. **7**(3), 223–247 (2015)

20. Rozenshtein, P., Anagnostopoulos, A., Gionis, A., Tatti, N.: Event detection in activity networks. In: Proceedings of the 20th ACM SIGKDD International Conference on Knowledge Discovery and Data Mining, KDD '14, pp. 1176–1185 (2014)
21. Sebastião, R., Silva, M.M., Rabiço, R., Gama, J., Mendonça, T.: Real-time algorithm for changes detection in depth of anesthesia signals. Evol. Syst. **4**(1), 3–12 (2013)
22. Wei, W., Carley, K.M.: Measuring temporal patterns in dynamic social networks. ACM Trans. Knowl. Discov. Data (TKDD) **10**(1), 9 (2015)

Large-Scale Learning from Data Streams with Apache SAMOA

Nicolas Kourtellis, Gianmarco De Francisci Morales, and Albert Bifet

Abstract Apache SAMOA (Scalable Advanced Massive Online Analysis) is an open-source platform for mining big data streams. Big data is defined as datasets whose size is beyond the ability of typical software tools to capture, store, manage, and analyze, due to the time and memory complexity. Apache SAMOA provides a collection of distributed streaming algorithms for the most common data mining and machine learning tasks such as classification, clustering, and regression, as well as programming abstractions to develop new algorithms. It features a pluggable architecture that allows it to run on several distributed stream processing engines such as Apache Flink, Apache Storm, and Apache Samza. Apache SAMOA is written in Java and is available at https://samoa.incubator.apache.org under the Apache Software License version 2.0.

1 Introduction

Big data are "data whose characteristics force us to look beyond the traditional methods that are prevalent at the time" [18]. For instance, social media are one of the largest and most dynamic sources of data. These data are not only very large due to their fine grain, but also being produced continuously. Furthermore, such data are nowadays produced by users in different environments and via a multitude of devices. For these reasons, data from social media and ubiquitous environments are perfect examples of the challenges posed by big data.

N. Kourtellis (✉)
Telefonica Research, Barcelona, Spain
e-mail: nicolas.kourtellis@telefonica.com

G. De Francisci Morales
Qatar Computing Research Institute, Doha, Qatar
e-mail: gdfm@acm.org

A. Bifet
LTCI, Télécom ParisTech, Paris, France
e-mail: albert.bifet@telecom-paristech.fr

© Springer International Publishing AG, part of Springer Nature 2019
M. Sayed-Mouchaweh (ed.), *Learning from Data Streams in Evolving Environments*, Studies in Big Data 41, https://doi.org/10.1007/978-3-319-89803-2_8

Currently, there are two main ways to deal with these challenges: streaming algorithms and distributed computing (e.g., MapReduce). Apache SAMOA aims at satisfying the future needs for big data stream mining by combining the two approaches in a single platform under an open source umbrella [9].

Data mining and machine learning are well-established techniques among Web and social media companies to draw insights from data coming from ubiquitous and social environments. Online content analysis for detecting aggression [6], stock trade volume prediction [5], online spam detection [7], recommendation [12], and personalization [10] are just a few of the applications made possible by mining the huge quantity of data available nowadays. Just think of Facebook's relevance algorithm for the news feed for a famous example.

In order to cope with Web-scale datasets, data scientists have resorted to *parallel and distributed computing*. MapReduce [11] is currently the de-facto standard programming paradigm in this area, mostly thanks to the popularity of Hadoop,[1] an open source implementation of MapReduce. Hadoop and its ecosystem (e.g., Mahout[2]) have proven to be an extremely successful platform to support the aforementioned process at web scale.

However, nowadays most data are generated in the form of a stream, especially when dealing with social media. Batch data are just a snapshot of streaming data obtained in an interval (window) of time. Researchers have conceptualized and abstracted this setting in the *streaming model*. In this model, data arrive at high speed, one instance at a time, and algorithms must process it in one pass under very strict constraints of space and time. Extracting knowledge from these massive data streams to perform dynamic network analysis [19] or to create predictive models [1], and using them, e.g., to choose a suitable business strategy, or to improve healthcare services, can generate substantial competitive advantages. Many applications need to process incoming data and react on-the-fly by using comprehensible prediction mechanisms (e.g., card fraud detection) and, thus, streaming algorithms make use of probabilistic data structures to give fast and approximate answers.

On the one hand, MapReduce is not suitable to express streaming algorithms. On the other hand, traditional sequential online algorithms are limited by the memory and bandwidth of a single machine. *Distributed stream processing engines* (DSPEs) are a new emergent family of MapReduce-inspired technologies that address this issue. These engines allow to express parallel computation on streams, and combine the scalability of distributed processing with the efficiency of streaming algorithms. Examples include Storm,[3] Flink,[4] Samza,[5] and Apex.[6]

---

[1] http://hadoop.apache.org.

[2] http://mahout.apache.org.

[3] http://storm.apache.org.

[4] http://flink.apache.org.

[5] http://samza.apache.org.

[6] https://apex.apache.org.

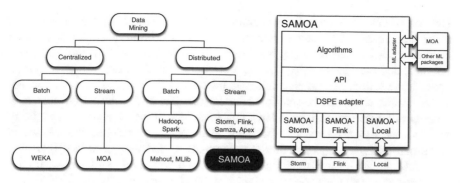

Fig. 1 (Left) Taxonomy of data mining and machine learning tools. (Right) High level architecture of Apache SAMOA

Alas, currently there is no common solution for mining big data streams, that is, for running data mining and machine learning algorithms on a distributed stream processing engine. The goal of Apache SAMOA is to fill this gap, as exemplified by Fig. 1 (left).

2 Description

Apache SAMOA (SCALABLE ADVANCED MASSIVE ONLINE ANALYSIS) is a platform for mining big data streams [8]. For a simple analogy, think of Apache SAMOA as Mahout for streaming. As most of the rest of the big data ecosystem, it is written in Java.

Apache SAMOA is both a framework and a library. As a framework, it allows the algorithm developer to abstract from the underlying execution engine, and therefore reuse their code on different engines. It features a pluggable architecture that allows it to run on several distributed stream processing engines such as Storm, Flink, Samza, and Apex. This capability is achieved by designing a minimal API that captures the essence of modern DSPEs. This API also allows to easily write new bindings to port Apache SAMOA to new execution engines. Apache SAMOA takes care of hiding the differences of the underlying DSPEs in terms of API and deployment.

As a library, Apache SAMOA contains implementations of state-of-the-art algorithms for distributed machine learning on streams. For classification, Apache SAMOA provides a Vertical Hoeffding Tree (VHT), a distributed streaming version of a decision tree. For clustering, it includes an algorithm based on CluStream. For regression, HAMR, a distributed implementation of Adaptive Model Rules. The library also includes meta-algorithms such as bagging and boosting [26]. The platform is intended to be useful for both research-oriented settings for the design and experimentation of new algorithms, and real-world deployments in production settings.

Related Work: We identify two frameworks that belong to the category of distributed streaming machine learning: Jubatus and StormMOA. Jubatus[7] is an example of distributed streaming machine learning framework. It includes a library for streaming machine learning such as regression, classification, recommendation, anomaly detection, and graph mining. It introduces the local ML model concept which means there can be multiple models running at the same time and they process different sets of data. Using this technique, Jubatus achieves scalability via horizontal parallelism in partitioning data. We test horizontal parallelism in our experiments, by implementing a horizontally scaled version of the Hoeffding tree. Jubatus establishes tight coupling between the machine learning library implementation and the underlying distributed stream processing engine (SPE). The reason is Jubatus builds and implements its own custom distributed SPE. In addition, Jubatus does not offer any tree learning algorithm, as all of its models need to be linear by construction.

StormMOA[8] is a project to combine MOA with Storm to satisfy the need of scalable implementation of streaming ML frameworks. It uses Storm's Trident abstraction and MOA library to implement OzaBag and OzaBoost [21]. Similarly to Jubatus, StormMOA also establishes tight coupling between MOA (the machine learning library) and Storm (the underlying distributed SPE). This coupling prevents StormMOA's extension by using other SPEs to execute the machine learning library. StormMOA only allows to run a single model in each Storm bolt (processor). This characteristic restricts the kind of models that can be run in parallel to ensembles.

3   High Level Architecture

We identify three types of Apache SAMOA users:

1. Platform users, who use available ML algorithms without implementing new ones.
2. ML developers, who develop new ML algorithms on top of Apache SAMOA and want to be isolated from changes in the underlying SPEs.
3. Platform developers, who extend Apache SAMOA to integrate more DSPEs into Apache SAMOA.

There are three important design goals of Apache SAMOA:

1. **Flexibility** in terms of developing new ML algorithms or reusing existing ML algorithms from other frameworks.
2. **Extensibility** in terms of porting Apache SAMOA to new DSPEs.
3. **Scalability** in terms of handling ever increasing amount of data.

---

[7]http://jubat.us/en.

[8]http://github.com/vpa1977/stormmoa.

Figure 1(right) shows the high-level architecture of Apache SAMOA which attempts to fulfill the aforementioned design goals. The *algorithm* layer contains existing distributed streaming algorithms that have been implemented in Apache SAMOA. This layer enables platform users to easily use the existing algorithm on any DSPE of their choice.

The *application programming interface* (API) layer consists of primitives and components that facilitate ML developers when implementing new algorithms. The *ML-adapter* layer allows ML developers to integrate existing algorithms in MOA or other ML frameworks into Apache SAMOA. The API layer and ML-adapter layer in Apache SAMOA fulfill the flexibility goal since they allow ML developers to rapidly develop algorithms.

Next, the *DSPE-adapter* layer supports platform developers in integrating new DSPEs into Apache SAMOA. To perform the integration, platform developers should implement the *samoa-SPE* layer as shown in Fig. 1(right). Currently Apache SAMOA is equipped with four adapters: the *samoa-Storm* adapter for Storm, the *samoa-Samza* adapter for Samza, the *samoa-Flink* adapter for Flink, and the *samoa-Apex* adapter for Apex. To satisfy the extensibility goal, the DSPE-adapter layer decouples DSPEs and ML algorithms implementations in Apache SAMOA, so that platform developers are able to easily integrate more DSPE platforms.

The last goal, scalability, implies that Apache SAMOA should be able to scale to cope ever increasing amount of data. To fulfill this goal, Apache SAMOA utilizes modern DSPEs to execute its ML algorithms. The reason for using modern DSPEs such as Storm, Flink, Samza, and Apex in Apache SAMOA is that they are designed to provide horizontal scalability to cope with Web-scale streams.

4   System Design

An algorithm in Apache SAMOA is represented by a directed graph of nodes that communicate via messages along streams which connect pairs of nodes. Borrowing the terminology from Storm, this graph is called a *Topology*. Each node in a Topology is a *Processor* that sends messages through a *Stream*. A Processor is a container for the code implementing the algorithm. A Stream can have a single source but several destinations (akin to a pub-sub system). A Topology is built by using a *TopologyBuilder*, which connects the various pieces of user code to the platform code and performs the necessary bookkeeping in the background. The following code snippet builds a topology that joins two data streams in Apache SAMOA:

```
TopologyBuilder builder = new TopologyBuilder();
Processor sourceOne = new SourceProcessor();
builder.addProcessor(sourceOne);
Stream streamOne = builder
    .createStream(sourceOne);
```

```
Processor sourceTwo = new SourceProcessor();
builder.addProcessor(sourceTwo);
Stream streamTwo = builder
    .createStream(sourceTwo);

Processor join = new JoinProcessor();
builder.addProcessor(join)
    .connectInputShuffle(streamOne)
    .connectInputKey(streamTwo);
```

A *Task* is an execution entity, similar to a job in Hadoop. A Topology is instantiated inside a Task to be run by Apache SAMOA. An example of a Task is *PrequentialEvaluation*, a classification task where each instance is used for testing first, and then for training.

A message or an event is called *Content Event* in Apache SAMOA. As the name suggests, it is an event which contains content that needs to be processed by the processors. Finally, a *Processing Item* is a hidden physical unit of the topology and is just a wrapper of Processor. It is used internally, and it is not accessible from the API.

5 Machine Learning Algorithms

In Apache SAMOA there are currently three types of algorithms performing basic machine learning functionalities such as classification via a decision tree (VHT), clustering (CluStream), and regression rules (AMR).

The Vertical Hoeffding Tree (VHT) [20] is a distributed extension of the VFDT [13]. VHT uses vertical parallelism to split the workload across several machines. Vertical parallelism leverages the parallelism across attributes in the same example, rather than across different examples in the stream. In practice, each training example is routed through the tree model to a leaf. There, the example is split into its constituting attributes, and each attribute is sent to a different Processor instance that keeps track of sufficient statistics. This architecture has two main advantages over one based on horizontal parallelism. First, attribute counters are not replicated across several machines, thus, reducing the memory footprint. Second, the computation of the fitness of an attribute for a split decision (via, e.g., entropy or information gain) can be performed in parallel. The drawback is that in order to get good performances, there must be sufficient inherent parallelism in the data. That is, the VHT works best for high-dimensional data.

Apache SAMOA includes a distributed version of CluStream, an algorithm for clustering evolving data streams. CluStream keeps a small summary of the data received so far by computing micro-clusters online. These micro-clusters are further refined to create macro-clusters by a micro-batch process, which is triggered periodically. The period is configured via a command line parameter (e.g., every 10,000 examples).

For regression, Apache SAMOA provides a distributed implementation of Adaptive Model Rules [23]. The algorithm, HAMR, uses a hybrid of vertical and horizontal parallelism to distribute AMRules on a cluster.

Apache SAMOA also includes adaptive implementations of ensemble methods such as bagging and boosting. These methods include state-of-the-art change detectors such as ADWIN, DDM, EDDM, and Page-Hinckley [15]. These meta-algorithms are most useful in conjunction with external single-machine classifiers, which can be plugged in Apache SAMOA in several ways. For instance, open-source connectors for MOA [4] are provided separately by the Apache SAMOA-MOA package.[9]

The following listing shows how to download, build, and run Apache SAMOA.

```
# download and build SAMOA
git clone http://git.apache.org/incubator-samoa.git
cd incubator-samoa
mvn package

# download the Forest Cover Type dataset
wget "http://downloads.sourceforge.net/project/moa-datastream/
    Datasets/Classification/covtypeNorm.arff.zip"
unzip "covtypeNorm.arff.zip"

# run SAMOA in local mode
bin/samoa local target/SAMOA-Local-0.4.0-SNAPSHOT.jar "
    PrequentialEvaluation -l classifiers.ensemble.Bagging -s
    (ArffFileStream -f covtypeNorm.arff) -f 100000"
```

6 Vertical Hoeffding Tree

We explain the details of the *Vertical Hoeffding Tree* [20], which is a data-parallel, distributed version of the Hoeffding tree. The *Hoeffding tree* [13] (a.k.a. VFDT) is a streaming decision tree learner with statistical guarantees. In particular, by leveraging the Chernoff-Hoeffding bound [16], it guarantees that the learned model is asymptotically close to the model learned by the batch greedy heuristic, under mild assumptions. The learning algorithm is very simple. Each leaf keeps track of the statistics for the portion of the stream it is reached by, and computes the best two attributes according to the splitting criterion. Let ΔG be the difference between the value of the functions that represent the splitting criterion of these two attributes. Let ϵ be a quantity that depends on a user-defined confidence parameter δ, and that decreases with the number of instances processed. When $\Delta G > \epsilon$, then the current best attribute is selected to split the leaf. The Hoeffding bound guarantees that this choice is the correct one with probability at least $1 - \delta$.

---

[9]https://github.com/samoa-moa/samoa-moa.

In this section, first, we describe the parallelization and the ideas behind our design choice. Then, we present the engineering details and optimizations we employed to obtain the best performance.

6.1 Vertical Parallelism

Data parallelism is a way of distributing work across different nodes in a parallel computing environment such as a cluster. In this setting, each node executes the same operation on different parts of the dataset. Contrast this definition with task parallelism (aka pipelined parallelism), where each node executes a different operator, and the whole dataset flows through each node at different stages. When applicable, data parallelism is able to scale to much larger deployments, for two reasons: (1) data has usually much higher intrinsic parallelism that can be leveraged compared to tasks, and (2) it is easier to balance the load of a data-parallel application compared to a task-parallel one. These attributes have led to the high popularity of the currently available DSPEs. For these reasons, we employ data parallelism in the design of VHT.

In machine learning, it is common to think about data in matrix form. A typical linear classification formulation requires to find a vector x such that $A \cdot x \approx b$, where A is the data matrix and b is a class label vector. The matrix A is $n \times m$-dimensional, with n being the number of data instances and m being the number of attributes of the dataset.

There are two ways to *slice* the data matrix to obtain data parallelism: by row or column. The former is called *horizontal parallelism*, the latter *vertical parallelism*. With horizontal parallelism, data instances are independent from each other, and can be processed in isolation while considering all available attributes. With vertical parallelism, instead, attributes are considered independently from each other.

The fundamental operation of the algorithm is to accumulate statistics n_{ijk} (i.e., counters) for triplets of attribute i, value j, and class k, for each leaf l of the tree. The counters for each leaf are independent, so let us consider the case for a single leaf. These counters, together with the learned tree structure, constitute the state of the VHT algorithm.

Different kinds of parallelism distribute the counters across computing nodes in different ways. With horizontal parallelism [3], the instances are distributed randomly, thus multiple instances of the same counter can exist on several nodes. On the other hand, when using vertical parallelism, the counters for one attribute are grouped on a single node.

This latter design has several advantages. First, by having a single copy of the counter, the memory requirements for the model are the same as in the sequential version. In contrast, with horizontal parallelism a single attribute may be tracked on every node, thus the memory requirements grow linearly with the parallelism level. Second, by having each attribute tracked independently, the computation of the split criterion can be performed in parallel by several nodes. Conversely, with horizontal partitioning the algorithm needs to (centrally) aggregate the partial counters before being able to compute the splitting criterion.

Of course, the vertically-parallel design has also its drawbacks. In particular, horizontal parallelism achieves a good load balance more easily, even though solutions for these problems have recently been proposed for vertical parallelism as well [24, 25]. In addition, if the instance stream arrives in row-format, it needs to be transformed in column-format, and this transformation generates additional CPU overhead at the source. Indeed, each attribute that constitutes an instance needs to be sent independently, and needs to carry the class label of its instance. Therefore, both the number of messages and the size of the data transferred increase. Nevertheless, the advantages of vertical parallelism outweigh its disadvantages for several real-world settings.

6.2   Algorithm Structure

We are now ready to explain the structure of the VHT algorithm. In general, there are two main parts to the Hoeffding tree algorithm: *sorting* the instances through the current model, and accumulating *statistics* of the stream at each leaf node. This separation offers a neat cut point to modularize the algorithm in two separate components. We call the first component *model aggregator*, and the second component *local statistics*. Figure 2 presents a visual depiction of the algorithm, and specifically, of its components and how the data flow among them. Also, Table 1 summarizes a list of components used in the rest of the algorithm description.

The model aggregator holds the current model (the tree) produced so far in a Processor node. Its main duty is to receive the incoming instances and sort them to the correct leaf. If the instance is unlabeled, the model predicts the label at the leaf and sends it downstream (e.g., for evaluation). Otherwise, if the instance is labeled, it is also used as training data. The VHT decomposes the instance into its constituent attributes, attaches the class label to each, and sends them independently

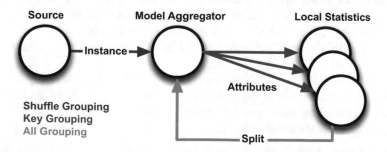

Fig. 2 High level diagram of the VHT topology

Table 1 Definitions and symbols

Description	Symbol
Source receiving/distributing instances	S
Training instance from S	E
Model aggregator	MA
Current state of the decision tree in MA	VHT_tree
Local statistic	LS
Counter for attribute i, value j, class k	n_{ijk}
Instances seen at leaf l	n_l
Information gain or entropy of attribute i in leaf l	$\overline{G}_l(X_i^{local})$

Algorithm 1 MA: VerticalHoeffdingTreeInduction(E, VHT_tree)

Require: E wrapped in `instance` content event
Require: VHT_tree in MA
1: Use VHT_tree to sort E into a leaf l
2: Send `attribute` content events to LSs
3: Increment n_l
4: **if** n_l *mod* $n_{min} = 0$ **and** not all instances seen at l belong to the same class **then**
5:     Add l into the list of splitting leaves
6:     Send `compute` content event with the id of leaf l to all LSs
7: **end if**

Algorithm 2 LS: UpdateLocalStatistic($attribute$, $local\_statistic$)

Require: $attribute$ is an `attribute` content event
Require: $local\_statistic$ is the LS in charge of $attribute$, could be implemented as
     $Table < leaf\_id, attribute\_id >$
1: Update $local\_statistic$ with data in $attribute$: attribute value, class value and instance
     weights

---

to the following stage, the *local statistics*. Algorithm 1 shows a pseudocode for the model aggregator.

The local statistics contain the counters n_{ijk} for a set of attribute-value-class triplets. Conceptually, the local statistics can be viewed as a large distributed table, indexed by leaf id (row), and attribute id (column). The value of the cell represents a set of counters, one for each pair of attribute value and class. The local statistics accumulate statistics on the data sent by the model aggregator. Pseudocode for the update function is shown in Algorithm 2.

In Apache SAMOA, we implement vertical parallelism by connecting the model to the statistics via key grouping. We use a composite key made by the leaf id and the attribute id. Horizontal parallelism can similarly be implemented via shuffle grouping on the instances themselves.

Leaf Splitting: Periodically, the model aggregator tries to see if the model needs to evolve by splitting a leaf. When a sufficient number of instances n_{min} have been sorted through a leaf, and not all instances that reached l belong to the same class

---

Algorithm 3 LS: ReceiveComputeMessage(*compute*, *local_statistic*)

Require: *compute* is a compute content event
Require: *local_statistic* is the LS in charge of *attribute*, could be implemented as
    $Table < leaf\_id, attribute\_id >$
1: Get leaf l ID from compute content event
2: For each attribute i that belongs to leaf l in local statistic, compute $\overline{G}_l(X_i)$
3: Find X_a^{local}, i.e., the attribute with highest $\overline{G}_l$ based on the local statistic
4: Find X_b^{local}, i.e., the attribute with second highest $\overline{G}_l$ based on the local statistic
5: Send X_a^{local} and X_b^{local} using local-result content event to model-aggregator PI via computation-result stream

---

(line 4, Algorithm 1), the aggregator sends a broadcast message to the statistics, asking to compute the split criterion for the given leaf id. The statistics Processor gets the table corresponding to the leaf, and for each attribute compute the splitting criterion in parallel (an information-theoretic function such as information gain or entropy). Each local statistic then sends back to the model the top two attributes according to the chosen criterion, together with their scores $(\overline{G}_l(X_i^{local}), i = a, b$; Algorithm 3).

Subsequently, the model aggregator (Algorithm 4) simply needs to compute the overall top two attributes received so far from the available statistics, apply the Hoeffding bound (line 4), and see whether the leaf needs to be split (line 5). The algorithm also computes the criterion for the scenario where no split takes places $(X_\emptyset)$. Domingos and Hulten [13] refer to this inclusion of a no-split scenario with the term *pre-pruning*. The decision to split or not is taken after a time has elapsed, as explained next.

By using the top two attributes, the model aggregator computes the difference of their splitting criterion values $\Delta \overline{G}_l = \overline{G}_l(X_a) - \overline{G}_l(X_b)$. To determine whether the leaf needs to be split, it compares the difference $\Delta \overline{G}_l$ to the Hoeffding bound $\epsilon = \sqrt{\frac{R^2 \ln(1/\delta)}{2n_l}}$ for the current confidence parameter δ (where R is the range of possible values of the criterion). If the difference is larger than the bound ($\Delta \overline{G}_l > \epsilon$), then X_a is the best attribute with high confidence $1 - \delta$, and can therefore be used to split the leaf. If the best attribute is the no-split scenario $(X_\emptyset)$, the algorithm does not perform any split. The algorithm also uses a tie-breaking τ mechanism to handle the case where the difference in splitting criterion between X_a and X_b is very small. If the Hoeffding bound becomes smaller than τ ($\Delta \overline{G}_l < \epsilon < \tau$), then the current best attribute is chosen regardless of the values of $\Delta \overline{G}_l$.

Two cases can arise: the leaf needs splitting, or it does not. In the latter case, the algorithm simply continues without taking any action. In the former case instead, the model modifies the tree by splitting the leaf l on the selected attribute, replacing l with an internal node (line 6), and generating a new leaf for each possible value of the branch (these leaves are initialized using the class distribution observed at the best attribute splitting at l (line 8)). Then, it broadcasts a drop message containing the former leaf id to the local statistics (line 10). This message is needed to release the resources held by the leaf and make space for the newly created leaves.

Algorithm 4 MA: Receive($local\_result$, VHT_tree)

Require: $local\_result$ is an `local-result` content event
Require: VHT_tree in MA
 1: Get correct leaf l from the list of splitting leaves
 2: Update X_a and X_b in the splitting leaf l with X_a^{local} and X_b^{local} from $local\_result$
 3: **if** $local\_results$ from all LSs received or time out reached **then**
 4:     Compute Hoeffding bound $\epsilon = \sqrt{\frac{R^2 \ln(1/\delta)}{2n_l}}$
 5:     **if** $X_a \neq X_\emptyset$ and $(\overline{G_l}(X_a) - \overline{G_l}(X_b) > \epsilon$ **or** $\epsilon < \tau)$ **then**
 6:         Replace l with a split-node on X_a
 7:         **for all** branches of the split **do**
 8:             Add new leaf with derived sufficient statistic from split node
 9:         **end for**
10:         Send `drop` content event with id of leaf l to all LSs
11:     **end if**
12: **end if**

Table 2 Types of content events used during the execution of VHT algorithm

Name	Parameters	From	To
`instance`	$<$ attr 1, ..., attr m, class C $>$	S	MA
`attribute`	$<$ attr id, attr value, class C $>$	MA	LS id =$<$ leaf id + attr id $>$
`compute`	$<$ leaf id $>$	MA	All LS
`local-result`	$< \overline{G_l}(X_a^{local}), \overline{G_l}(X_b^{local}) >$	LS_{id}	MA
`drop`	$<$ leaf id $>$	MA	All LS

Subsequently, the tree can resume sorting instances to the new leaves. The local statistics creates a new table for the new leaves lazily, whenever they first receive a previously unseen leaf id. In its simplest version, while the tree adjustment is performed, the algorithm drops the new incoming instances. We show in the next section an optimized version that buffers them to improve accuracy.

Messages: During the VHT execution several types of events are sent and received from the different parts of the algorithm, as summarized in Table 2.

6.3 Evaluation

In our experimental evaluation of the VHT method, we aim to study the following questions:

Q1: How does a centralized VHT compare to a centralized hoeffding tree with respect to accuracy and throughput?
Q2: How does the vertical parallelism used by VHT compare to the horizontal parallelism?
Q3: What is the effect of number and density of attributes?
Q4: How does discarding or buffering instances affect the performance of VHT?

Experimental Setup: In order to study these questions, we experiment with five datasets (two synthetic generators and three real datasets), five different versions of the hoeffding tree algorithm, and up to four levels of computing parallelism. We measure classification accuracy during, and at the end of the execution, and throughput (number of classified instances per second). We execute each experimental configuration ten times and report the average of these measures.

Synthetic Datasets: We use synthetic data streams produced by two random generators: one for dense and one for sparse attributes.

- **Dense attributes** are extracted from a random decision tree. We test different number of attributes, and include both categorical and numerical types. The label for each configuration is the number of categorical-numerical used (e.g, 100–100 means the configuration has 100 categorical and 100 numerical attributes). We produce 10 differently seeded streams with 1M instances for each tree, with one of two balanced classes in each instance, and take measurements every 100k instances.
- **Sparse attributes** are extracted from a random tweet generator. We test different dimensionalities for the attribute space: 100, 1k, 10k. These attributes represent the appearance of words from a predefined bag-of-words. On average, the generator produces 15 words per tweet (size of a tweet is Gaussian), and uses a Zipf distribution with skew $z = 1.5$ to select words from the bag. We produce 10 differently seeded streams with 1M tweets in each stream. Each tweet has a binary class chosen uniformly at random, which conditions the Zipf distribution used to generate the words.

Real Datasets: We also test VHT on three real data streams to assess its performance on benchmark data.[10]

- (*elec*) Electricity: 45312 instances, 8 numerical attributes, 2 classes.
- (*phy*) Particle Physics: 50000 instances, 78 numerical attributes, 2 classes.
- (*covtype*) CovertypeNorm: 581012 instances, 54 numerical attributes, 7 classes.

Algorithms: We compare the following versions of the hoeffding tree algorithm.

- **moa**: This is the standard Hoeffding tree in MOA.
- **local**: This algorithm executes VHT in a local, sequential execution engine. All split decisions are made in a sequential manner in the same process, with no communication and feedback delays between statistics and model.
- **wok**: This algorithm discards instances that arrive during a split decision. This version is the vanilla VHT.
- **wk(z)**: This algorithm sends instances that arrive during a split decision downstream. It also adds instances to a buffer of size z until full. If the split decision is taken, it replays the instances in the buffer through the new tree model.

---

[10]http://moa.cms.waikato.ac.nz/datasets/,
http://osmot.cs.cornell.edu/kddcup/datasets.html.

Otherwise, it discards the buffer, as the instances have already been incorporated in the statistics downstream.

- **sharding**: Splits the incoming stream horizontally among an ensemble of Hoeffding trees. The final prediction is computed by majority voting. This method is an instance of horizontal parallelism applied to Hoeffding trees. It creates an ensemble of hoeffding trees, but each tree is built with a subset of instances split horizontally, while using all available attributes.

Experimental Configuration: All experiments are performed on a Linux server with 24 cores (Intel Xeon X5650), clocked at 2.67 GHz, L1d cache: 32 kB, L1i cache: 32 kB, L2 cache: 256 kB, L3 cache: 12288 kB, and 65 GB of main memory. On this server, we run a Storm cluster (v0.9.3) and zookeeper (v3.4.6). We use Apache SAMOA v0.4.0 (development version) and MOA v2016.04 available from the respective project websites.

We use several parallelism levels in the range of $p = 2, \ldots, 16$, depending on the experimental configuration. For dense instances, we stop at $p = 8$ due to memory constraints, while for sparse instances we scale up to $p = 16$. We disable model replication (i.e., use a single model aggregator), as in our setup the model is not the bottleneck.

6.3.1 Accuracy and Time of VHT Local vs. MOA

In this first set of experiments, we test if VHT is performing as well as its counterpart hoeffding tree in MOA. This is mostly a sanity check to confirm that the algorithm used to build the VHT does not affect the performance of the tree when all instances are processed sequentially by the model. To verify this fact, we execute VHT local and MOA with both dense and sparse instances. Figure 3 shows that VHT local achieves the same accuracy as MOA, even besting it at times. However, VHT local always takes longer than MOA to execute. Indeed, the local execution engine of Apache SAMOA is optimized for simplicity rather than speed. Therefore, the additional overhead required to interface VHT to DSPEs is not amortized by scaling the algorithm out. Future optimized versions of VHT and the local execution engine should be able to close this gap.

6.3.2 Accuracy of VHT Local vs. Distributed

Next, we compare the performance of VHT local with VHT built in a distributed fashion over multiple processors for scalability. We use up to $p = 8$ parallel statistics, due to memory restrictions, as our setup runs on a single machine. In this set of experiments we compare the different versions of VHT, **wok** and **wk(z)**, to understand what is the impact of keeping instances for training after a model's split. Accuracy of the model might be affected, compared to the local execution, due to delays in the feedback loop between statistics and model. That is, instances

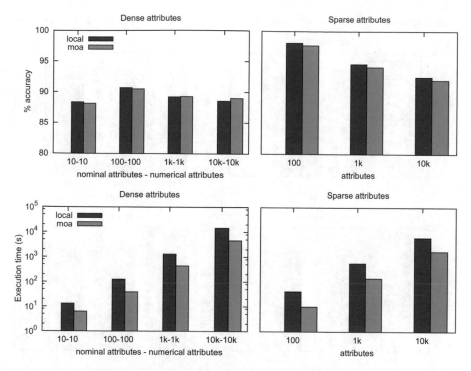

Fig. 3 Accuracy and execution time of VHT executed in local mode on Apache SAMOA compared to MOA, for dense and sparse datasets

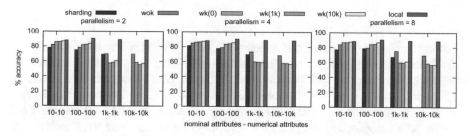

Fig. 4 Accuracy of several versions of VHT (local, **wok**, **wk(z)**) and sharding, for dense datasets

arriving during a split are classified using an older version of the model compared to the sequential execution. As our target is a distributed system where independent processes run without coordination, this delay is a characteristic of the algorithm as much as of the distributed SPE we employ.

We expect that buffering instances and replaying them when a split is decided would improve the accuracy of the model. In fact, this is the case for dense instances with a small number of attributes (i.e., around 200), as shown in Fig. 4. However, when the number of available attributes increases significantly, the load imposed on the model seems to outweigh the benefits of keeping the instances for replaying. We

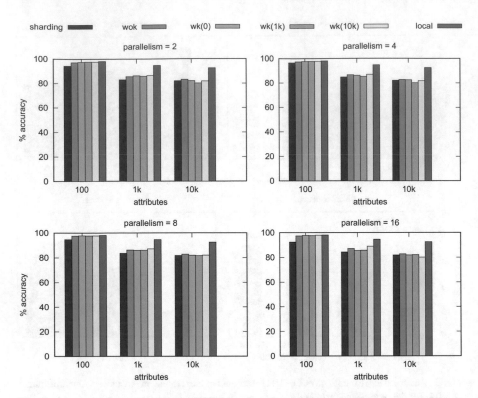

Fig. 5 Accuracy of several versions of VHT (local, **wok**, **wk(z)**) and sharding, for sparse datasets

conjecture that the increased load in computing the splitting criterion in the statistics further delays the feedback to compute the split. Therefore, a larger number of instances are classified with an older model, thus, negatively affecting the accuracy of the tree. In this case, the additional load imposed by replaying the buffer further delays the split decision. For this reason, the accuracy for VHT **wk(z)** drops by about 30% compared to VHT local. Conversely, the accuracy of VHT **wok** drops more gracefully, and is always within 18% of the local version.

VHT always performs approximatively 10% better than sharding. For dense instances with a large number of attributes (20k), sharding fails to complete due to its memory requirements exceeding the available memory. Indeed, sharding builds a full model for each shard, on a subset of the stream. Therefore, its memory requirements are p times higher than a standard hoeffding tree.

When using sparse instances, the number of attributes per instance is constant, while the dimensionality of the attribute space increases. In this scenario, increasing the number of attributes does not put additional load on the system. Indeed, Fig. 5 shows that the accuracy of all versions is quite similar, and close to the local one. This observation is in line with our conjecture that the overload on the system is the cause for the drop in accuracy on dense instances.

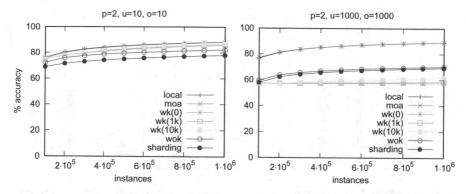

Fig. 6 Evolution of accuracy with respect to instances arriving, for several versions of VHT (local, **wok**, **wk(z)**) and sharding, for dense datasets

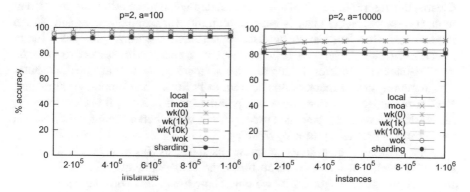

Fig. 7 Evolution of accuracy with respect to instances arriving, for several versions of VHT (local, **wok**, **wk(z)**) and sharding, for sparse datasets

We also study how the accuracy evolves over time. In general, the accuracy of all algorithms is rather stable, as shown in Figs. 6 and 7. For instances with 10 to 100 attributes, all algorithms perform similarly. For dense instances, the versions of VHT with buffering, **wk(z)**, outperform **wok**, which in turn outperforms sharding. This result confirms that buffering is beneficial for small number of attributes. When the number of attributes increases to a few thousand per instance, the performance of these more elaborate algorithms drops considerably. However, the VHT **wok** continues to perform relatively well and better than sharding. This performance coupled with good speedup over MOA (as shown next) makes it a viable option for streams with a large number of attributes and a large number of instances.

Speedup of VHT distributed vs. MOA Since the accuracy of VHT **wk(z)** is not satisfactory for both types of instances, next we focus our investigation on VHT **wok**. Figure 8 shows the speedup of VHT for dense instances. VHT **wok** is about 2–10 times faster than VHT local and up to 4 times faster than MOA.

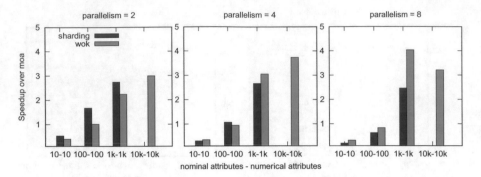

Fig. 8 Speedup of VHT **wok** executed on Apache SAMOA compared to MOA for dense datasets

Clearly, the algorithm achieves a higher speedup when more attributes are present in each instance, as (1) there is more opportunity for parallelization, and (2) the implicit load shedding caused by discarding instances during splits has a larger effect. Even though sharding performs well in speedup with respect to MOA on small number of attributes, it fails to build a model for large number of attributes due to running out of memory. In addition, even for a small number of attributes, VHT **wok** outperforms sharding with a parallelism of 8. Thus, it is clear from the results that the vertical parallelism used by VHT offers better scaling behavior than the horizontal parallelism used by sharding.

When testing the algorithms on sparse instances, as shown in Fig. 9, we notice that VHT **wok** can reach up to 60 times the throughput of VHT local and 20 times the one of MOA (for clarity we only show the results with respect to MOA). Similarly to what observed for dense instances, a higher speedup is observed when a larger number of attributes are present for the model to process. This very large, superlinear speedup ($20\times$ with $p = 2$) is due to the aggressive load shedding implicit in the **wok** version of VHT. The algorithm actually performs consistently less work than the local version and MOA.

However, note that for sparse instances the algorithm processes a constant number of attributes, albeit from an increasingly larger space. Therefore, in this setup, **wok** has a constant overhead for processing each sparse instance, differently from the dense case. VHT **wok** outperforms sharding in most scenarios and especially for larger numbers of attributes and larger parallelism.

Increased parallelism does not impact accuracy of the model (see Figs. 4 and 5), but its throughput is improved. Boosting the parallelism from 2 to 4 makes VHT **wok** up to 2 times faster. However, adding more processors does not improve speedup, and in some cases there is a slowdown due to additional communication overhead (for dense instances). Particularly for sparse instances, parallelism does not impact accuracy which enables handling large sparse data streams while achieving high speedup over MOA.

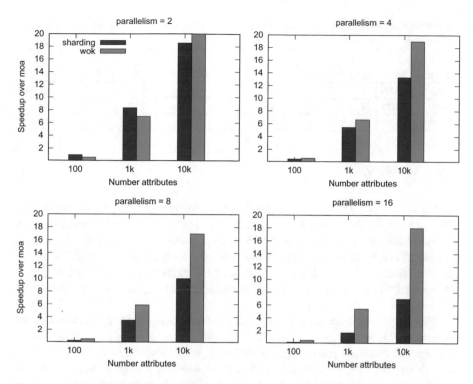

Fig. 9 Speedup of VHT **wok** executed on Apache SAMOA compared to MOA for sparse datasets

Table 3 Average accuracy (%) for different algorithms, with parallelism level (p), on the real-world datasets

| Dataset | MOA | VHT | | | | | Sharding | |
		Local	**wok** $p = 2$	**wok** $p = 4$	wk(0) $p = 2$	wk(0) $p = 4$	$p = 2$	$p = 4$
elec	75.4	75.4	75.0	75.2	75.4	75.6	74.7	74.3
phy	63.3	63.8	62.6	62.7	63.8	63.7	62.4	61.4
covtype	67.9	68.4	68.0	68.8	67.5	68.0	67.9	60.0

Performance on Real-World Datasets: Tables 3 and 4 show the performance of VHT, either running in a local mode or in a distributed fashion over a storm cluster of a few processors. We also test two versions of VHT: **wok** and wk(0). In the same tables we compare VHT's performance with MOA and sharding.

The results from these real datasets demonstrate that VHT can perform similarly to MOA with respect to accuracy and at the same time process the instances faster. In fact, for the larger dataset, covtypeNorm, VHT **wok** exhibits 1.8 speedup with respect to MOA, even though the number of attributes is not very large (54 numeric attributes). VHT **wok** also performs better than sharding, even though the latter is faster in some cases. However, the speedup offered by sharding decreases when the parallelism level is increased from 2 to 4 shards.

Table 4 Average execution time (seconds) for different algorithms, with parallelism level (p), on the real-world datasets

Dataset	MOA	VHT					Sharding	
		Local	**wok** $p = 2$	**wok** $p = 4$	wk(0) $p = 2$	wk(0) $p = 4$	$p = 2$	$p = 4$
elec	1.09	1	2	2	2	2	2	2.33
phy	5.41	4	3.25	4	3	3.75	3	4
covtype	21.77	16	12	12	13	12	9	11

6.4 Summary

In conclusion, our VHT algorithm has the following performance traits. We learned that for a small number of attributes, it helps to buffer incoming instances that can be used in future decisions of split. For larger number of attributes, the load in the model can be high and larger delays can be observed in the integration of the feedback from the local statistics into the model. In this case, buffered instances may not be used on the most up-to-date model and this can penalize the overall accuracy of the model.

With respect to a centralized sequential tree model (MOA), it processes dense instances with thousands of attributes up to 4× faster with only 10–20% drop in accuracy. It can also process sparse instances with thousands of attributes up to 20× faster with only 5–10% drop in accuracy. Also, its ability to build the tree in a distributed fashion using tens of processors allows it to scale and accommodate thousands of attributes and parse millions of instances. Competing methods cannot handle these data sizes due to increased memory and computational complexity.

7 Distributed AMRules

Decision rule learning is a category of machine learning algorithms whose goal is to extract a set of decision rules from the training data. These rules are later used to predict the unknown *label* values for test data. A rule is a logic expression of the form:

$$\textbf{IF } antecedent \textbf{ THEN } consequent$$

or, equivalently, *head* ← *body*, where *head* and *body* correspond to the *consequent* and *antecedent*, respectively.

The body of a rule is a conjunction of multiple clauses called *features*, each of which is a condition on an attribute of the instances. Such conditions consist of the identity of an attribute, a threshold value and an operator. For instance, the feature "$x < 5$" is a condition on attribute x, with threshold value 5 and operator *less-than* (<). An instance is said to be *covered* by a rule if its attribute values satisfy all the features in the rule body. The head of the rule is a function to be applied on the covered instances to determine their label values. This function can be a constant or a function of the attributes of the instances, e.g., $ax + b \leftarrow x < 5$.

AMRules is an algorithm for learning regression rules on streaming data. It incrementally constructs the rule model from the incoming data stream. The rule model consists of a set of *normal* rules (which is empty at the beginning), and a *default* rule. Each normal rule is composed of 3 parts: a *body* which is a list of features, a *head* with information to compute the prediction for those instance covered by the rule, and *statistics* of past instances to decide when and how to add a new feature to its body. In fact, the default rule is a rule with an empty *body*.

For each incoming instance, AMRules searches the current rule set for those rules that cover the instance. If an instance is not covered by any rule in the set, it is considered as being covered by the default rule. The heads of the rules are first used to compute the prediction for the instance they cover. Later, their statistics are updated with the attribute values and label value of the instance. There are two possible modes of operation: ordered and unordered. In ordered-rules mode, the rules are checked according to the order of their creation, and only the first rule is used for prediction and then updated. In unordered-rules mode, all covering rules are used and updated. In this work, we focus on the former which is more often used albeit more challenging to parallelize.

Each rule tries to expand its body after it receives N_m updates. In order to decide on the feature to expand, the rule incrementally computes a standard deviation reduction (SDR) measure [17] for each potential feature. Then, it computes the *ratio* of the second-largest SDR value over the largest SDR value. This *ratio* is used with a high confidence interval ϵ computed using the Hoeffding bound [16] to decide to expand the rule or not: if $ratio + \epsilon < 1$, the rule is expanded with the feature corresponding to the largest SDR value. Besides, to avoid missing a good feature when there are two (or more) equally good ones, rules are also expanded if the Hoeffding bound ϵ falls below a threshold. If the default rule is expanded, it becomes a *normal* rule and is added to the rule set. A new default rule is initialized to replace the previous one.

Each rule records its prediction error and applies a modified version of the Page-Hinkley test [22] for streaming data to detect changes. If the test indicates that the cumulative error has exceeded a threshold, the rule is evicted from the rule set. The algorithm also employs outlier detection to check if an instance, although being covered by a rule, is an anomaly. If an instance is deemed as an anomaly, it is treated as if the rule does not cover it and is checked against other rules. The following sections describe two possible strategies to parallelize AMRules that are implemented in Apache SAMOA.

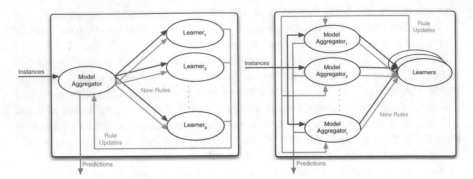

Fig. 10 (Left) Vertical AMRules (VAMR). (Right) AMRules with multiple horizontally parallelized model aggregators

7.1 Vertical Parallelism

In AMRules, each rule can evolve independently, as its expansion is based solely
on the statistics of instances it covers. Also, searching for the best feature among all
possible ones in an attempt to expand a rule is computationally expensive.

Given these observations, we decide to parallelize AMRules by delegating the
training process of rules to multiple *learner processors*, each of which handles
only a subset of the rules. Besides the learners, a *model aggregator processor* is
also required to filter and redirect the incoming instances to the correct learners.
The aggregator manages a set of simplified rules that have only head and body,
i.e., do not keep statistics. The bodies are used to identify the rules that cover
an instance, while the heads are used to compute the prediction. Each instance is
forwarded to the designated learners by using the ID of the covering rule. At the
learners, the corresponding rules' statistics are updated with the forwarded instance.
This parallelization scheme guarantees that the rules created are the same as in the
sequential algorithm. Figure 10(left) depicts the design of this vertically parallelized
version of AMRules, or Vertical AMRules (VAMR for brevity).

The model aggregator also manages the statistics of the default rule, and updates
it with instances not being covered by any other rule in the set. When the default
rule is expanded and adds a new rule to the set, the model aggregator sends a
message with the newly added rule to one of the learners, which is responsible for its
management. The assignment of a rule to a learner is done based on the rule's ID via
key grouping. All subsequent instances that are covered by this rule are forwarded
to the same learner.

At the same time, learners update the statistics of each corresponding rule with
each processed instance. When enough statistics have been accumulated and a rule
is expanded, the new feature is sent to the model aggregator to update the body of
the rule. Learners can also detect changes and remove existing rules. In such an
event, learners inform the model aggregator with a message containing the removed
rule ID.

As each rule is replicated in the model aggregator and in one of the learners, their bodies in model aggregator might not be up-to-date. The delay between rule expansion in the learner and model update in the aggregator depends mainly on the queue length at the model aggregator. The queue length, in turn, is proportional to the volume and speed of the incoming data stream. Therefore, instances that are in the queue before the model update event might be forwarded to a recently expanded rule which no longer covers the instance.

Coverage test is performed again at the learner, thus the instance is dropped if it was incorrectly forwarded. Given this additional test, and given that rule expansion can only increase the selectivity of a rule, when using unordered rules the accuracy of the algorithm is unaltered. However, in ordered-rules mode, these temporary inconsistencies might affect the statistics of other rules because the instance should have been forwarded to a different rule.

7.2 Horizontal Parallelism

A bottleneck in VAMR is the centralized model aggregator. Given that there is no straightforward way to vertically parallelize the execution of the model aggregator while maintaining the order of the rules, we explore an alternative based on horizontal parallelism. Specifically, we introduce multiple replicas of the model aggregator, so that each replica maintains the same copy of the rule set but processes only a portion of the incoming instances.

Horizontally Parallelized Model Aggregator: The design of this scheme is illustrated in Fig. 10(right). The basic idea is to extend VAMR and accommodate multiple model aggregators into the design. Each model aggregator still has a rule set and a default rule. The behavior of this scheme is similar to VAMR, except that each model aggregator now processes only a portion of the input data, i.e., the amount of instances each of them receives is inversely proportional to the number of model aggregators. This affects the prediction statistics and, most importantly, the training statistics of the default rules.

Since each model aggregator processes only a portion of the input stream, each default rule is trained independently with different portions of the stream. Thus, these default rules evolve independently and potentially create overlapping or conflicting rules. This fact also introduces the need for a scheme to synchronize and order the rules created by different model aggregators. Additionally, at the beginning, the scheme is less reactive compared to VAMR as it requires more instances for the default rules to start expanding. Besides, as the prediction function of each rule is adaptively constructed based on attribute values and label values of past instances, having only a portion of the data stream leads to having less information and potentially lower accuracy. We show how to address these issues next.

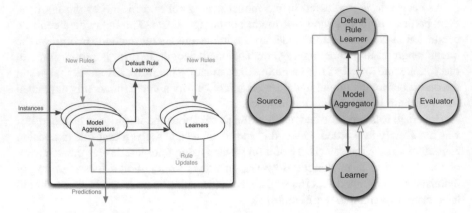

Fig. 11 (Left) Hybrid AMRules (HAMR) with multiple model aggregators and separate default rule learner. (Right) Prequential evaluation task for HAMR. Single lines represent single messages (*key grouping, shuffle grouping*) while double lines represent broadcast messages (*all grouping*)

Centralized Rule Creation: In order to address the issues with distributed creation of rules, we move the default rule in model aggregators to a specialized default rule learner processor. With the introduction of this new component, some modifications are required in the model aggregators, but the behavior of the learners is still the same as in VAMR. However, as a result, all the model aggregators are in synch.

As the default rule is now moved to the designated learner, those instances that are not covered by any rules are forwarded from the model aggregators to this learner. This specialized learner updates its statistics with the received instances and, when the default rule expands, it broadcasts the newly created rule to the model aggregators. The new rule is also sent to the assigned learner, as determined by the rule's ID.

The components of this scheme are shown in Fig. 11(left), where this scheme is referred to as Hybrid AMRules (HAMR), as it is a combination of vertical and horizontal parallelism strategies.

7.3 Evaluation

We evaluate the performance of the 2 distributed implementations of AMRules, i.e., VAMR and HAMR, in comparison to the centralized implementation in MOA[11] (MAMR).

Evaluation Methodology: We plug VAMR and HAMR into a *prequential evaluation task* [14], where each instance is first used to test and then to train the model.

---

[11]http://moa.cms.waikato.ac.nz.

This evaluation task includes a *source* processor which provides the input stream and an *evaluator* processor which records the rate and accuracy of prediction results. The final task for HAMR is depicted in Fig. 11(right). The parallelism level of the model is controlled by setting the number of learners p and the number of model aggregators r. The task for VAMR is similar but the default rule learner is excluded and model aggregator's parallelism level is always 1. Each task is repeated for five runs.

Datasets: We perform the same set of experiments with 3 different datasets, i.e., *electricity, airlines*, and *waveform*.

- *Electricity*: A dataset from the UCI Machine Learning Repository [2] which records the electricity power consumption (in watt-hour) of a household from December 2006 to November 2010. The dataset contains more than 2 millions 12-attribute records.
- *Airlines*: A dataset recording the arrival delay (in seconds) of commercial flights within the USA in year 2008.[12] It contains more than 5.8 millions records, each with 10 numerical attributes.
- *Waveform*: A dataset generated using an artificial random generator. To generate an instance, it picks a random waveform among the 3 available ones, and 21 attribute values for this instance are generated according to the chosen waveform. Another 19 noise attributes are generated and included in the new instance, making a total of 40 attributes for each instance. The label value is the index of the waveform (0, 1, or 2). Although this dataset does not fit perfectly to the definition of a regression task, it allows us to test our implementations with a high number of numerical attributes.

Setup: The evaluation is performed on an OpenStack[13] cluster of 9 nodes, each with 2 Virtual CPUs @ 2.3 GHz and 4 GB of RAM. All the nodes run Red Hat Enterprise Linux 6. The distributed implementations are evaluated on a Samza[14] cluster with Hadoop YARN 2.2[15] and Kafka 0.8.1.[16] The Kafka brokers coordinate with each other via a single-node ZooKeeper 3.4.3[17] instance. The replication factor of Kafka's streams in these experiments is 1. The performance of MAMR is measured on one of the nodes in the cluster.

Throughput: The throughput of several variants of AMRules is shown in Fig. 12 for each dataset examined. HAMR-1 and HAMR-2 stand for *HAMR with 1 learner* and *HAMR with 2 learners*. The parallelism levels of VAMR represent the number

---

[12]http://kt.ijs.si/elena_ikonomovska/data.html.

[13]http://www.openstack.org.

[14]http://samza.incubator.apache.org.

[15]http://hadoop.apache.org.

[16]http://kafka.apache.org.

[17]http://zookeeper.apache.org.

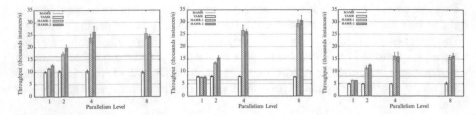

Fig. 12 Throughput of distributed AMRules with *electricity* (left), *airlines* (mid), and *waveform* (right)

Table 5 Statistics for features and rules (with MAMR) for different datasets

	Electricity	Airlines	Waveform
Instances	2,049,280	5,810,462	1,000,000
# Attributes	12	10	40
Result message size (B)	891	764	1446
# Rules created	1203	2501	270
# Rules removed	1103	1040	51
Avg. # rules	100	1461	219
# Features created	1069	10,606	1245

of its learners, while in the case of HAMR it represents the number of model aggregators.

With *electricity* and *waveform*, the communication overhead in VAMR (to send instances from model aggregator to learners and evaluator) exceeds the throughput gained from delegating the training process to the learners and results in a lower overall throughput compared to MAMR's. However, with *airlines*, the performance of VAMR is better than MAMR. To verify that the training process for *airlines* is more computationally intensive than the other two datasets and, thus, VAMR is more effective for this dataset, we compare the statistics of rules and predicates creation for the 3 datasets. The number of rules created, rules removed, and features created (when a rule is expanded) by MAMR with the 3 datasets are presented in Table 5. By subtracting the total number of rules removed from the total number of rules created, we can have an estimation of the average number of rules in the model for each dataset. A higher number of rules in the model and a higher number of features created suggest that the model is more complex and it takes more processing power to search for the best new feature when a rule attempts to expand.

Although VAMR can perform better than MAMR for computationally intensive datasets, its throughput does not change with different parallelism level. This is due to the bottleneck at the model aggregator. The processing of each instance at the model aggregator consists of three steps: finding the covering rules, forwarding the instance to the corresponding learners, and applying covering rules to predict the label value of the instance. Since the complexity of these three steps for an instance is constant, the final throughput is unaffected by the parallelism level.

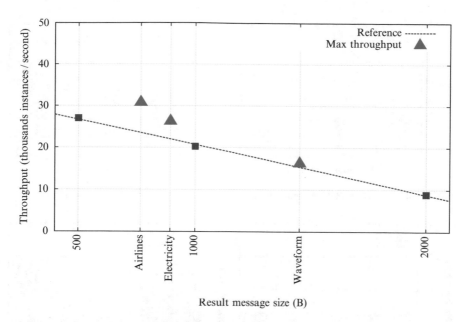

Fig. 13 Maximum throughput of HAMR vs message size

The throughput of HAMR-1 and HAMR-2 exhibits a better scalability compared to VAMR. Up to parallelism level of 4, the throughput increases almost linearly with the number of model aggregators. However, there is no or little improvement when this number is increased from 4 to 8. As we measure this throughput at a single evaluator, we suspect that the bottleneck is in the maximum rate the evaluator can read from the output streams of the model aggregators and default rule learner. To investigate this issue, we plot the maximum throughput of HAMR against the size of messages from model aggregators and default rule learner to evaluator in Fig. 13. The values of throughput of a single-partition Samza stream with messages of size 500 B, 1000 B, and 2000 B are used to compute the linear regression line (reference line) in the figure. The message size for different datasets is shown in Table 5.

As reading is expected to be faster than writing in Samza and Kafka, the maximum rate the evaluator in HAMR can read from multiple stream partitions is expected to be higher than the throughput represented by the reference line. This fact is reflected in Fig. 13 as the maximum throughput of HAMR for the 3 datasets constantly exceeds the reference line. However, the difference between them is relatively small. This result is a strong indicator that the bottleneck is the maximum reading rate of the evaluator. If there is no need to aggregate the result from different streams, this bottleneck can be eliminated.

Accuracy: We evaluate accuracy of the different implementations of AMRules in terms of *Mean Absolute Error* (MAE) and *Root Mean Square Error* (RMSE). Figures 14, 15, and 16 show the MAE and RMSE for the three datasets, normalized by the range of label values in each dataset.

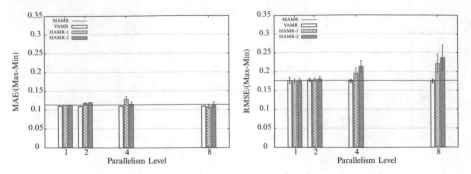

Fig. 14 MAE (left) and RMSE (right) of distributed AMRules with the *electricity* dataset

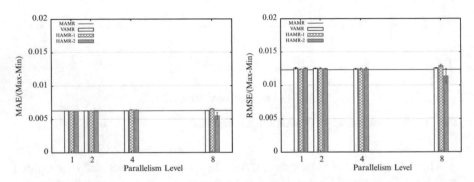

Fig. 15 MAE (left) and RMSE (right) of distributed AMRules with the *airlines* dataset

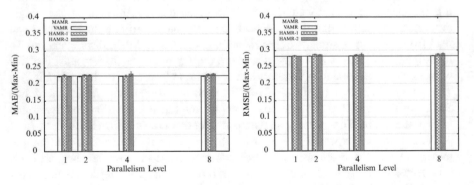

Fig. 16 MAE (left) and RMSE (right) of distributed AMRules with the *waveform* random generator

Most of the figures show that error values of distributed implementations present very small fluctuations around the corresponding MAMR error line. However, there is a significant increment in the value of RMSE in HAMR with *electricity* dataset when the number of model aggregators is increased to 4 or 8. Moreover, larger variances, i.e., *standard error* is greater than 5% of the average MAE or RMSE,

Table 6 Memory consumption of MAMR for different datasets

Dataset	Memory consumption (MB)	
	Avg.	Std. dev.
Electricity	52.4	2.1
Airlines	120.7	51.1
Waveform	223.5	8

Table 7 Memory consumption of VAMR for different datasets and parallelism levels (AVG: average; SD: standard deviation)

Dataset	Parallelism	Memory consumption (MB)			
		Model aggregator		Model learner	
		AVG	SD	AVG	SD
Electricity	1	266.5	5.6	40.1	4.3
	2	264.9	2.8	23.8	3.9
	4	267.4	6.6	20.1	3.2
	8	273.5	3.9	34.7	2.9
Airlines	1	337.6	2.8	83.6	4.1
	2	338.1	1.0	38.7	1.8
	4	337.3	1.0	38.8	7.1
	8	336.4	0.8	31.7	0.2
Waveform	1	286.3	5.0	171.7	2.5
	2	286.8	4.3	119.5	10.4
	4	289.1	5.9	46.5	12.1
	8	287.3	3.1	33.8	5.7

are also observed in the case of higher parallelism levels ($p \geq 4$) of HAMR. The probable cause is that when a rule in the model aggregators is out-of-sync with the corresponding one in the learners, i.e., model aggregators are not yet updated with a newly created rule, the number of instances that use an outdated model is multiplied by the throughput.

Memory: The reference memory consumption of MAMR for different datasets is presented in Table 6. This table shows that *waveform* consumes the largest amount of memory, followed by *airlines* and then *electricity*.

Table 7 reports the memory consumption (*average* and *standard deviation*) of model aggregator and learner processors of VAMR for the 3 datasets with different parallelism levels. First of all, we notice a high memory consumption at the model aggregator. However this amount does not differ much for different datasets. This suggests that, at the model aggregator, there is a constant overhead in memory usage, but the memory consumption due to the growing decision model is small.

Second, the memory consumption per learner decreases as more learners are used. As there is definitely some memory overhead for each learner instance, there is no significant reduction of memory usage per learner when the parallelism level goes from 4 to 8. However, the result indicates that we can spread a large decision model over multiple learners and make it possible to learn very large models whose size exceeds the memory capacity of a single machine.

8 Conclusions

We presented the Apache SAMOA platform for mining big data streams. The platform supports the most common machine learning tasks such as classification, clustering, and regression. It also provides a simple API for developers that allows to implement distributed streaming algorithms easily. Apache SAMOA is already available and can be found online at:

https://samoa.incubator.apache.org

The website includes a wiki, an API reference, and a developer's manual. Several examples of how the software can be used are also available. The code is hosted on GitHub. Finally, Apache SAMOA is released as open source software under the Apache Software License v2.0.

References

1. Aggarwal, C.C.: Data Streams: Models and Algorithms. Springer, Berlin (2007)
2. Bache, K., Lichman, M.: UCI machine learning repository (2013). http://archive.ics.uci.edu/ml
3. Ben-Haim, Y., Tom-Tov, E.: A streaming parallel decision tree algorithm. J. Mach. Learn. Res. **11**, 849–872 (2010). ISSN 1532–4435. http://dl.acm.org/citation.cfm?id=1756006.1756034
4. Bifet, A., Holmes, G., Kirkby, R., Pfahringer, B.: MOA: Massive online analysis. J. Mach. Learn. Res. **11**, 1601–1604 (2010)
5. Bordino, I., Kourtellis, N., Laptev, N., Billawala, Y.: Stock trade volume prediction with Yahoo Finance user browsing behavior. In: 30th International Conference on Data Engineering (ICDE), pp. 1168–1173. IEEE, New York (2014)
6. Chatzakou, D., Kourtellis, N., Blackburn, J., De Cristofaro, E., Stringhini, G., Vakali, A.: Mean birds: detecting aggression and bullying on Twitter. In: 9th International Conference on Web Science (WebSci). ACM, New York (2017)
7. Chen, C., Zhang, J., Chen, X., Xiang, Y., Zhou, W.: 6 million spam tweets: a large ground truth for timely Twitter spam detection. In: International Conference on Communications (ICC). IEEE, New York (2015)
8. De Francisci Morales, G.: SAMOA: a platform for mining big data streams. In: RAMSS: 2nd International Workshop on Real-Time Analysis and Mining of Social Streams @WWW (2013)
9. De Francisci Morales, G., Bifet, A.: SAMOA: scalable advanced massive online analysis. J. Mach. Learn. Res. **16**, 149–153 (2015)
10. De Francisci Morales, G., Gionis, A., Lucchese, C.: From chatter to headlines: harnessing the real-time web for personalized news recommendation. In: 5th ACM International Conference on Web Search and Data Mining (WSDM), pp. 153–162. ACM, New York (2012)
11. Dean, J., Ghemawat, S.: MapReduce: simplified data processing on large clusters. In: 6th Symposium on Operating Systems Design and Implementation (OSDI), pp. 137–150. USENIX Association, Berkeley (2004)

12. Devooght, R., Kourtellis, N., Mantrach, A.: Dynamic matrix factorization with priors on unknown values. In: 21st International Conference on Knowledge Discovery and Data Mining (SIGKDD), pp. 189–198. ACM, New York (2015)
13. Domingos, P., Hulten, G.: Mining high-speed data streams. In: 6th International Conference on Knowledge Discovery and Data Mining (KDD), pp. 71–80 (2000)
14. Gama, J., Sebastião, R., Rodrigues, P.P.: On evaluating stream learning algorithms. Mach. Learn. **90**(3), 317–346 (2013)
15. Gama, J., Zliobaite, I., Bifet, A., Pechenizkiy, M., Bouchachia, A.: A survey on concept drift adaptation. ACM Comput. Surv. **46**(4), 13–30 (2014)
16. Hoeffding, W.: Probability inequalities for sums of bounded random variables. J. Am. Stat. Assoc. **58**(301), 13–30 (1963). http://amstat.tandfonline.com/doi/abs/10.1080/01621459.1963.10500830
17. Ikonomovska, E., Gama, J., Džeroski, S.: Learning model trees from evolving data streams. Data Min. Knowl. Disc. **23**(1), 128–168 (2011)
18. Jacobs, A.: The pathologies of big data. Commun. ACM **52**(8), 36–44 (2009)
19. Kourtellis, N., Bonchi, F., De Francisci Morales, G.: Scalable online betweenness centrality in evolving graphs. IEEE Trans. Knowl. Data Eng. **27**(9), 2494–2506 (2015)
20. Kourtellis, N., De Francisci Morales, G., Bifet, A.: VHT: vertical hoeffding tree. In: 4th IEEE International Conference on Big Data (BigData) (2016)
21. Oza, N.C., Russell, S.: Online bagging and boosting. In: Artificial Intelligence and Statistics, pp. 105–112. Morgan Kaufmann, Los Altos (2001)
22. Page, E.: Continuous inspection schemes. Biometrika **41**(1–2), 100–115 (1954)
23. Thu Vu, A., De Francisci Morales, G., Gama, J., Bifet, A.: Distributed adaptive model rules for mining big data streams. In: 2nd IEEE International Conference on Big Data (BigData) (2014)
24. Uddin Nasir, M.A., De Francisci Morales, G., Garcia-Soriano, D., Kourtellis, N., Serafini, M.: The power of both choices: practical load balancing for distributed stream processing engines. In: 31st International Conference on Data Engineering (ICDE) (2015)
25. Uddin Nasir, M.A., De Francisci Morales, G., Kourtellis, N., Serafini, M.: When two choices are not enough: balancing at scale in distributed stream processing. In: 32nd International Conference on Data Engineering (ICDE) (2016)
26. Vasiloudis, T., Beligianni, F., De Francisci Morales, G.: BoostVHT: boosting distributed streaming decision trees. In: 26th ACM International Conference on Information and Knowledge Management (CIKM) (2017)

Process Mining for Analyzing Customer Relationship Management Systems: A Case Study

Ahmed Fares, João Gama, and Pedro Campos

Abstract Process Mining aims to discover and evaluate As-Is processes from sets of sequential events, by examining different instances of the same process and building models that can detect patterns and behaviors. In the meanwhile, organizational perspective is being considered in Process Mining by taking advantage of the ability to extract social networks that represent different kinds of relations between resources performing the process. The case study tries to describe how Process Mining could be applied in order to detect and improve "Customer Relationship Management" process and extract some kind of social networks that represent the relations between the employees(resources) of National Institute of Statistics of Portugal (INE) using event logs.

1 Introduction

Process Mining is the link between traditional process model analyses and data-oriented analysis like Data Mining and Machine Learning, focused on end to end process using the real data [1]. Nowadays almost all enterprise information systems store relevant events in more or less structured form using several software systems like Enterprise Resource Planning systems (ERP) and Customer Relationship Management (CRM) systems. There is a standard definition of the structure of the saved event log, whatever the name it is being referred to, e.g. transaction log, audit trial, history, etc. The high-level structure of any of these logs could be explained in the form of "Case" and "Activity." The case is the process instance which has

A. Fares (✉)
LIAAD-INESC TEC, Porto, Portugal
e-mail: ahmed.a.fares@inesctec.pt

J. Gama · P. Campos
LIAAD-INESC TEC, Porto, Portugal

Faculty of Economics, University of Porto, Porto, Portugal
e-mail: jgama@fep.up.pt; pcampos@fep.up.pt

© Springer International Publishing AG, part of Springer Nature 2019
M. Sayed-Mouchaweh (ed.), *Learning from Data Streams in Evolving Environments*, Studies in Big Data 41, https://doi.org/10.1007/978-3-319-89803-2_9

Table 1 Sample of event log

Case	Activity	Timestamp	Performer
Case 1	**Record the request**	**1-1-2015**	**Agent01**
Case 2	Record the request	1-1-2015	Agent02
Case 3	Record the request	2-1-2015	Agent01
Case 1	**Response from internal department**	**2-1-2015**	**Cons02**
Case 1	**Sending final answer**	**2-1-2015**	**Agent01**
Case 2	Response from internal department	2-1-2015	Cons01
Case 2	Sending final answer	2-1-2015	Agent02
Case 3	Response from internal department	3-1-2015	Cons01
Case 2	Close the request	3-1-2015	Agent02
Case 1	**Close the request**	**3-1-2015**	**Agent01**
Case 4	Record the request	4-1-2015	Agent03
Case 5	Record the request	4-1-2015	Agent03
Case 4	Response from internal department	4-1-2015	Agent03
Case 3	Sending final answer	4-1-2015	Agent01
Case 3	Close the request	4-1-2015	Agent01
Case 5	Sending clarification or link	4-1-2015	Agent03
Case 5	Close the request	4-1-2015	Agent03
Case 4	Sending final answer	5-1-2015	Cons01
Case 4	Close the request	6-1-2015	Agent03

begging, intermediate, and ending activities. Moreover, as most of the activities are performed by resources, event logs also would contain information about these resources who are initiating or executing activities.

A sample of the event log for a CRM system is presented in Table 1 involving 19 events, 5 process instances(cases), and 6 performers. Considering Case 1 as an example. It has four unique activities that started with activity "Record the request" on 1-1-2015 by Agent01, followed by activities "Response from internal department" and "sending final answer" on 2-1-2015 by Consultant02 and Agent01, respectively, finally it ends by activity "Close the request" on 3-1-2015 by Agent01. Using a similar event log, Process Mining techniques are able to construct process models like the one in Fig. 1 where most of patterns and sequences of the work are represented. For example, Case 1 is perfectly fit in the model as we can trace the flow of its events going through the model. Also, the social network in Fig. 2 shows the relationships among the performers based on the handover of work, e.g. Agent01 and Cons02 have a two-way relation as each one is handing the work over to each other. On the other hand, Cons01 has relations with all agents, while Cons02 is only receiving work from Agent01.

Of course, using a sample of five cases, the information is so clear to be caught, but considering the full data set with thousands of cases, it will not be that easy to capture the real process model and the relations between performers.

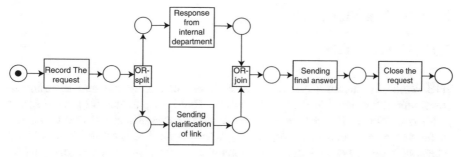

Fig. 1 Potential process model

Fig. 2 Social network based on handover-of-work

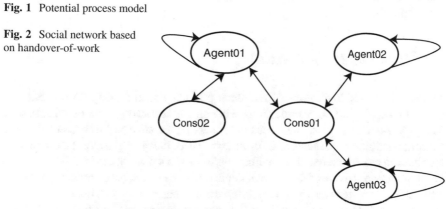

2 Related Work

Some theoretical researches and practical case studies had been done in order to prove the applicability of Process Mining and Mining Social Networks in real life. Also, tools have been developed in order to aggregate enhance Process Mining algorithms. ProM [2] is the most powerful and extensible framework that is entirely pluggable environment used mainly for discovering, enhancing, and doing conformance checking for process models.

Aalst and Song [3] were pioneers in extracting social structure from event logs. Also, they defined the basic metrics and developed a tool to mine social networks from event logs (MiSoN) which have been embedded in ProM software.

Later, case studies have been done especially in healthcare [4] and industrial [5] domains considering (a)process perspective, (b)organizational perspective, and (c)case perspective. This paper's focus on the applicability of Process Mining in the customer service domain considering organizational perspective.

3   INE Case Study

3.1   What Is INE?

INE stands for Instituto Nacional de estatstica in Portuguese and the English translation is "National Institute of Statistics." INE is responsible for preparing and publishing official statistical information regarding several sectors in Portugal, e.g. health sector, environment sector, economic sector, etc. It was created in 1935 with a head office in Lisbon and four delegations in the main cities Évora, Faro, Coimbra, and Porto [6].

3.2   Data and Pre-processing

The data is related to Customer Relationship Management (CRM) system that INE uses in recording all communications with clients requesting data or information. The process of CRM starts when customer service resource receives a request from a client regarding some kind of information. Then, he/she replies to the client with feedback in case the request was within his/her knowledge, otherwise, the request is being forwarded to one of the internal department resources based on the type of the request. Later, whenever the internal department has a response, customer service resource receives the response from the internal department and forward it to the client.

The event log file has all requests received during the year 2015. There were 5811 requests with 18 unique activities and 81 originators (resources execute that activities). The numbers of activities per case vary from 3 to 18 with the mean value of 4. All cases started with activity "Record the request," but 98% of cases ended with activity "Close the request." We can categorize the cases with the irregular behavior of closing the request into two categories. (a) When a client asks for clarification after INE considering the request is closed, which could be a minor problem. (b) When receiving a response from the internal department after closing the request, which is a major problem that needs a further investigation.

3.3   Questions

Our goal is to answer the following questions:

- How is the real process looks like based on the actual data?
- What activities have to be included in the model?
- What activities could be ignored from the model without causing a significant impact on the quality of the model?
- Does the process take so much time? And what are the causes?
- Which performers are deeply involved? And how are they related to each other?

3.4 Process Discovery

The primary function of Process Mining is to discover the process model using the real data. We had tried several algorithms in order to know the best of them to describe our process model taking into consideration the balance between the four quality forces: fitness, simplicity, precision, and generalization, as most of the observations in the log should fit in the model but at the same time the model should be represented in a simple way. Also, the model should be precise enough and not allow too many behaviors that were not observed in the log, but at the same time should not be too restricted and to be more general to accept more behaviors.

Before considering the quality forces, the generated model should be reliable by guaranteeing the soundness (bug-free). In order to consider a model as a sound model the following three properties should be satisfied: loops of length 1, option to complete, and proper completion [1, 7].

Alpha Algorithm is the first algorithm that has fit the gap between event logs and process models [1]. The major flaw of the Alpha algorithm is that it does not consider frequencies, so all connections between activities are presented even if it just happened only one time, which produces very complicated models as shown in Fig. 3a where all the 18 activities were presented while 9 of them occurred in just 2% of the cases. Because of that we had regenerated the model after filtering out the less frequent activities from the original log file. The new model is presented in Fig. 3b. The model was less complicated but it was not sound as it was not able to figure out Loops of length one (cannot detect unary relationship) so it has represented these relations as an isolated transition i.e. "Response from Internal Department."

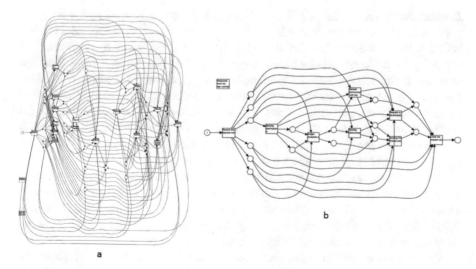

a   b

Fig. 3 Process models discovered with Alpha algorithm miner

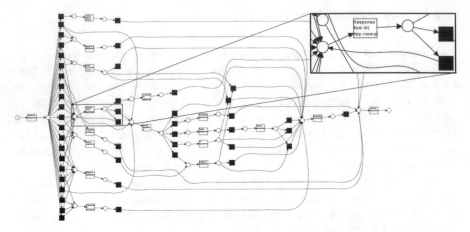

Fig. 4 Process models discovered with Heuristics miner

Heuristics Miner is an improvement of the Alpha algorithm, especially in three issues [8]. (a) It takes frequencies into account so that it can filter out noisy or infrequent behaviors, (b) it's able to detect short loops (like loops of length one), and (c) it allows skipping of single activities. However, it does not guarantee sound process models. The model generated using Heuristics miner had overcome loops of length 1 (see Fig. 4) with focus on transition "Response from internal department," but the model is not sound yet because the proper completion problem still exists as some transitions produce two tokens, one of them was stuck in the model and couldn't reach the end stat.

Evolutionary Tree Mine ETM algorithm always guarantees to generate sound models as it uses process trees which reduce the search space, so unsound models will not be considered [7]. On the other hand, it gives users the flexibility of choosing the preferred balance between quality criteria through two ways. The first way is to define relative weights for the quality criteria, and ETM miner will search for the optimal model satisfying the balance between quality criteria according to user-defined weights. The second way is to define a set of constraints for the desired qualities for each criterion independently, so ETM will return a set of models satisfying these constraints. Later the user can preview and select the best model in case he is not able to define relative weights between quality criteria.

The first way, we had defined the relative weights according to the importance of each criterion as following: Fitness = 10, Precision = 5, Simplicity = 1, Generalization = 1. The algorithm has reached the best model with overall fitness of 0.967. The process model (Petri net) in Fig. 5 represents the best model, after coding activities in letters for the readability as mentioned in Table 2.

The process model has nine transitions or activities, starts with "Record the request" activity, then having a non-exclusive choice (OR) between three activities on the first hand that either "Request is not feasible" or exclusive choice loop with

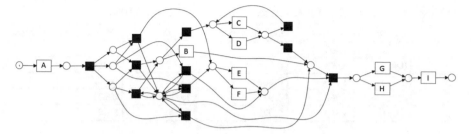

Fig. 5 Petri net produced by ETM miner

Table 2 Activities legend

Code	Activity
A	Record the request
B	Request is not feasible
C	Response from internal department
D	Budget accepted
E	Sending clarification or link
F	General communication with the client
G	Sending final answer
H	Request is duplicated
I	Close the request
S	Silent transition

"Budget accepted" or "Response from Int.dep.". Normally the process instant has one "Budget accepted" event and one or more "Response from Int.dep." which in most of the cases followed itself because the system does not record the event of sending the request to an internal department and just record the response from the internal department. Also, there could be several responses from the internal department for the same request as the client could not be satisfied by the response or he had changed his requirements during the same request. On the other hand of the non-exclusive choice, there could be one of two activities "Sending clarification of link" or "General communication with the client." Then the process goes to an exclusive choice (XOR) between either "Request is duplicated" or "Sending final answer" as the last step before reaching the end of the process by "Close the request" activity.

The second way, we had used the following constraints: Minimum 0.85 for fitness and precision, minimum 0.75 for simplicity and generalization and 1000 maximum number of generations. As the primary quality dimension is replay fitness, we had chosen the best five models with the highest fitness values but also considering the business requirements of the desired model as some models return unacceptable results, e.g. having the option to start with activities other than "Record the request" which is illogical for the business. The results are shown in Table 3, ordered by the average quality assuming equal weights for all dimensions.

Table 3 Quality dimensions for the best 5 models in Pareto front with ETM

Seq.	Fitness	Precision	Generalization	Simplicity	Overall average quality
1	0.962	0.907	0.967	1.000	0.959
2	0.962	0.895	0.974	1.000	0.958
3	0.966	0.895	0.967	1.000	0.957
4	0.968	0.862	0.975	1.000	0.951
5	0.971	0.853	0.972	1.000	0.949

Although numbers in Table 3 are very close to each other, the models represented in process trees are quite different, e.g. models 1, 3, and 5 have ten nodes but models 2 and 4 have just nine nodes, and some nodes are not in common. Another example of the differences between models is that models 1, 3, and 4 do not have a parallel relation (AND) but models 2 and 5 have a parallel relation between activities "General communication with the client" and "Budget accepted" that two activities are represented in the rest of models by non-exclusive relation. These differences prove that quality dimensions are helping to filter the candidate models but the opinion of business experts and business process requirements will always have the final decision.

The major drawback of ETM miner is not in the quality of the generated models but in time consumed in generating these models. As ETM is a genetic algorithm, it is building models based on repeating the following cycle until one or more stopping criteria are satisfied, it could be a maximum number of generations, a maximum time consumed and/or the number of steady states. The cycle starts with generating random models, evaluates each model by computing overall quality based on the user defined weights, selects the best candidates and adds them to the new generated random models [9], and then repeats the cycle. After satisfying the stopping criteria, the best model for the last generation will be selected.

3.5 Conformance Checking

The alignment-based approach is the most advanced conformance checking approach [1]. It starts with aligning observed behavior from the log with the modeled traces (by choosing the trace in the model which is the best-matched to the behavior from the log). In the case of not having a perfect match (Synchronous move), the most similar trace would be chosen based on the cost function. (Cost function: giving a cost to each move happening in log only and does not have a similar move in the model and another cost for move in the model only which does not have a similar move in the log). The total cost defines which alignment should be considered (which trace should be considered for a particular behavior). There could be more than one alignment solution (equals cost), so fitness and other quality measures will be used to decide the best alignment solution.

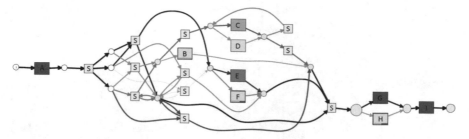

Fig. 6 Petri net after projected with alignment

First, we had to convert process tree into a Petri net and coding activities in letters for the readability as mentioned in Table 2, also note that silent transition is not an activity but it is being used to handle multiple paths between activities [1].

Later, we had projected the Petri net with alignment as shown in Fig. 6, where each rectangle (Transition) represents an activity and each circle (Place) used to control firing transitions. The color of transitions and the thickness of arrows show the frequency of cases going through this transition. The size of places shows moves on log frequency (traces in the log which couldn't be aligned to the model).

For example, transition E has a high number of traces going through it, and all of them are entirely aligned with the model (100% Synchronous) with a total number of cases went through it 4083. On the other hand, Transition F have a total of 1035 moves in the model, 425 of them moved in the model only without occurrence in the original log, and 610 moves were Synchronous in both model and log. Finally, most activities have a perfect match between log and model except for activities B, F, and H. Where cases had deviated in these three activities and algorithm had to make some moves on the model in order to align these traces to the closest path.

3.6 Performance Analysis

We had generated the same process model (Petri net) but now it is projected with time consumed in each transition as shown in Fig. 7 which allows us to detect the bottlenecks in the process by discovering which activity takes much time to be performed.

Using the alignments, we know exactly how to relate events to the process model, and we can annotate activities with the times at which it has been observed. For now, we only know the completion times, and we had no clue when they had started. So the time has been calculated for each activity as the difference between its completion time and the previous activity's completion time (which leads to the current activity). If we also had the starting time, we could distinguish between waiting time and execution time of activity. The color of transitions shows the relative time consumed in order to perform each activity, e.g. activity C has the

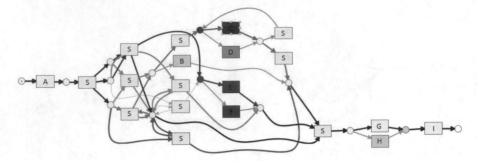

Fig. 7 Performance analysis using time perspective

longest time as it appears with very dark color which means it is taking so much time to be accomplished comparing with the rest of activities. Also, we had discovered that it takes on average 5.59 days to be performed while the whole cycle takes on average 5.41 days, so we can say that there is a potential bottleneck in activity C which needs further analysis with the cooperation of business owners in order to know the origin of the problem.

3.7   Building Social Network

All organizations establish a formal organizational structure where all the hierarchical relationships between employees are defined. However, in most cases, the workflow in the organizations has some gaps with the predefined structure [10].

Our focus here was on organizational perspective, more precisely in extracting social network from event logs and makes it available for further social network analysis. As events are being executed and/or initiated by resources, so it is natural to collect information about these resources while recording the activities details [11]. The social network in Fig. 8 has been inferred from the event log, where each node represents a resource, connection represents a handover of work between two resources in the same case, and the arrow shows the direction of handover. Also, there are two types of resources, customer service resources (Agentxx) and internal department resources (Consxx). Agents with very high number of connections with Cons. are centralized, while resources on the periphery (most of them are Cons.) are only communicating with particular agents, e.g. in Fig. 9, Cons39 is connected with just four resources, but always receiving tasks from 3 of them (Agent01, Agent03, Agent08) and handover tasks to two resources (Agent01, Cons41). It is not normal to have two resources with the same type connected to each other (Cons39 and Cons41) which needs a further investigation.

There are some drawbacks in ProM social network package such as the results are just presented in a graphical form, so the relations between nodes cannot be extracted in a numerical way to be used in further analysis.

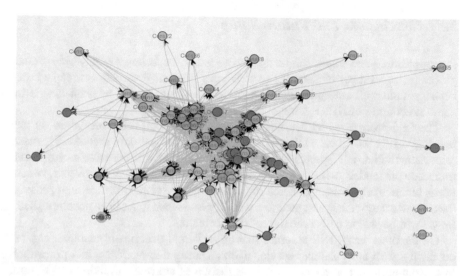

Fig. 8 Social network based on handover-of-work

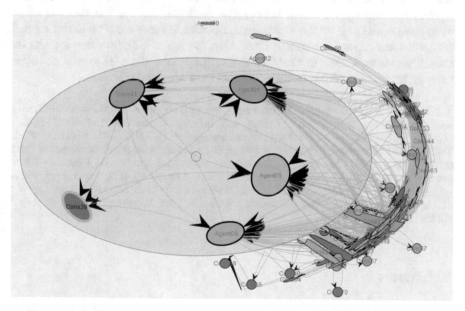

Fig. 9 Zoom in social network

3.8   Conclusions and Future Study

Process Mining techniques can be applied to analyze and improve any operational processes in a variety of domains as it is very relevant for today's organizations because almost all enterprise information systems are storing relevant events in some structured form (ERP, CRM, WFMS, etc.).

The final process model that has been extracted during this work based on the real log history would help INE to have a clear model that only includes the most relevant activities, and ignores the rest of activities that will not have a significant impact on the model quality. Also, a potential bottleneck has been discovered, which needs further analysis using detailed row data including some information regarding the time when the task is assigned to a resource and when it has been accomplished. So we will be able to define the root of the problem.

On the other hand, INE assumes that any internal department resource can be contacting with any customer service agent. During this work, we had proof that this assumption is not correct and some resources are only receiving tasks from specific resources. Also, it has been discovered that some resources have a dual role and can act as a customer service and internal department resource at the same time.

For a future study, we will need to extract more detailed row data that will allow us to analyze the time perspective deeper, in order to focus more on detecting real bottleneck and recourses response time. Also, we are waiting for some updates on Social Network package in ProM software that would allow us to extract social network in a numerical form which could be used to extend this work.

Acknowledgements This work was supported by the research project TEC4Growth—Pervasive Intelligence, Enhancers and Proofs of Concept with Industrial Impact/NORTE-01-0145-FEDER-000020, , North Portugal Regional Operational Programme (NORTE 2020), under the POR-TUGAL 2020 Partnership Agreement, and through the European Regional Development Fund (ERDF) and the ERDF European Regional Development Fund through the Operational Programme for Competitiveness and Internationalization—COMPETE 2020 Programme within project POCI-01-0145-FEDER-006961, and by National Funds through the FCT Fundao para a Cincia e a Tecnologia (Portuguese Foundation for Science and Technology) as part of project UID/EEA/50014/2013.

References

1. Van der Aalst, W.M.P.: Process Mining: Discovery, Conformance and Enhancement of Business Processes. Springer, Berlin (2011)
2. ProM: Prom 6 tutorial. http://www.promtools.org/prom6/downloads/prom-6.0-tutorial.pdf (2010)
3. Van der Aalst, W., Song, M.: Mining social networks: uncovering interaction patterns in business processes. In: International Conference on Business Process Management. Springer, Berlin (2004)
4. Mans, R.S., Schonenberg, M.H., Song, M., et al.: Application of process mining in healthcare : a case study in a Dutch hospital. Biomed. Eng. Syst. Technol. **25**, 425–438 (2009)

5. van der Aalst, W.M.P., Reijers, H.A., Weijters, A., et al.: Business process mining: an industrial application. Inf. Syst. **32**(5), 713–732 (2007)
6. INE: Statistics Portugal. https://www.ine.pt/xportal/xmain?xpgid=ine_main&xpid=INE (2015)
7. Buijs, J.C.A.M., van Dongen, B.F., van der Aalst, W.M.P.: On the role of fitness, precision, generalization and simplicity in process discovery. In: OTM 2012: On the Move to Meaningful Internet Systems: OTM 2012, pp. 305–322. Springer, Berlin (2012)
8. Weijters, A.J.M.M., Ribeiro, J.T.S.: Flexible heuristics miner (fhm). In: IEEE Symposium on Computational Intelligence and Data 565 Mining (CIDM), pp. 310–317 (2011)
9. Van Eck, M. L., Buijs, J.C.A.M., van Dongen, B.F.: Genetic process mining: alignment-based process model mutation. In: Business Process Management Workshops, pp. 291–303. Springer International Publishing, Cham (2015)
10. CROSS, R.: Knowing what we know: supporting knowledge creation and sharing in social networks. Organ. Dyn. **30**, 100120 (2001)
11. Song, M., van der Aalst, W.: Towards comprehensive support for organizational mining. Decis. Support. Syst. **46**(1), 300317 (2008)

Detecting Smooth Cluster Changes in Evolving Graph Structures

Sohei Okui, Kaho Osamura, and Akihiro Inokuchi

Abstract Graph mining is a set of techniques for finding useful patterns in various types of structured data. Many effective algorithms for mining static graphs have been proposed. However, graphs of human relationships and evolving genes change over time, and such evolving graphs require different algorithms for analysis. In this chapter, we explain a method called O2I for clustering in evolving graphs that can detect changes in clusters over time. O2I partitions the graph sequence into smooth clusters, even when the numbers of clusters and vertices vary. It first constructs a graph from the graph sequence, then uses spectral clustering and the RatioCut to apply k partitioning to this graph. O2I is compared in detail with the preserving clustering membership (PCM) algorithm, which is a conventional online graph-sequence clustering algorithm in which the numbers of clusters and vertices must remain constant. We further show that, in contrast to PCM, the performance of O2I is not dependent on the clustering of the initial graph in the graph sequence. Experiments on synthetic evolving graphs show that O2I is practical to calculate and addresses the main disadvantages of PCM. Further tests on real-world data show that O2I can obtain reasonable clusters. This method is hence a flexible clustering solution and will be useful on a wide range of graph-mining applications in which the connections, number of clusters, and number of vertices of the graphs evolve over time.

S. Okui
Graduate School of Science and Technology, Kwansei Gakuin University, Sanda, Japan

K. Osamura · A. Inokuchi (✉)
School of Science and Technology, Kwansei Gakuin University, Sanda, Japan
e-mail: osamura.kaho.oe5@is.naist.jp; inokuchi@kwansei.ac.jp

© Springer International Publishing AG, part of Springer Nature 2019
M. Sayed-Mouchaweh (ed.), *Learning from Data Streams in Evolving Environments*, Studies in Big Data 41, https://doi.org/10.1007/978-3-319-89803-2_10

223

1   Introduction

Studies on graph mining have established many approaches for finding useful patterns in various types of structured data. Although the major algorithms for graph mining are quite effective in practice, most of them focus on *static* graphs, whose structures do not change over time. However, evolving graphs are used to model many real-world applications [12]. For example, a human network can be represented as a graph in which each human and each relationship between two humans correspond to a vertex and an edge, respectively. If a human joins (or leaves) a community in the human network, the numbers of vertices and edges in the graph can change. Similarly, the evolution of a gene network, which consists of genes and their interactions, produces a graph sequence when genes are added, deleted, or mutated. Recently, much attention has been given to graph mining from evolving graphs [6, 19]. Figure 1 shows an example of an evolving graph with four steps and ten unique IDs, indicated by the numbers attached to the vertices. In addition, edge weights are represented by line thickness. For example, humans in a human network correspond to vertices, each of the humans has a unique ID. The strength of the friendship between two humans is represented as an edge weight between two vertices. The current human network is represented as a weighted graph, and the network evolves over time. To represent the evolving network, we use a graph sequence consisting of a series of graphs.

In this chapter, we tackle the problem of clustering in evolving graphs to detect changes in the clusters. In an evolving graph, the number of clusters increases when a cluster divides or decreases when two clusters merge. Although most conventional clustering algorithms focus on partitioning a set of points in a vector space into k clusters, von Luxburg [18] notes that the k partition problem in a vector space can be reduced to the k partition problem in a graph, where each vertex corresponds to a point in the set to be partitioned and an edge indicates the similarity between two vertices. Therefore, the methods in this chapter are applicable to both evolving graphs and points arriving over time.

In [19], problems involving clustering points arriving over time are categorized into four types. Let D_t be a set of points at time t in the vector space, and let $D = \langle D_1, D_2, \ldots, D_T \rangle$ be a series of sets of points. The first type of clustering

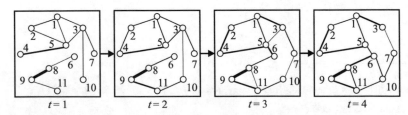

Fig. 1 Example of a weighted graph sequence with four steps (numbers attached to vertices represent vertex IDs)

problem is where only one point arrives at each time t [1–3, 7, 9, 11, 15]. This type of problem focuses on online data processing. The second type of problem is clustering n sequences into k clusters and is applicable to clustering DNA sequences or protein sequences in bioinformatics [4, 14, 17]. This differs from the first type of problem as it does not require online data processing and D_t consists of n points. The third type of problem is to cluster n data streams into k clusters [5, 8]. Although this is the same as the first type in terms of the online analysis of data, each set D_t contains n points, unlike the first type. Preserving cluster membership in Sect. 2.2 tackles this type of problem.

The fourth type of problem, which is addressed in this chapter, analyzes a series D of sets D_t, each of which contains at most n points and is given before the analysis [19]. Although each point clustered in the second type of problem is a sequence, each point clustered in the fourth type of problem is a point in D_t. In addition, while a set of k clusters is returned in the second type of problem, T sets of k clusters are returned in the fourth type. While the predecessors D_τ ($\tau < t$) of D_t are used to cluster D_t in the third type of problem, its successors D_τ ($\tau \geq t$) are also used to cluster D_t in the fourth type.

In this chapter, we explain an algorithm called O2I that partitions the vertices of a graph sequence into smooth clusters, even when the number of vertices is allowed to vary over time [16]. O2I uses spectral clustering and relies on applying the k partition problem to a graph constructed from a graph sequence. Several experiments demonstrate the performance of O2I and its advantages over existing methods.

The remainder of this chapter is organized as follows: In Sect. 2, we formalize the graph sequence clustering problem that we consider in this chapter and explain the conventional method called PCM and its some drawbacks. In Sect. 3, we explain a method called O2I that overcomes the drawbacks of PCM and discuss the relationship between the performance of OI2 and connectivities of graphs in a graph sequence. In Sect. 4, we compare O2I with PCM in terms of clustering accuracy using artificially generated datasets, and verify the practicality of O2I on a real-world dataset. Finally, we conclude the chapter in Sect. 5.

2 Clustering a Graph Sequence

2.1 Problem Definition

In this chapter, to model an evolving graph, we use a weighted graph sequence. A weighted graph at time t is represented by $G^{(t)} = \left(V^{(t)}, E^{(t)}, w^{(t)}\right)$, where $V^{(t)}$ is a set of vertices, each of which has a unique ID, $E^{(t)} = V^{(t)} \times V^{(t)}$ is the set of all edges, and $w^{(t)}$ is a function that assigns nonnegative real values to the edges at time t. A series of T graphs is called a weighted graph sequence with T steps and is denoted by $\langle G^{(1)}, G^{(2)}, \ldots, G^{(T)} \rangle$. Although we assume that the value of

$|V^{(t)}| = n$ is unchanged in the graph sequence in this section, the value of $|V^{(t)}|$ in O2I explained in the next section changes over time.

Figure 1 shows an example of a graph sequence with four steps. In the figure, edge weights are represented by line thickness. For the sake of simplicity, we do not show edges with weight 0 in the figures in this chapter.[1]

The vertices $V^{(t)}$ in a graph at time t are partitioned into k disjoint subsets $P^{(t)} = \left\{ C_1^{(t)}, C_2^{(t)}, \ldots, C_k^{(t)} \right\}$, where $\bigcup_{j=1}^{k} C_j^{(t)} = V^{(t)}$. Using this notation, a cluster sequence can be written as $\left\langle C_j^{(1)}, C_j^{(2)}, \ldots, C_j^{(T)} \right\rangle$ for $1 \le j \le k$.

Given a graph sequence $\left\langle G^{(1)}, G^{(2)}, \ldots, G^{(T)} \right\rangle$ and the number of clusters k as inputs, the problem addressed in this chapter is how to determine cluster sequences $\left\{ \left\langle C_j^{(1)}, C_j^{(2)}, \ldots, C_j^{(T)} \right\rangle \mid 1 \le j \le k \right\}$ that satisfy the following two requirements:

1. Vertices connected by high-weight edges in a graph at time t should appear in the same cluster for each graph in the sequence.

2. Clusters $C_j^{(t)}$ and $C_j^{(t+1)}$ should be almost the same. This requirement is called cluster smoothness.

Figures 2 and 3 show cluster sequences obtained from the graph sequence in Fig. 1. When we do not take requirement (2) into account, vertices 5 and 6 appear in the same cluster because the edge $(5, 6)$ at time 3 has a high weight, as shown in Fig. 2. In contrast, when we take requirement (2) into account, vertex 6 is assigned to $C_2^{(t)}$ before and after time 3, and hence both clusters $C_2^{(2)}$ and $C_2^{(3)}$ are the same.

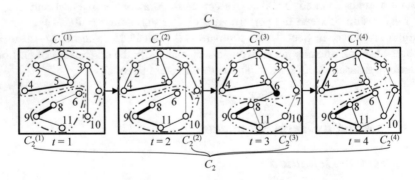

Fig. 2 Cluster sequences (1) obtained from the graph sequence in Fig. 1

---

[1]The weight 0 means that there is no connection between two vertices. We need these zeros to create Laplacian matrices in Sect. 2.2.

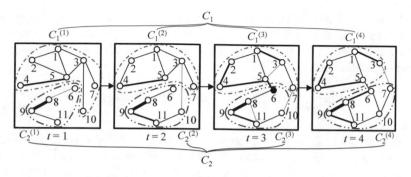

Fig. 3 Cluster sequences (2) obtained from the graph sequence in Fig. 1

2.2 Preserving Cluster Membership

The k partition problem for a graph $G = (V, E, w)$ is defined as the problem of finding non-empty sets $C_1, C_2, \ldots, C_k$ that partition V and that minimize

$$\sum_{j=1}^{k} \frac{1}{|C_j|} \sum_{e \in E(C_j, V \setminus C_j)} w(e),$$

where $E(S, V \setminus S)$ is the set of edges (v, u) with $v \in S$ and $u \in V \setminus S$. This optimization problem minimizes the function called RatioCut. According to [18], the above minimization problem is equivalent to

$$\min_{X \in \mathcal{R}^{n \times k}} tr(X^T L X) \quad s.t. \ X^T X = I, \,^2 \tag{1}$$

where the n-by-k matrix X indicates the cluster to which each vertex belongs, with element x_{ij} of the matrix given by

$$x_{ij} = \begin{cases} \frac{1}{\sqrt{|C_j|}} & if \ v_i \in C_j, \\ 0 & otherwise, \end{cases}$$

where $X^T X = I$ indicates that each vertex in graph G belongs to one cluster, where I is the identity matrix of size n. In addition, L is the Laplacian matrix of G, defined as follows. Let A be an adjacency matrix for G, where the (i, j)th element a_{ij} is weight $w((i, j))$ if an edge exists between v_i and v_j in G. Otherwise, a_{ij} is 0. Setting $D = diag(\sum_{i=1}^{n} a_{i1}, \sum_{i=1}^{n} a_{i2}, \ldots, \sum_{i=1}^{n} a_{in})$, the Laplacian matrix is $L = D - A$. Equation (1) is called spectral clustering.

---

[2]Because of space limitations, we omit $X^T X = I$ henceforth.

One online algorithm for clustering a graph sequence is called preserving cluster membership (PCM) [10]. In this algorithm, the matrix X_{t-1} corresponding to $G^{(t-1)}$ is known and we are given graph $G^{(t)}$. The algorithm obtains cluster sequences by iteratively optimizing

$$\min_{X_t \in \mathscr{R}^{n \times k}} tr(X_t^T L_t X_t) + \alpha ||X_t X_t^T - X_{t-1} X_{t-1}^T||^2, \qquad (2)$$

where $\alpha \geq 0$ and L_t is the Laplacian matrix of $G^{(t)}$. If the ith and jth vertices belong to the same cluster at time t, then the (i, j)th element of $X_t X_t^T$ is a positive real number. Otherwise, the (i, j)th element is 0. Minimizing the second term of the objective function in Eq. (2) under the Frobenius norm, where $||W||^2 = tr(W^T W)$, satisfies requirement (2). The objective function is transformed as follows:

$$tr(X_t^T L_t X_t) + \alpha ||X_t X_t^T - X_{t-1} X_{t-1}^T||^2$$
$$= tr(X_t^T L_t X_t) + \alpha \, tr(X_t X_t^T - X_{t-1} X_{t-1}^T)^T (X_t X_t^T - X_{t-1} X_{t-1}^T)$$
$$= tr(X_t^T L_t X_t) + \alpha \, tr(X_t X_t^T X_t X_t^T - 2 X_t X_t^T X_{t-1} X_{t-1}^T + X_{t-1} X_{t-1}^T X_{t-1} X_{t-1}^T)$$
$$= tr(X_t^T L_t X_t) + 2\alpha k - 2\alpha \, tr(X_t^T X_{t-1} X_{t-1}^T X_t)$$
$$= 2\alpha k + tr(X_t^T L_t X_t - 2\alpha X_t^T X_{t-1} X_{t-1}^T X_t)$$
$$= 2\alpha k + tr[X_t^T (L_t - 2\alpha X_{t-1} X_{t-1}^T) X_t].$$

Therefore, Eq. (2) is equivalent to

$$\min_{X_t \in \mathscr{R}^{n \times k}} tr \left[X_t^T \left(L_t - 2\alpha X_{t-1} X_{t-1}^T \right) X_t \right]. \qquad (3)$$

In [10], an offline algorithm was also proposed as an extension to PCM. To demonstrate the offline algorithm, the authors introduced an optimization problem for clustering $G^{(t)}$ using known X_{t-1} and X_{t+1} corresponding to clusters $P^{(t-1)}$ and $P^{(t+1)}$, respectively:

$$\min_{X_t \in \mathscr{R}^{n \times k}} tr \left[X_t^T \left(L_t - \alpha X_{t-1} X_{t-1}^T - \alpha X_{t+1} X_{t+1}^T \right) X_t \right]. \qquad (4)$$

We define the functions $func_1(L)$, $func_2(L_t, X_{t-1}, \alpha)$, and $func_3(L_t, X_{t-1}, X_{t+1}, \alpha)$ to be the minimum values of Eqs. (2), (3), and (4), respectively. Using these functions, the PCM offline algorithm is shown in Algorithm 1. First, the PCM offline algorithm clusters $G^{(1)}$, and then it clusters $G^{(2)}$ using the results from time 1. This process is repeated for the series of graphs. Next, it clusters $G^{(1)}$ using the results from time 2. Then, it clusters $G^{(2)}$ using the results from times 1 and 3. The process repeats until convergence.

Algorithm 1: PCM_offline

     Data: $\langle G^{(1)}, G^{(2)}, \ldots, G^{(T)} \rangle, k$
     Result: $X_1, X_2, \ldots, X_T$
 1 **for** $t \in [1, T]$ **do**
 2   **if** $t = 1$ **then**
 3 | $X_1 = func_1(L_1)$;
 4   **else**
           $X_t = func_2(L_t, X_{t-1}, \alpha)$;

 5 **repeat**
 6   **for** $t \in [1, T]$ **do**
 7   **if** $t = 1$ **then**
 8 | $X_1 = func_2(L_1, X_2, \alpha)$;
 9   **else**
                 if $t = T$ **then**
 10 | $X_T = func_2(L_T, X_{T-1}, \alpha)$;
 11   **else**
                      $X_t = func_3(L_t, X_{t-1}, X_{t+1}, \alpha)$;

     until $X_1, X_2, \ldots, X_T$ *converge*;
 12 **return** $X_1, X_2, \ldots, X_T$;

When α is decreased, each graph in a graph sequence is clustered independently because the first term in Eq. (2), which relates to requirement (1), is emphasized over requirement (2). This results in the cluster sequences in Fig. 2. On the one hand, when α is increased, the smooth cluster sequences in Fig. 3 are obtained because the second term in Eq. (2), which relates to requirement (2), is emphasized. In concrete terms, vertex 6 at time 3 belongs to cluster sequence C_1 in Fig. 2, while it belongs to the other cluster sequence C_2 at times 2 and 4. On the other hand, placing vertex 6 in C_1 in Fig. 3 decreases the second term of Eq. (2), and hence vertex 6 belongs to C_1 at all times.

2.3 Drawbacks of PCM

We point out three drawbacks of the PCM offline algorithm. First, the performance of PCM is dependent on $G^{(1)}$. If the vertices in each latent cluster of $G^{(1)}$ are strongly connected, then the problem of obtaining cluster sequences from a graph sequence is relatively easy because the algorithm uses X_1 to cluster $G^{(2)}$ and then uses X_t to cluster $G^{(t+1)}$ for $t > 1$. However, if the clusters for $G^{(1)}$ are not suitable, then this unsuitability propagates to clusters in $P^{(t)}$ for $t > 1$ because of the second term in Eq. (2).

The second drawback comes from having k clusters at all times. For this reason, PCM cannot determine suitable cluster sequences from a graph sequence when the number of clusters increases after one cluster divides or when the number of clusters decreases after two clusters merge. To detect suitable cluster sequences from a graph sequence, we should allow the number of clusters to vary over time.

The third drawback of PCM is related to the second drawback. The number of vertices in the graph is the same at all times in PCM. However, the members of a social network are not constant, but change over time. Therefore, we should allow members to join and leave the network and develop a clustering algorithm for data in which the number of vertices is not constant.

3   Detecting Smooth Cluster Changes in a Graph Sequence

3.1   Clustering a Graph Sequence Using Smoothness Between Two Successive Graphs

Okui et al. have proposed a method called O2I that overcomes the first and second drawbacks of PCM that are explained in the previous section [16]. To explain the method, we discuss the problem of obtaining the $X_1, X_2, \ldots, X_T \in \mathscr{R}^{n \times k}$ that minimize

$$\sum_{t=1}^{T} tr(X_t^T L_t X_t) + \alpha' \sum_{t=1}^{T-1} ||X_t - X_{t+1}||^2, \tag{5}$$

where $\alpha' > 0$. Minimizing the first term in Eq. (5) corresponds to clustering each graph $G^{(t)}$ in a graph sequence according to requirement (1). To show that minimizing the second term in Eq. (5) corresponds to requirement (2), we consider a graph sequence consisting of only two graphs. The objective function for the sequence is given by

$$tr(X_1^T L_1 X_1) + tr(X_2^T L_2 X_2) + \alpha' ||X_1 - X_2||^2. \tag{6}$$

From Eq. (6), we derive the equation in Fig. 4. Similarly, the equation shown in Fig. 5 is derived from Eq. (5). When D' is the underlined matrix in Fig. 5 and W' is the double underlined matrix in Fig. 5, matrix $L' = D' - W'$ is the Laplacian matrix for a graph G' that satisfies the following:

- The number of vertices in G' is $n \times T$. Henceforth, the ith vertex of G' at time t is represented by $v_{t,i}$.
- If $G^{(t)}$ contains an edge (i, j) of weight $w((i, j))$, then G' also contains an edge $(v_{t,i}, v_{t,j})$ of $w((i, j))$.
- Graph G' contains an edge $(v_{t,i}, v_{t+1,i})$ of weight α' for $1 \le t \le T - 1$ and $1 \le i \le n$.

Therefore, the problem of minimizing Eq. (5) is reduced to the k partition problem for G'. The edges between vertices $v_{t,i}$ and $v_{t+1,i}$ have weight α'. Cutting some

$$tr\left[X_1^T L_1 X_1\right] + tr\left[X_2^T L_2 X_2\right] + \alpha' \,||X_1 - X_2||^2$$
$$= tr\left[X_1^T L_1 X_1 + X_2^T L_2 X_2\right] + \alpha' \, tr[X_1^T X_1 + X_2^T X_2 - X_2^T X_1 - X_1^T X_2]$$
$$= tr\left[\binom{X_1}{X_2}^T \begin{pmatrix} L_1 & 0 \\ 0 & L_2 \end{pmatrix} \binom{X_1}{X_2}\right] + \alpha' \, tr\left[\binom{X_1}{X_2}^T \begin{pmatrix} I & -I \\ -I & I \end{pmatrix} \binom{X_1}{X_2}\right]$$
$$= tr\left[\binom{X_1}{X_2}^T \begin{pmatrix} L_1 + \alpha' I & -\alpha' I \\ -\alpha' I & L_2 + \alpha' I \end{pmatrix} \binom{X_1}{X_2}\right]$$
$$= tr\left[\binom{X_1}{X_2}^T \left\{ \begin{pmatrix} D_1 + \alpha' I & 0 \\ 0 & D_2 + \alpha' I \end{pmatrix} - \begin{pmatrix} W_1 & \alpha' I \\ \alpha' I & W_2 \end{pmatrix} \right\} \binom{X_1}{X_2}\right]$$

Fig. 4 Objective function for a graph sequence with two steps

$$tr\left[\begin{pmatrix} X_1 \\ X_2 \\ \vdots \\ \vdots \\ X_T \end{pmatrix}^T \left\{ \begin{pmatrix} D_1 + \alpha' I & 0 & \cdots & \cdots & 0 \\ 0 & D_2 + 2\alpha' I & \ddots & & \vdots \\ \vdots & \ddots & \ddots & \ddots & 0 \\ \vdots & & \ddots & D_{T-1} + 2\alpha' I & 0 \\ 0 & \cdots & \cdots & 0 & D_T + \alpha' I \end{pmatrix} - \begin{pmatrix} W_1 & \alpha' I & 0 & \cdots & 0 \\ \alpha' I & W_2 & \ddots & \ddots & \vdots \\ 0 & \ddots & \ddots & \ddots & 0 \\ \vdots & \ddots & \ddots & W_{T-1} & \alpha' I \\ 0 & \cdots & 0 & \alpha' I & W_T \end{pmatrix} \right\} \begin{pmatrix} X_1 \\ X_2 \\ \vdots \\ \vdots \\ X_T \end{pmatrix}\right]$$

Fig. 5 Objective function for a graph sequence with T steps

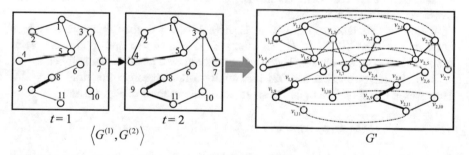

Fig. 6 Conversion of a graph sequence with two steps into a graph G'

of these edges increases the value of the objective function in Eq. (5) when G' is partitioned to k subgraphs. For this reason, $v_{t,i}$ and $v_{t+1,i}$ are likely to appear in the same cluster, so these edges may not be cut. Therefore, minimizing Eq. (5) satisfies requirement (2). Figure 6 shows an example of transforming a graph sequence with two steps $\langle G^{(1)}, G^{(2)} \rangle$ to a graph G'. In this figure, broken lines represent edges of weights α'.

The problem of minimizing Eq. (5) has hence been reduced to the k partition problem for G'. The cluster sequences obtained by O2I is not dependent on clustering $G^{(1)}$, unlike in PCM, which first partitions $G^{(1)}$ and then iteratively partitions the other graphs. Thus, the first drawback of PCM is overcome. In

---

Algorithm 2: O2I

Data: $\langle G^{(1)}, G^{(2)}, \ldots, G^{(T)} \rangle, k$
Result: $X_1, X_2, \ldots, X_T$
1 Construct G' from $\langle G^{(1)}, G^{(2)}, \ldots, G^{(T)} \rangle$;
2 $\ell = n \times T$;
3 Compute the Laplacian matrix L' of G';
4 Compute the first k eigenvectors $\mathbf{u}_1, \mathbf{u}_2, \ldots, \mathbf{u}_k$ of L';
5 Let $U \in \mathscr{R}^{\ell \times k}$ be the matrix that has $\mathbf{u}_q$ as its qth column;
6 For $i = 1, \ldots, \ell$, let $\mathbf{y}_i \in \mathscr{R}^k$ be the vector corresponding to the ith row of Γ;
7 Use the k-means algorithm to cluster the points $\{\mathbf{y}_1, \mathbf{y}_2, \ldots, \mathbf{y}_\ell\}$ in $\mathscr{R}^k$ into clusters
   $P_1, P_2, \ldots, P_k$;
8 **for** $t \in [1, T]$ **do**
9 $\quad \lfloor \quad X_t = 0$;
10 **for** $j \in [1, k]$ **do**
11 $\quad$ **for** $v_{t,i} \in P_j$ **do**
12 $\quad \quad \lfloor \quad x_{i,j}^{(t)} = \frac{1}{\sqrt{|\{v_{t',i'} \in P_j | t = t'\}|}}$;

13 **return** $X_1, X_2, \ldots, X_T$;

---

addition, some clusters obtained using O2I may not contain any vertex from time t. Therefore, O2I does not guarantee a partition of each graph $G^{(t)}$ into exactly k clusters, but instead partitions the graph into k or fewer clusters. Hence, O2I overcomes the second drawback.

Note that O2I requires $\alpha' > 0$ in Eq. (5). If $\alpha' = 0$, then there are no edges between $G^{(t)}$ and $G^{(t+1)}$ in G', so when G' is partitioned into T clusters, each graph $G^{(t)}$ becomes a cluster. Therefore, α' should be a positive real number.

The objective function for O2I is similar to the objective function for PCM. However, it is impossible to derive an equation in the form of Fig. 4 from

$$tr(X_1^T L_1 X_1) + tr(X_2^T L_2 X_2) + \alpha||X_1^T X_1 - X_2^T X_2||^2.$$

Thus, it is impossible to reduce the objective function for PCM to the k partition problem for a graph.

Algorithm 2 shows the pseudo code for O2I. The method uses the spectral clustering algorithm [18] in lines 3–6. It then initializes X_t to the zero matrix in lines 7–8. In lines 9–11, if the jth cluster P_j contains $v_{t,i}$, then $\frac{1}{\sqrt{|\{v_{t',i'} \in P_j | t=t'\}|}}$ replaces $x_{i,j}^{(t)}$ in X_t.

Using Algorithm 2, the number of vertices does not have to be n at all times. The third drawback is hence resolved by replacing $\ell = n \times T$ in line 2 with $\ell = \sum_{t=1}^{T} |V^{(t)}|$.

3.2 Clustering Using the Forgetting Rate

We considered the smoothness of clusters between two consecutive timesteps in the previous subsection. In this section, we extend O2I to consider cluster smoothness between timesteps separated by distance τ. This is formulated in the following equation:

$$\sum_{t=1}^{T} tr(X_t^T L_t X_t) + \alpha' \sum_{\tau=1}^{T-1} \gamma^{\tau-1} \sum_{t=1}^{T-\tau} ||X_t - X_{t+\tau}||^2, \qquad (7)$$

where γ is called the forgetting rate and $0 \leq \gamma \leq 1$. Equation (7) is a generalization of Eq. (5), as they are equivalent when $\gamma = 0$. The equation shown in Fig. 7 is derived in a manner similar to that in the previous section, where $B_t = \alpha' \sum_{\tau=1, \tau \neq t}^{T} \gamma^{|t-\tau|-1} I$ is a diagonal matrix. The underlined part in Fig. 7 is the Laplacian matrix L'' of graph G'' that satisfies the following:

- The number of vertices in G'' is $n \times T$.
- If $G^{(t)}$ contains an edge (i, j) of weight $w((i, j))$, then G' also contains an edge $(v_{t,i}, v_{t,j})$ of $w((i, j))$.
- Graph G'' contains an edge of weight $\alpha' \gamma^{(t'-t-1)}$ between $v_{t,i}$ and $v_{t',i}$ for $1 \leq t < t' \leq T$ and $1 \leq i \leq n$.

Graph G' is a subgraph of G''. When $\gamma = 0$, G' is isomorphic to G''. Algorithm 2 is applicable to G'' by replacing G' and L' with G'' and L'', respectively. In the previous section, we explained that O2I overcomes the third drawback of PCM. However, when v_i is contained in $G^{(t)}$ but not in both $G^{(t-1)}$ and $G^{(t+1)}$, we cannot consider the smoothness of clusters for this vertex. In contrast, O2I using the forgetting rate further overcomes the third drawback by introducing edges with weights that decrease exponentially with the distance between graphs.

$$tr\left[\begin{pmatrix} X_1 \\ X_2 \\ \vdots \\ \vdots \\ X_T \end{pmatrix}^T \left\{ \begin{pmatrix} D_1+B_1 I & 0 & \cdots & \cdots & 0 \\ 0 & D_2+B_2 & \ddots & & \vdots \\ \vdots & & \ddots & \ddots & 0 \\ \vdots & & & \ddots & \ddots & 0 \\ 0 & \cdots & & 0 & D_T+B_T \end{pmatrix} - \begin{pmatrix} W_1 & \alpha' I & \alpha' \gamma I & \cdots & \alpha' \gamma^{T-2} I \\ \alpha' I & W_2 & \ddots & & \vdots \\ \alpha' \gamma I & \ddots & \ddots & & \alpha' \gamma I \\ \vdots & & \ddots & W_{T-1} & \alpha' I \\ \alpha' \gamma^{T-2} I & \cdots & \alpha' \gamma I & \alpha' I & W_T \end{pmatrix} \right\} \begin{pmatrix} X_1 \\ X_2 \\ \vdots \\ \vdots \\ X_T \end{pmatrix} \right]$$

Fig. 7 Objective function with forgetting rate for a graph sequence with T steps

3.3 Connectivities of Graphs

In this section, we discuss the effect of the connectivity of each graph in a graph sequence on the clustering result. For the sake of simplicity, we consider the simple example of a sequence of sets of points, as shown in Fig. 8. Each timestep consists of three points whose coordinates are given in the figure. We convert each of the sets of the points into a graph where the vertices and edge weights are the points and $\exp(-\frac{d^2}{2})$, respectively, where d is the Euclidean distance between two points. We then obtain a graph sequence with two steps. We assume that $\langle\{v_1\}, \{v_1, v_2\}\rangle$ and $\langle\{v_2, v_3\}, \{v_3\}\rangle$ are desirable cluster sequences obtained from the graph sequence for $k = 2$.

Figure 9 shows the graph $G'_1 = (V'_1, E'_1, w'_1)$ created from the graph sequence with two steps. Because G'_1 is partitioned by RatioCut, we obtain the following solutions depending on the value of α'.

$$\min \sum_{j=1}^{2} \frac{1}{|C_j|} \sum_{e \in E'_1(C_j, V'_1 \setminus C_j)} w'_1(e)$$

$$= \begin{cases} 2\alpha' & \text{if } 0 < \alpha' \le 0.56 \ (C_1 = \{v_{1,1}, v_{1,2}, v_{1,3}\}), \text{ and} \\ 1.11 & \text{otherwise} \qquad (C_1 = \{v_{1,1}, v_{2,1}\}). \end{cases} \tag{8}$$

Fig. 8 Example of a sequence of sets of points

Fig. 9 Graph G'_1 created from the sequence of points in Fig. 8

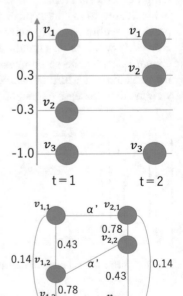

Fig. 10 Solutions for G_1' and various α'

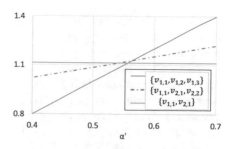

Fig. 11 Solutions for G_2' and various α'

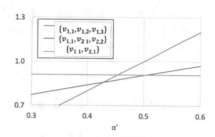

Equation (8) is represented by Fig. 10 for various α'. When α' is less than 0.56, G_1' is partitioned into $\{v_{1,1}, v_{1,2}, v_{1,3}\}$ and $\{v_{2,1}, v_{2,2}, v_{2,3}\}$, which does not satisfy cluster smoothness. In contrast, when α' is greater than 0.56, G_1' is partitioned into $\{v_{1,1}, v_{2,1}\}$ and $\{v_{1,2}, v_{1,3}, v_{2,2}, v_{2,3}\}$, which means that O2I cannot detect any changes in the clusters because requirement (2) is oversatisfied. Thus, O2I has no chance to obtain the desirable cluster sequences from G_1' shown in Fig. 9, even if α' is tuned to the optimal value.

In the previous example, we created the complete graph from each set of points in a timestep. In [18], the ϵ-neighborhood graph and κ-nearest neighbor graph are introduced as an alternative to a complete graph created from a set of points. In the ϵ-neighborhood graph, two points are connected by an edge if a distance d between the points is less than ϵ. In the κ-nearest neighbor graph, two points are connected if one of them is among the κ-nearest neighbors of the other. When ϵ-neighborhood graph $G_2' = (V_2', E_2', w_2')$ of ϵ 1 is created from a sequence of points, we obtain the following solutions, depending on the value of α' after partitioning G_2' using RatioCut.

$$\min \sum_{j=1}^{2} \frac{1}{|C_j|} \sum_{e \in E_2'(C_j, V_2' \setminus C_j)} w_2'(e)$$

$$= \begin{cases} 2\alpha' & \text{if } 0 < \alpha' \le 0.43 \quad (C_1 = \{v_{1,1}, v_{1,2}, v_{1,3}\}) \\ 0.67\alpha' + 0.57 & \text{if } 0.43 < \alpha' \le 0.50 \ (C_1 = \{v_{1,1}, v_{2,1}, v_{2,2}\}) \\ 1.11 & \text{otherwise} \quad\quad\quad (C_1 = \{v_{1,1}, v_{2,1}\}). \end{cases} \quad (9)$$

Similarly to Eq. (8) and Fig. 10, Eq. (9) is illustrated by Fig. 11 for various α'. In contrast to the case for G_1', Fig. 11 indicates that O2I can obtain the desirable cluster sequence when $0.43 < \alpha' \leq 0.50$.

When G_1' consists of two complete graphs, each of which is created from a set of points, vertices coming from the same timestep are connected densely with each other, while every pair of vertices coming from different timesteps is rarely connected. In this case, $\{v_{1,1}, v_{1,2}, v_{1,3}\}$ coming from the same timestep minimizes RatioCut rather than $\{v_{1,1}, v_{2,1}, v_{2,2}\}$. In contrast, when a sparse graph is created from a set of points instead of a dense graph, the connectivities among vertices coming from the same timestep and among vertices coming from the different timesteps are balanced, and O2I can select $\{v_{1,1}, v_{2,1}, v_{2,2}\}$ to minimize RatioCut for G' by tuning α'.

4 Experimental Evaluation

4.1 Experimental Setup

In this section, we compare O2I with the PCM offline algorithm using the adjusted Rand index (ARI). The ARI measures the similarity of two sets of clusters, and is calculated using the number of vertices common to each pair of clusters. We assume that a set of n vertices is partitioned both into r disjoint subsets $\mathcal{U} = \{U_1, U_2, \ldots, U_r\}$ and into c disjoint subsets $\mathcal{V} = \{V_1, V_2, \ldots, V_c\}$, so that $\sum_{i=1}^{r} |U_i| = \sum_{j=1}^{c} |V_j| = n$. The number of vertices common to U_i and V_j is denoted by n_{ij}, as shown in Table 1 , where $n_{i.}$ and $n_{.j}$ are the numbers of vertices in clusters U_i and V_j, respectively. The number of pairs of vertices commonly contained in U_i and V_j is calculated by $\binom{n_{ij}}{2}$. The ARI is hence calculated as

$$\frac{\sum_{i,j} \binom{n_{ij}}{2} - \left[\sum_i \binom{n_{i.}}{2} \sum_j \binom{n_{.j}}{2}\right] / \binom{n}{2}}{\frac{1}{2}\left[\sum_i \binom{n_{i.}}{2} + \sum_j \binom{n_{.j}}{2}\right] - \left[\sum_i \binom{n_{i.}}{2} \sum_j \binom{n_{.j}}{2}\right] / \binom{n}{2}}.$$

The ARI takes a value between 0 and 1. Larger ARI values indicate that the obtained clusters are more suitable because $\mathcal{U}$ and $\mathcal{V}$ correspond to the original partition of

Table 1 Contingency table comparing partitions $\mathcal{U}$ and $\mathcal{V}$

$\mathcal{U}$	$\mathcal{V}$				
	V_1	V_2	$\ldots$	V_c	Total
U_1	n_{11}	n_{12}	$\ldots$	n_{1c}	$n_{1.}$
U_2	n_{21}	n_{22}	$\ldots$	n_{2c}	$n_{2.}$
$\vdots$	$\vdots$	$\vdots$	$\ddots$	$\vdots$	$\vdots$
U_r	n_{r1}	n_{r2}	$\ldots$	n_{rc}	$n_{r.}$
Total	$n_{.1}$	$n_{.2}$	$\ldots$	$n_{.c}$	$n_{..} = n$

the data and the partition obtained by the algorithm, respectively. The ARI values given in this chapter are averages for 20 trials.

4.2 Results

4.2.1 Dependence on the Initial Graph of the Graph Sequence

To compare O2I with PCM, we generated artificial datasets using the following procedures with the parameters shown in Table 2. The means of the k Gaussian distributions were placed at equal intervals on a circle of radius r whose center is the origin. A set of points was generated under a Gaussian distribution for each of the means. In this experiment, the sizes of the three sets of points were set to 600, 300, and 200. The set of generated points corresponds to a latent cluster of vertices. The sets of points move toward and away from the origin as time advances, as shown in Fig. 12. The means of the Gaussian distributions are on a sine curve whose amplitude, angular frequency, and initial phase are denoted by A, ω, and φ, respectively. Therefore, the mean of the Gaussian distribution corresponding to the jth cluster at time t is given by

Table 2 Parameters of the artificial data

Parameters	Default values
Number of cluster sequences	$k = 3$
Radius	$r = 3$
Variance of each Gaussian distribution	$var = 1.0$
Number of vertices	$n = 1100 \ (=600+300+200)$
Amplitude	$A = 1.0$
Angular frequency	$\omega = \frac{\pi}{4}$
Initial phase	$\varphi = 0$
Steps	$T = 10$
Number of moving vertices	$m = 5$
Connectivity	$\kappa = 10$

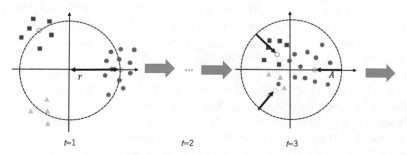

Fig. 12 Generation of the artificial datasets

Fig. 13 ARIs for O2I for various values of α'

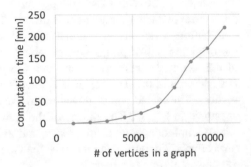

Fig. 14 Computation time for various values of n

$$\begin{pmatrix} x \\ y \end{pmatrix} = R\left(\frac{2\pi(j-1)}{k}\right)\left[A\begin{pmatrix} \sin(\omega(t-1)+\varphi) \\ 0 \end{pmatrix} + \begin{pmatrix} r \\ 0 \end{pmatrix}\right], \qquad (10)$$

where $R(\theta)$ is a rotation matrix for angle θ, with $1 \le j \le k$ and $1 \le t \le T$. In addition, m points in the largest cluster move to the second-largest cluster at each step. The sets of points at time t are converted into the κ-nearest neighbor graph $G^{(t)}$ with edges with weights $\exp\left(-\frac{d^2}{2}\right)$, where d is the Euclidean distance between two of the points.

Figure 13 shows the ARIs for O2I as α' increases from 1 to 100,000. When $\alpha' = 40$, the two requirements are balanced. When $\alpha' > 40$, the ARI decreases because requirement (2) is oversatisfied because of the influence of the second term in Eq. (5). In contrast, when $\alpha' < 40$, the ARI decreases because requirement (1) is oversatisfied because of the influence of the first term in Eq. (5). Moreover, when α' is decreased to less than 1, the ARI decreases drastically because the weights of the edges between two successive graphs in the graph sequence are less than the weights of the edges in each graph $G^{(t)}$ and O2I partitions G' into clusters by cutting the edges between successive graphs. Therefore, these results confirm that O2I obtains suitable cluster sequences satisfying the requirements by tuning α'. Henceforth, α' is set to 40. Because a similar result was obtained for PCM, α was set to 4.

Figure 14 shows the computation time for O2I when the number of vertices in each graph in a graph sequence increases. The computation time is proportional to the cube of the number of the vertices because the most time-consuming procedure in O2I is the calculation of the eigenvectors of the Laplacian matrix of G'. Figure 14 shows that O2I is practical for graphs G' with more than $10,000 \times 10$ vertices.

Fig. 15 ARIs for various values of φ

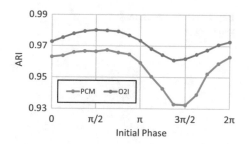

Figure 15 shows the ARIs for O2I and PCM when φ increases from 0 to 2π. The ARI for PCM substantially decreases around $\varphi = \frac{3}{2}\pi$. When $\varphi = \frac{3}{2}\pi$, the distributions of the points significantly overlap at $t = 1$, 5, and 9, because the means of the distributions approach one another. In particular, when the clusters for $G^{(1)}$ are not suitable, the ARI value decreases substantially because the unsuitability propagates through the clusters for $G^{(t)}$ with $t > 1$. In contrast, the ARI for O2I is better than the ARI for PCM for all φ, although the ARI decreases slightly around $\varphi = \frac{3}{2}\pi$. Hence, the results confirm that O2I overcomes the first drawback of PCM.

4.2.2 Varying Cluster Numbers

In this experiment, the means of the two small latent clusters out of the three clusters are located at

$$\begin{pmatrix} x \\ y \end{pmatrix} = R\left(\frac{2\pi(j-1)}{k}\right) r \begin{pmatrix} \exp\left[\beta(t-T)\right] \\ 0 \end{pmatrix}, \tag{11}$$

rather than the points in Eq. (10). Although the means of the two clusters are close to each other at time 1, they diverge exponentially and move toward a circle of radius r whose center is the origin at time T. By assuming that the points generated from distributions whose means are closer than $2var$ belong to the same cluster, we generate an artificial graph sequence where one of the latent clusters divides into two latent clusters. In this experiment, m is set to 0.

Figure 16 shows the ARIs for O2I and PCM as β increases from 0 to 0.4. When β is low, the ARI for PCM is high because the three latent clusters are separate from one another in $G^{(1)}$ and remain separate until time T. However, when β is high, the ARI value decreases because PCM partitions each graph $G^{(t)}$ into three clusters for small t even though the number of latent clusters is 2. In contrast, O2I partitions each graph into k or fewer clusters because it first converts the graph sequence into G' and then solves the k partition problem for G'. Thus, O2I partitions each graph $G^{(t)}$ into two clusters for small t and into three clusters for large t. Hence, it detects suitable clusters for various values of β.

Figure 17 shows three sets of points: One of the sets is derived from Eq. (10) and the others are derived from Eq. (11). Each point is colored according to cluster

Fig. 16 ARIs for various
values of β

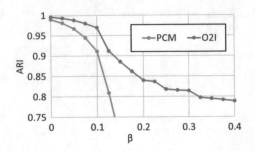

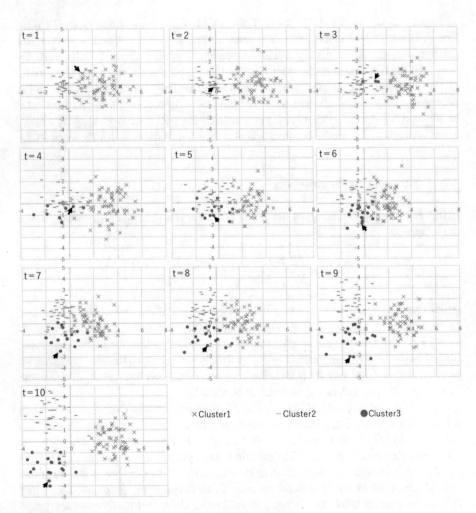

×Cluster1   −Cluster2   ●Cluster3

Fig. 17 Distribution of vertices in detected cluster sequences

sequences detected by O2I. This result was obtained for $n = 110$ and $\beta = 0.35$. The points with arrows are points with the same ID. Because cluster 1 vibrates in a sinusoid, the number of points is almost the same over time. In contrast, although the number of points in detected cluster 3 is 0 at time 1, the number of vertices increases and becomes 18 at the last timestep.[3] It is a difficult task to detect the three different clusters at time 2 by conventional methods because the point detected as a point in cluster 3 at time 2 exists near the centroid of cluster 2. However, the second term of Eq. (5) enables O2I to detect the three clusters. This figure indicates that O2I detects cluster sequences in which one cluster divides into two clusters.

In this experiment, we generated artificial graph sequences where one of the latent clusters divides into two latent clusters. O2I does not depend on the direction of the temporal axis because it converts the graph sequence into graph G' and solves the k partition problem for G'. If Eq. (12) is used instead of Eq. (11), we can generate an artificial graph sequence in which two latent clusters merge.

$$\begin{pmatrix} x \\ y \end{pmatrix} = R\left(\frac{2\pi(j-1)}{k}\right) r \begin{pmatrix} \exp\left[-\beta(t-1)\right] \\ 0 \end{pmatrix} \tag{12}$$

In this case, the same result is obtained for O2I as in Fig. 16. Therefore, we have confirmed that O2I overcomes the second drawback of PCM.

4.2.3 Varying Numbers of Vertices

Figure 18 shows the ARIs for O2I for artificial data with $ratio\%$ of the vertices removed from a graph sequence generated using Eq. (10). In this experiment, we measure the ARIs for forgetting rates γ equal to 0, 0.1, 0.2, 0.3, 0.4, and 0.5, as $ratio$ increases from 0 to 30. The setting for $ratio = 0$ and $\gamma = 0$ is the same as in the experiments for Fig. 13. Figure 18 does not contain any results for PCM because PCM cannot be applied to this data. When γ is increased, then ARI increases except

Fig. 18 ARIs for various values of $ratio$ and γ

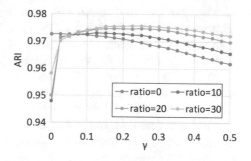

---

[3]The numbers of vertices in the third detected cluster sequence increases with time as $\langle 0, 1, 4, 8, 13, 16, 16, 18, 18, 18 \rangle$.

Fig. 19 AIRs for various
values of α' and κ (1)

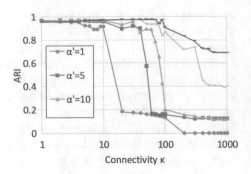

for $ratio = 0$ because vertices with the same ID that are Δ timesteps apart are connected by an edge weight $\alpha' \gamma^{\Delta-1}$ and requirement (2) is satisfied. In particular, increasing γ from 0 to 0.025 is the most effective. When γ is increased further, the ARI decreases because requirement (2) is oversatisfied. This result is consistent with Sect. 4.2.1 In contrast, because α' for $ratio = 0$ is sufficiently tuned in the experiment of Sect. 4.2.1. The ARI decreases when γ is increased. Hence, these results confirm that O2I overcomes the third drawback of PCM.

For $\gamma \geq 0.1$, the reason why ARI for large $ratio$ is more than for small $ratio$ is as follows. Because the graph $G^{(t)}$ for $ratio = 0$ and $\gamma = 0$ has about $\kappa |V^{(t)}|/2$ edges and it connects to $G^{(t+1)}$ with $|V^{(t)}|$ edges, the former connectivity is higher than the latter. Their connectivities are not balanced. In contrast, when $ratio$ or γ is increased, both connectivities are balanced and the ARI increases. In the next subsections, we further investigate effect of other parameters on connectivities of graphs in graph sequences.

4.2.4 Graph Connectivities

Figure 19 shows the result of ARIs when α' and κ are changed. When κ is increased for a certain α', the ARI decreases. This is because the connectivity among vertices coming from the same timestep becomes dense by increasing κ, and cutting edges with weights α' in G' minimizes the objective function Eq. (5) of O2I compared with cutting edges among the vertices coming from the same timestamp. Figure 20 shows the maximum ARIs and their corresponding α' obtained by setting α' to 1, 5, 10, 50, 100, 500, 1000, 5000, and 10,000 for each κ. Because vertices in the latent clusters in G' rarely connect with one another for small κ, the ARI increases when κ is increased. When κ is further increased, the ARI decreases drastically. When κ is greater than 100, the tuned α' is greater than 500. In this case, most of the clusters in the obtained cluster sequences satisfy $C_j^{(t)} = C_j^{(t+1)}$, which means that the cluster sequences do not change over time. Thus, although the effectiveness of O2I decreases when κ becomes too large, we can improve its effectiveness by making the graphs in the graph sequences sparse.

Fig. 20 AIRs for various values of α' and κ (2)

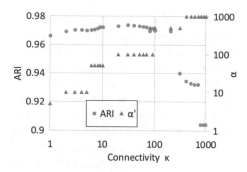

Fig. 21 Distribution of $\mathbf{y}_i$ for $\alpha' = 50$ and $\kappa = 10$

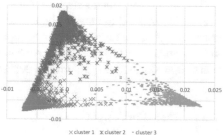

Fig. 22 Distribution of $\mathbf{y}_i$ for $\alpha' = 50$ and $\kappa = 1000$

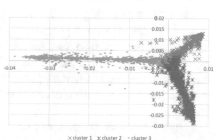

As mentioned in Algorithm 2, O2I contains spectral clustering. In general, when the spectral clustering is applied to a certain graph G_1 consisting of k connected components, all vertices in the jth ($1 \leq j \leq k$) connected component are mapped to the same point $\mathbf{p}_j$ in the k-dimensional space in Line 6 of Algorithm 2. In addition, when the spectral clustering is applied to another connected graph G_2 created from G_1 by adding some edges with small weights, all of the vertices that formed the jth ($1 \leq j \leq k$) connected component in G_1 are mapped to points $\mathbf{y}_i$ around the point $\mathbf{p}_j$ in k-dimensional space [18].[4] Because k-means is applied to points $\mathbf{y}_i$ around this point $\mathbf{p}_j$, the spectral clustering adequately detects clusters in which the vertices are connected with large weights to one another in G_2. Figures 21 and 22 show the distributions of $\mathbf{y}_i$ derived from artificially generated graph sequences for $\kappa = 10$ and $\kappa = 1000$, respectively. Although $\mathbf{y}_i$ are 3-dimensional vectors, we use their second and third elements to plot the distributions because their first

---

[4]Vector $\mathbf{y}_i$ is the same symbol used in Algorithm 2.

elements are the same. Each point is colored according to the latent clusters to which the point belongs. In Fig. 21, because points in each cluster are distributed around either $(0.00, 0.015)$, $(-0.005, -0.005)$, or $(-0.02, 0.005)$, k-means detects the latent clusters accurately. In this figure, the reason why the distributions for clusters 1 and 2 overlap is because m points at each timestep move from cluster 1 to cluster 2. In contrast, in Fig. 22, the distributions of the three clusters overlap around the origin. This is because vertices coming from the same timestamps connect to each other, vertices coming from the different timestamps connect with large weights α' and G' becomes a large connected component. In this case, k-means cannot partition the points around the origin accurately, and the ARI of O2I decreases.

The above experiments confirm that κ is an important hyperparameter of O2I that enables the method to cluster graph sequences accurately. When a graph sequence consisting of dense graphs is given as input, we must select edges with large weights in the graphs to make the graphs sparse before line 1 of Algorithm 2. Making graphs sparse is easier than making graphs dense.

4.2.5   Real-World Data

To assess the practicality of O2I, we applied it to the Enron e-mail dataset [13]. We divided the dataset into T periods according to timestamps of e-mails, assigned a unique ID to the e-mail address for each person participating in the communication, and assigned an edge to a pair of individuals if they communicate via e-mail within each period, assigned weight $\log(c+1)$ to the edge between two vertices if c e-mails are sent between the corresponding individuals, and obtained graphs $G^{(t)}$ for each period t.

Figure 23 shows the number of vertices in the clusters detected by O2I at each timestep. This result was obtained from a graph sequence with eight steps in which the vertices correspond to about 150 senior managers. The hyperparameters for O2I were set as $k = 24$, $\alpha' = 12$, and $\gamma = 0.1$, and O2I took about 5 s to obtain this result. For the sake of visibility, we omitted any

Fig. 23 Number of vertices in each cluster at time t

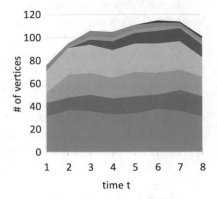

detected outliers having only a few vertices from the figure. The reason why O2I detects the outliers is that there are many senior managers who each contacted a particular senior manager in the dataset. This figure shows that O2I is applicable to a graph sequence where the number of vertices varies over time, with $(|V^{(1)}|, |V^{(2)}|, \ldots, |V^{(8)}|) = (79, 99, 113, 113, 114, 123, 124, 109)$. In addition, the number of clusters detected by O2I also varies over time, with $(|P^{(1)}|, |P^{(2)}|, \ldots, |P^{(8)}|) = (8, 9, 12, 14, 9, 15, 17, 16)$. As shown in Fig. 23, the cluster represented by dark blue appears at time 3 and then gradually grows larger, although it does not exist at times 1 and 2.

We also applied O2I to the same graph sequence for the approximately 150 managers, with $\gamma = 0$ and the other hyperparameters set as before. In the detected cluster sequences, more than 90% of the vertices belong to a particular cluster sequence at each time and the rest of the vertices belong to $k - 1$ outliers. Therefore, the forgetting rate γ in O2I is beneficial for obtaining suitable cluster sequences.

5 Conclusion

In this chapter, we explained O2I for clustering in evolving graphs that can detect changes in clusters over time. In O2I, the graph sequence is partitioned into smooth clusters, even when the numbers of clusters and vertices vary. The method first constructs a graph from the graph sequence, then uses spectral clustering and the RatioCut to apply k partitioning to this graph. The method approach was compared in detail with the preserving clustering membership (PCM) algorithm, which is a conventional online graph-sequence clustering algorithm in which the numbers of clusters and vertices must remain constant. We further showed that, in contrast to PCM, the performance of O2I is not dependent on the clustering of the initial graph in the graph sequence. Experiments on synthetic evolving graphs showed that O2I is practical to calculate and addresses the main disadvantages of PCM. Further tests on real-world data showed that O2I can obtain reasonable clusters. It is hence a flexible clustering solution and will be useful on a wide range of graph-mining applications in which the connections, number of clusters, and number of vertices of the graphs evolve over time.

References

1. Aggarwal, C.C., Han, J., Wang, J., Philip S.Y.: A framework for clustering evolving data streams. In: Proceedings of International Conference on Very Large Data Bases (VLDB), pp. 81–92 (2003)
2. Aggarwal, C.C., Han, J., Wang, J., Philip S.Y.: A framework for projected clustering of high dimensional data streams. In: Proc. of International Conference on Very Large Data Bases (VLDB), pp. 852–863 (2004)

3. Aggarwal, C.C., Han, J., Wang, J., Philip S.Y.: On demand classification of data streams. In: Proceedings of International Conference on Knowledge Discovery and Data Mining (KDD), pp. 503–508 (2004)
4. Bar-Joseph, Z., Gerber, G.K., Gifford, D.K., Jaakkola, T.S., Simon, I.: A new approach to analyzing gene expression time series data. In: Proceedings of International Conference on Computational Biology (RECOMB), pp. 39–48 (2002)
5. Beringer, J., Hüllermeier, E.: Online clustering of parallel data streams. Data Knowl. Eng. **58**(2), 180–204 (2006)
6. Berlingerio, M., Bonchi, F., Bringmann, B., Gionis, A.: Mining graph evolution rules. In: Proceedings of European Conference on Machine Learning and Knowledge Discovery in Databases (ECML/PKDD), pp. 115–130 (2009)
7. Cao, F., Ester, M., Qian, W., Zhou, A.: Density-based clustering over an evolving data stream with noise. In: Proceedings of SIAM International Conference on Data Mining (SDM), pp. 328–339 (2006)
8. Chakrabarti, D., Kumar, R., Tomkins, A.: Evolutionary clustering. In: Proceedings of International Conference on Knowledge Discovery and Data Mining (KDD), pp. 554–560 (2006)
9. Charikar, M., O'Callaghan, L., Panigrahy, R.: Better streaming algorithms for clustering problems. In: Proceedings of Annual ACM Symposium on Theory of Computing (STOC), pp. 30–39 (2003)
10. Chi, Y., Song, X., Zhou, D., Hino, K., Tseng, B.L.: On evolutionary spectral clustering. ACM Trans. Knowl. Discov. Data **3**(4), 17:1–17:30 (2009)
11. Domingos, P.M., Hulten, G.: A general method for scaling up machine learning algorithms and its application to clustering. In: Proceedings of International Conference on Machine Learning (ICML), pp. 106–113 (2001)
12. Inokuchi, A., Washio, I.: Mining frequent graph sequence patterns induced by vertices. In: Proceedings of SIAM International Conference on Data Mining (SDM), pp. 466–477 (2010)
13. Klimmt, B., Yang, Y.: Introducing the Enron corpus. In: CEAS Conference (2004)
14. Möller-Levet, C.S., Klawonn, F., Cho, K.-H., Yin, H., Wolkenhauer, O.: Clustering of unevenly sampled gene expression time-series data. Fuzzy Sets Syst. **152**(1), 49–66 (2005)
15. O'Callaghan, L., Meyerson, A., Motwani, R., Mishra, N., Guha, S.: Streaming-data algorithms for high-quality clustering. In: Proceedings of International Conference on Data Engineering (ICDE), pp. 685–694 (2002)
16. Okui, S., Osamura, K., Inokuchi, A.: Detecting smooth cluster changes in evolving graphs. In: Proceedings of International Conference on Machine Learning and Applications (ICMLA), pp. 369–374 (2016)
17. van Wijk, J.J., van Selow, E.R.: Cluster and calendar based visualization of time series data. In: Proceedings of IEEE Symposium on Information Visualization (INFOVIS), pp. 4–9 (1999)
18. von Luxburg, U.: A tutorial on spectral clustering. Stat. Comput. **17**(4), 395–416 (2007)
19. Wang, Y., Liu, S.-X., Feng, J., Zhou, L.: Mining naturally smooth evolution of clusters from dynamic data. In: Proceedings of SIAM International Conference on Data Mining (SDM), pp. 125–134 (2007)

Efficient Estimation of Dynamic Density Functions with Applications in Data Streams

Abdulhakim Qahtan, Suojin Wang, and Xiangliang Zhang

Abstract Recently, many applications such as network monitoring, traffic management and environmental studies generate huge amount of data that cannot fit in the computer memory. Data of such applications arrive continuously in the form of streams. The main challenges for mining data streams are the high speed and the large volume of the arriving data. A typical solution to tackle the problems of mining data streams is to learn a model that fits in the computer memory. However, the underlying distributions of the streaming data change over time in unpredicted scenarios. In this sense, the learned models should be updated continuously and rely more on the most recent data in the streams.

In this chapter, we present an online density estimator that builds a model called KDE-Track for characterizing the dynamic density of the data streams. KDE-Track summarizes the distribution of a data stream by estimating the Probability Density Function (PDF) of the stream at a set of resampling points. KDE-Track is shown to be more accurate (as reflected by smaller error values) and more computationally efficient (as reflected by shorter running time) when compared with existing density estimation techniques. We demonstrate the usefulness of KDE-Track in visualizing the dynamic density of data streams and change detection.

A. Qahtan
Qatar Computing Research Institute (QCRI), HBKU, Doha, Qatar
e-mail: aqahtan@hbku.edu.qa

S. Wang
Department of Statistics, TAMU, College Station, TX, USA
e-mail: sjwang@stat.tamu.edu

X. Zhang (✉)
CEMSE, King Abdullah University of Science and Technology (KAUST), Thuwal, Kingdom of Saudi Arabia
e-mail: xiangliang.zhang@kaust.edu.sa

© Springer International Publishing AG, part of Springer Nature 2019
M. Sayed-Mouchaweh (ed.), *Learning from Data Streams in Evolving Environments*, Studies in Big Data 41, https://doi.org/10.1007/978-3-319-89803-2_11

1 Introduction

Recent advances in computing technology allow for collecting vast amount of data
that arrive continuously in data streams. Examples of data streams can be found
in fields such as sensor networks, mobile data collection platform, and network
traffic. The data need to be processed and analyzed once they arrive. However, the
unbounded, rapid, and continuous arrival of data streams disallows the usage of
traditional data mining techniques. Therefore, the development of online algorithms
for processing data streams becomes highly important.

Density estimation has been widely used in various applications. Estimating
the Probability Density Function (PDF) for a given data set provides knowledge
about the underlying distribution of the data. Consequently, dense regions can be
recognized as clusters and quantities such as medians and centers of clusters can be
computed [1]. By contrast, sparse regions are reported as outliers that can be used
for fault detection, for example, in sensor networks [2]. Moreover, the estimated
dynamic density can be visualized to help on placing taxicabs in places with high
pickup rate at certain period of time [3], reducing ambulance emergencies response
time [4] and reflecting people's interest at a particular location for specific seasons
[5].

Estimating the dynamic density that comes with evolving streams needs to
address the following challenges. First, the data distribution changes dynamically in
an unpredictable fashion. Second, an anytime-available model should be efficiently
updated to allow real-time monitoring of the density. Third, the spatial nonuniformity of data distribution requires higher/lower resolutions in dense/sparse areas,
respectively, so that the estimation is accurate to catch the details.

Most of the existing approaches for estimating the density of data streams are
based on the Kernel Density Estimation (KDE) method due to its advantages for
estimating the true density [6]. Given a set of samples, $S = \{x_1, x_2, \ldots, x_n\}$ with
$x_j \in R^d$, KDE estimates the density at a point x as:

$$\hat{f}(x) = \frac{1}{n} \sum_{j=1}^{n} K_h\left(x, x_j\right), \tag{1}$$

where $K_h\left(x, x_j\right)$ is a kernel function, which is usually a radially symmetric
unimodal function that integrates to 1. Equation (1) shows that KDE uses all the
data samples to estimate the PDF at any given point. For online density estimation
of data stream, that is, estimating the density of every arriving data sample, KDE
has quadratic time complexity with respect to (w.r.t.) the stream size. Also, the space
requirement for KDE significantly increases with the dataset size.

In Sect. 3 of this chapter, we introduce our model that is called KDE-Track, to
model the data distribution as a set of resampling points with their estimated PDF. To
guarantee the estimation accuracy and to lighten the load on the model, an adaptive
resampling strategy is employed to control the number of resampling points, that is,

more points are resampled in the areas where the PDF has a larger curvature, while a low number of points are resampled in the areas where the function is approximately linear. In order to overcome the quadratic time complexity of KDE when evaluating the PDF for each new observation, linear interpolation is used with KDE for online density estimation. It therefore has advantages of evaluating the PDF for any new observation in linear time complexity and space complexity w.r.t. the number of resampling points. Evaluating the PDF for all received observations will then take linear time w.r.t. the stream size compared with the quadratic time complexity of KDE. To timely track the evolving density, we use a sliding window strategy in KDE-Track to estimate the density using the most recent data samples.

The KDE-Track has unique properties as follows:

1. It generates density functions that are available to visualize the dynamic density of data streams at any time. After receiving one streaming data sample x_t, KDE-Track updates the PDF of the data stream and also estimates $f(x_t)$
2. It has linear time and space complexities w.r.t. the model size for maintaining the dynamic PDF of data stream upon the arrival of every new sample. It is thus 8–85 times faster than traditional KDE depending on the window size.
3. The estimation accuracy is achieved by adaptive resampling and optimized bandwidth (h), which also address the spatial nonuniformity issue of data streams.

In Sect. 4 of this chapter, we evaluate the most popular density estimators, as well as KDE-Track in different scenarios of data streams. Advantages and disadvantages of these evaluated methods are discussed based on the comparison of their performance. The obtained dynamic density can be applied to diverse problems. Section 5 of this chapter firstly presents a density *visualization* example, which displays the real-time dynamic traffic distribution in the New York Taxi streams where interesting patterns of community behavior are discovered. The second application is for *unsupervised change detection*, where changes are usually detected by comparing the distribution in a current (test) window sliding over the data stream with a fixed reference window that contains data arrived after the last detected change. The third application is *outlier detection* in data streams, which is a straightforward application of estimated density. Data samples, which have small PDF values compared to other points, are more likely to be outliers. Compared with other outlier detection methods, the density-based approach is shown to report less false positives. This application is omitted in this chapter due to space limitation. Interested readers are referred to [7].

2   Related Work

2.1   Dynamic Density

The design of a dynamic density estimator should not only take into consideration the constraints of using limited memory and processing the data in real time [8, 9] due to the nature of streams, but also the dynamic changes of the underlying density function over time. To reduce the computational cost and space requirement of KDE, methods have been proposed based on *kernel merging, sampling*, or *space partitioning*. *Kernel merging* is used in [1, 10] and [11] where a specific number of kernels are maintained through merging two or more kernels. Each kernel summarizes a cluster of similar samples. A new arriving sample can either fall into one existing kernel or trigger a new kernel. Two kernels are merged if the number of kernels exceeds the specified number. Methods that utilize this concept include M-Kernel [1], Cluster Kernels (CK) [10], and AKDE [11]. Another kernel method emerges based on clustering by Self-Organizing Maps (SOM) [12]. Only trained SOM neurons are utilized in density estimation, rather than the whole set of kernels. In order to train the neurons and to minimize the time complexity, a data stream is considered as a sequence of disjoint windows where data in each window are assumed to have the same distribution.

Sampling was used in [13] to reduce the number of kernels while guaranteeing an ϵ-approximation of the density function. The authors studied random sampling and proposed group-selection and sort-selection, which achieve the same accuracy as random sampling but with a smaller number of samples.

Space partitioning is also used to reduce the computational cost of KDE. A *kd-tree* structure is used in [14] and [15] where the leaves contain a small number of kernels and each internal node contains a statistical summary about the subspace represented by that node. Estimating the density at any given point involves depth-first traversal of the tree where only close-by nodes will be visited. Grid-based methods were presented in [16, 17] for static datasets. They concentrate on the best setting of the bin width using a fixed number of resampling points. This approach will not work for data streams as the data are not available in advance, and the range of the data is changing over time.

KDE-Track [18] differs from the above-mentioned methods by updating the estimated density with the contribution of each new arriving sample. Hence, it provides any time available density values, which can be used for visualizing the density. The model size is controlled by adaptive resampling, rather than reducing the number of used kernels in [1, 10–14]. The estimation error is thus minimized. In addition, it is deployed with an accurate bandwidth selection method, which improves the density estimation significantly (Table 1).

Table 1 Summary of the key characteristics of the density estimators

Method	MV	Data streams	Bandwidth selection	Bandwidth of each dim in MV	Data points reduction	MV kernel function	Online update
CK [10]	No	Yes	Normal rule	N/A	Merge kernels	N/A	Yes
M-Kernels [1]	No	Yes	Normal rule	N/A	Merge kernels	N/A	Yes
SOMKE [12]	Yes	Yes	Normal rule	Different	Trained neurons	MV	BU
FFT-KDE [19]	Yes	Yes	Normal rule	Different	None	MV	No
KDE [20]	Yes	Yes	Normal rule	Different	None	Product	No
kd-tree [15]	Yes	No	Cross validation	Fixed	None	RI	No
RS, GS, SS [13]	Yes	No	User input	Fixed	Sampling	RI	No
MPLKernels [14]	Yes	Yes	Normal rule	Different	Sampling	Product	Yes
KDE-Track [18]	Yes	Yes	Plug in	Different	None	Product	Yes

MV multivariate data, *RS* random sampling, *GS* group selection, *SS* sort selection, *RI* rotation invariant, *BU* batch update

2.2 Change Detection

Change detection is relevant to a wide range of applications such as intrusion detection in computer networking [21], suspicious motion detection in vision systems [22], studying the effects of nuclear radiation on the environment [23], and online clustering and classification. For example, change detection can be used in online classification for reporting when the classifier should be retrained (only if a change in the data stream is observed). The problem of change detection has been widely studied and referred to as data evolution [24], event detection [25], or change-point detection [23, 26].

Unsupervised window-based change detection is based on comparing the distribution in a current stream window with a reference distribution [23, 27–30], where *density estimation* techniques and divergence metrics are essential to model and compare the distributions. Dasu et al. [28] used the *kdq-tree* data structure to model data distribution and presented a 3-step process for change detection: (1) updating the test distribution over the current window, (2) computing the change-score between the test and reference distributions, and (3) emitting an alarm signal if the change-score reaches the threshold specified using bootstrap sampling and the permutation test used in [23]. Kawahara and Sugiyama [26] used the density-ratio estimation that is based on the Kullback-Leibler Importance Estimation Procedure (KLIEP) to model data distribution. The method's complexity is quadratic w.r.t. the window size. An online version of KLIEP was studied in [26] and [31].

Table 2 Summary of change detection methods

Technique	MV	Data streams	Compare distribution/prediction	Threshold settings
kdq-tree [28]	Yes	Yes	Compare distribution	Bootstrap
PCA-SPLL [29]	Yes	No	Compare distribution	Fixed
KLIEP [26]	Yes	Yes	Compare distribution	Fixed
Kifer [23]	No	Yes	Compare distribution	Bootstrap
ADWIN [27]	No	Yes	Other	Dynamic
Density test [30]	Yes	No	Compare distribution	Bootstrap
CF [32]	No	Yes	Prediction	Fixed
CD-Area [33]	Yes	Yes	Compare distribution	Dynamic

MV multivariate

A statistical test, called the *density test* [30], determines whether the newly observed data S' are sampled from the same underlying distribution as the reference dataset S. The change detection summarized above has the limitation of user-based settings of key parameters, which requires users knowledge or they have high computational cost that makes them unusable for online change detection in data streams.

Table 2 summarizes the key characteristics of the existing methods for change detection. In Sect. 5, a PCA-based change detection framework will be presented [33]. The framework compares densities estimated by KDE-Track on selected principal components and dynamically adjusts the threshold for reporting changes, such that the false alarms are reduced and detection rate is improved.

3 KDE-Track: Dynamic Density Estimation

3.1 Theoretical Bases of Density Estimation

We first discuss the traditional KDE and its related issues, for example, selection of kernel functions and smoothing parameter (bandwidth), and complexity.

KDE estimates the density $\hat{f}(x)$ by Eq. (1). For the case of univariate data, Eq. (1) is written as

$$\hat{f}(x) = \left(\frac{1}{nh}\right) \sum_{j=1}^{n} K\left(\frac{x - x_j}{h}\right). \tag{2}$$

For the two-dimensional samples, where $x_j = (x_{1j}, x_{2j})^T \in R^2$, kernel functions $K_h(x, x_j)$ are defined as $\frac{1}{h_1 h_2} K\left(\frac{x_1 - x_{1j}}{h_1}, \frac{x_2 - x_{2j}}{h_2}\right)$, where h_i is the smoothing parameter, called the bandwidth, on dimension i [6].

A popular kernel function in case of multivariate data is called the multiplicative (product) kernel [6], which uses the product of univariate kernel functions on each dimension, and in the two-dimensional case computes $\hat{f}(x)$ as

$$\hat{f}(x) = \frac{1}{n} \sum_{j=1}^{n} \prod_{i=1}^{2} \left\{ \frac{1}{h_i} K \left(\frac{x_i - x_{ji}}{h_i} \right) \right\}. \tag{3}$$

Another option is to use the orientation-invariant kernel function [13] and [15], which is

$$\hat{f}(x) = \left(\frac{1}{nh^2} \right) \sum_{j=1}^{n} K \left(\frac{\|x - x_j\|}{h} \right). \tag{4}$$

This kernel function assumes that the data variation along all the dimensions is the same, which may fail to capture densities of arbitrary shapes.

KDE-Related Issues The choice of a kernel function is relatively unimportant provided that a kernel function is continuous with finite support [20]. It is recommended that the selected kernel is smooth, clearly unimodal and symmetric about the origin [6]. In the study of this chapter, we choose the multiplicative Epanechnikov kernel where the same univariate kernel function $K(x) = \frac{3}{4}(1 - x^2)I_{[-1,1]}(x)$ is used in each dimension with a different bandwidth value. We use the Epanechnikov kernel because of its asymptotically-optimal efficiency among all other kernel functions [34].

The estimation accuracy of KDE is mainly affected by the bandwidth value [6, 20]. A large bandwidth value over-smooths the density function curve and hides a lot of useful information, while a small bandwidth value makes the density function's curve too fluctuated. A general rule for selecting the bandwidth is to decrease the bandwidth value ($h \to 0$) as the number of samples used in the estimation increases ($n \to \infty$). However, the rate at which h is approaching 0 is much slower such that ($nh \to \infty$).

The Bandwidth h Equations (1) and (3) use different bandwidth values to capture the spread of the data on each dimension. This suggests using the same analysis of estimating the bandwidth for the case of univariate data on the marginal distribution of the data on each dimension. Typically, bandwidth setting should minimize the deviation between the true and the estimated densities. This deviation is measured by the Mean Integrated Square Error (MISE) [35].

Let $\mu_k(x)$, $R(f)$ be defined as $\mu_k(x) = \int x^k K(x)dx$ and $R(f) = \int f^2(x)dx$. The MISE of the estimator using a bandwidth value h is

$$\text{MISE}(h) = \int E \left[\hat{f}(x) - f(x) \right]^2 dx,$$

which has the asymptotic expansion $\text{MISE}(h) = \text{AMISE}(h) + O(n^{-1} + h^5)$ under suitable regularity conditions on K and f. The minimizer of the $\text{AMISE}(h) = \frac{1}{nh}R(K) + h^4 \left(\frac{\mu_2(K)}{4}\right)^2 R\left(f''\right)$ is considered a good approximation for the optimal bandwidth value, which can be estimated as

$$\hat{h} = \left(\frac{R(K)}{\mu_2^2(K)R(f'')n}\right)^{\frac{1}{5}}. \tag{5}$$

However, this minimizer cannot be computed as it depends on the unknown density f.

Many methods have been introduced to estimate $R(f'')$ in Eq. (5). The normal rule [20] is the most popular method for estimating the bandwidth, which assumes the unknown density f as a normal density and scales it according to the sample standard deviation. The bandwidth value selected using the normal rule is computed as

$$\hat{h} = c\hat{\sigma}n^{-1/5}, \tag{6}$$

where c is a constant that depends on the used kernel function K, $\hat{\sigma}$ is the sample standard deviation, and n is the number of kernels. This method is computationally efficient but it does not work well when the density deviates significantly from normality. Other methods based on cross-validation have been proposed in the literature [36, 37]. These methods require performing density estimation for each candidate of the bandwidth values, which multiplies the computational cost by another factor equal to the number of candidates.

Plug-in methods [38, 39] estimate an approximation of $R(f'')$ and plug it in Eq. (5) to compute the optimal bandwidth. Estimating $R(f'')$ requires also making assumptions about the density function but it becomes more accurate than using the normal rule. Sheather and Jones [37] estimate $R(f'')$ by estimating $f^{(4)}$, which in turn is estimated using $R(f^{(6)})$. The value $R(f^{(6)})$ is computed by assuming that $f^{(8)}$ is the eighth derivative of a normal density. After estimating $R(f^{(6)})$, a backward substitution is performed to estimate $R(f'')$. Shimazaki and Shinomoto [40] assume that the true density follows a Poisson distribution and use Estimation-Maximization (EM) method to find the optimal bandwidth value. The method requires estimating the density for each estimation step of the EM optimization procedure, which will be very expensive in the case of streaming data where the density is changing dynamically and the bandwidth value needs to be estimated frequently.

In the case of multidimensional data, most of the bandwidth selection methods either consider a fixed bandwidth value for all the dimensions [13, 41] or use the marginal distribution to estimate the bandwidth on each dimension [14, 20]. KDE fails to capture densities of arbitrary shapes when using the same bandwidth value for all dimensions. Methods with a different bandwidth value for each dimension rely on the marginal distribution of the data on that dimension.

The KDE-Track introduced next minimizes the effect of the normality assumption of f by using the data samples to estimate f''. The numerical integration technique is then used to compute $R(f'')$, which is plugged in Eq. (5) to estimate the bandwidths.

3.2 KDE-Track Method

KDE-Track models the distribution of the streaming data as a grid of resampling points and their corresponding estimated density values. For example in the 2-dim sample space, let $\mathcal{U}^1 = \{u_0^1, u_1^1, \ldots, u_{U^1-1}^1\}$ and $\mathcal{U}^2 = \{u_0^2, u_1^2, \ldots, u_{U^2-1}^2\}$ be the set of points that discretize the range of the data on the first and the second dimensions, respectively. The KDE-Track model $\mathcal{M}$ is defined as the set of the grid points from $\mathcal{U}^1 \times \mathcal{U}^2$ with their estimated densities. That is, $\mathcal{M} = \{M_0, M_1, \ldots, M_{q-1}\}$, where $q = U^1 U^2$ is the number of the resampling points and M_s is an ordered pair representing a grid point and its estimated PDF ($M_s = (\boldsymbol{m}_s, \hat{f}(\boldsymbol{m}_s))$). Here $\boldsymbol{m}_s = (u_k^1, u_l^2) \in \mathcal{U}^1 \times \mathcal{U}^2$ is the s-th resampling point with l, k being the quotient and the remainder of the division of s by U^1 and $\hat{f}(\boldsymbol{m}_s)$ is the density estimated using KDE at $\boldsymbol{m}_s$.

Density Estimation by Interpolation For a data sample $\boldsymbol{a}$ in 2-dim, the PDF at $\boldsymbol{a}$ can be efficiently estimated by bilinear interpolation of the resampling points, following two steps:

1. Fetch the estimated PDF values at resampling points $\boldsymbol{m}_{s1}, \boldsymbol{m}_{s1+1}, \boldsymbol{m}_{s2}$ and $\boldsymbol{m}_{s2+1}$ that surround the point $\boldsymbol{a}$ (as in Fig. 1). Let $\boldsymbol{y}^{(i)}$ be the projection of vector $\boldsymbol{y}$ on i-axis, then $m_{s1}^{(1)} = m_{s2}^{(1)} \leq a^{(1)} < m_{s1+1}^{(1)} = m_{s2+1}^{(1)}$ and $m_{s1}^{(2)} = m_{s1+1}^{(2)} \leq a^{(2)} < m_{s2}^{(2)} = m_{s2+1}^{(2)}$. Then linearly interpolate the density at $\boldsymbol{m}_{s1}, \boldsymbol{m}_{s1+1}$ to estimate the density at $\boldsymbol{r}_{s1}$ and interpolate the density at $\boldsymbol{m}_{s2}, \boldsymbol{m}_{s2+1}$ to compute the density at $\boldsymbol{r}_{s2}$.
2. Estimate the density at $\boldsymbol{a}$ by interpolating the density at $\boldsymbol{r}_{s1}, \boldsymbol{r}_{s2}$.

Fig. 1 Computing the density at $\boldsymbol{a}$ by interpolation given the KDE estimation $\hat{f}$ at $\boldsymbol{m}_{s1}, \boldsymbol{m}_{s1+1}, \boldsymbol{m}_{s2}$ and $\boldsymbol{m}_{s2+1}$

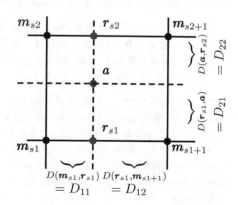

Let $D(b, c)$ be the Euclidean distance between b and c. The density at a will be computed as follows:

$$\tilde{f}(a) = \frac{D(a, r_{s2}) \tilde{f}(r_{s1}) + D(r_{s1}, a) \tilde{f}(r_{s2})}{D(r_{s1}, r_{s2})}, \tag{7}$$

where

$$\tilde{f}(r_{s1}) = \frac{D(r_{s1}, m_{s1+1}) \hat{f}(m_{s1}) + D(m_{s1}, r_{s1}) \hat{f}(m_{s1+1})}{D(m_{s1}, m_{s1+1})},$$

and

$$\tilde{f}(r_{s2}) = \frac{D(r_{s2}, m_{s2+1}) \hat{f}(m_{s2}) + D(m_{s2}, r_{s2}) \hat{f}(m_{s2+1})}{D(m_{s2}, m_{s2+1})}.$$

KDE interpolation is efficient as it stores only the function at the resampling points whose total number is in the constant order and is small compared to the stream size. The running time for estimating the PDF for all n arriving data samples will be in $O(n|\mathcal{M}|)$.

Error Analysis Three types of errors may be introduced by KDE-Track: the estimation error inherited from KDE, the interpolation error, and the rounding error. Since rounding error (occurring when an infinite number of digits after the decimal point are squeezed in a finite number of bits) is machine dependent, we focus on the interpolation error and propose an adaptive resampling model to minimize this type of error. The error inherited from KDE will be minimized by selecting the optimal bandwidth values for the KDE.

In [7], we studied the interpolation error for the case of univariate data. Let $\hat{f}(a)$ and $\tilde{f}(a)$ be the estimated PDFs using the traditional KDE and the KDE-Track, respectively, and D_m be the maximum distance between two consecutive resampling points. The error is $\tilde{f}(a) - \hat{f}(a) = \frac{\{D_m\}^2}{8} \hat{f}''(a) + O_p(\{D_m\}^3)$. The interpolation error will increase in the case of multidimensional data. In [18], we derived the interpolation error in two-dimension as $\tilde{f}(a) - \hat{f}(a) = \frac{D_m^2}{8} \left\{ \hat{f}_{x_1x_1}(a) + \hat{f}_{x_2x_2}(a) \right\} + O_p(\{D_m\}^3)$, where D_m is the distance between two consecutive resampling points in two-dimension. Note that the term $O_p(D_m^3)$ includes also the terms with higher-order derivatives. When using the Epanechnikov kernel function, the second partial derivative will be constant and partial derivatives of higher order will be zeros.

Adaptive Resampling Model From the above error analysis, we know that the accuracy of the linear interpolation depends on (1) the distance between two adjacent resampling points, and (2) the second derivative of the density function. To minimize the error while keeping the number of resampling points within a reasonable margin, more resampling points can be added in the regions where the density function has high curvature, as shown in Fig. 2. By contrast, in the regions with approximately linear function, less resampling points are used.

Fig. 2 Example of adaptive
resampling: more resampling
points are used in regions
with high curvature of the
function

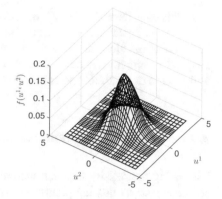

In multidimensional data streams, the distribution is usually spatial, nonuniform, and dynamic. Therefore, high resolution with sufficient resampling points is required (1) in dense areas with high PDF values to catch the details; and (2) in sensitive areas which are the boundary between dense and sparse areas to catch dynamic changes. Adaptive resampling meets the requirement perfectly.

Bandwidth Selection KDE-Track uses different bandwidth values for each dimension. The marginal distribution of the data is used to compute the bandwidth value on that dimension. From Eq. (5), the bandwidth value on dimension i can be computed as

$$\hat{h}_i = \left(\frac{R(K)}{\mu_2^2(K) R(f_i'') n} \right)^{\frac{1}{5}}, \tag{8}$$

where f_i'' is the second derivative of the density function on the i-th dimension. Initially, a pilot bandwidth is estimated using the normal rule $\tilde{h}_i = c\hat{\sigma}_i n^{-1/5}$, where c is a constant that depends on the kernel function K and $\hat{\sigma}_i$ is the standard deviation of the projection of the data on axis i. This pilot bandwidth is used to estimate the second derivative of the marginal distribution as

$$\hat{f}_i''(x) = \frac{1}{n\tilde{h}_i} \sum_{j=1}^{n} K'' \left(\frac{x - x_j}{\tilde{h}_i} \right). \tag{9}$$

In this case, $\hat{f}_i''$ will be a better approximation of f_i'' than considering f_i to be a normal density. Using KDE-Track on the one-dimensional data will speed up the computation of $R(\hat{f}_i'')$ and $\hat{h}_i$. Since the distribution of the data will change over time with the arrival of new samples from the stream, the bandwidth values $\hat{h}_i$ should be updated accordingly to represent the variation of the data along the different dimensions. Using KDE-Track will also allow for updating the values of $\hat{h}_i$ online and efficiently.

In this way, the estimated $\hat{f}_i''$ will serve two roles. First, it is used to approximate $R(f_i'')$ to compute the bandwidth value. Second, it is used as a more accurate indicator of the high curvature of the density function's curve, which facilitates the adaptive resampling in KDE-Track for obtaining more accurate estimation as we will discuss in the following subsections.

3.3 KDE-Track Implementation

Estimating the density for each incoming data sample using KDE-Track requires access to four resampling points only as discussed in Sect. 3.2. The key step is thus the maintenance of the resampling model (resampling points and their PDF values). Algorithm 1 shows the maintenance of KDE-Track's model and using it for online density estimation. Lines starting with the # sign represent comments.

Initializing the Resampling Model The resampling model is initialized by the beginning part of streaming data, for example, the first 5000 points.[1] The resampling points, m_s, $s = 1, \ldots, q$, are defined as the cartesian product of the set of equidistant points selected on each dimension within the range of initial points received so far.[2] The second derivative of the marginal density on each dimension is then estimated at the initial resampling points and used for selecting the bandwidth value. Moreover, the second derivative is used to add more resampling points in the regions with high curvature of the density. Using the estimated bandwidth value, the density values $\hat{f}(m_s)$ of these resampling points are computed using the traditional KDE on the initial batch of points.

Estimation Based on the Resampling Model Once all M_i have been initialized, the density at each arriving point x can be estimated. Due to the usage of interpolation method, the density at x can be estimated by (1) calculating, on each dimension, the index of one resampling point who and whose successive neighbor will contribute, (2) fetching the four resampling points in the cell surrounding x and their densities, and (3) computing $\tilde{f}(x)$ using Eq. (7).

Updating the Resampling Model The resampling model is the basis of the density estimator. The resampling points and their PDF values should be updated after receiving a new data point. As discussed in Sect. 3.2, the resampling points are adaptively maintained according to the curvature of density function. An interval $[u_j^i, u_{j+1}^i]$ is divided into two equal subintervals when

---

[1]This first batch of data is used for initializing the resampling points and setting bandwidth values. KDE-Track is not sensitive to how many points are used in this batch, as the resampling model and bandwidth are updated online with new arriving data after initialization.

[2]The initial model is defined such that the distance between any two consecutive points on the x_i axis is $\tilde{h}_{i_1}$, where $\tilde{h}_{i_1}$ is the pilot bandwidth estimated using the first batch of data points.

Algorithm 1 KDE-TRACK MODEL

Parameters: w (window size)
Online flow in: streaming data $S = \{x_1, \ldots, x_t, \ldots\}$
Online output: $\tilde{f}(x_t)$ (the PDF value at x_t)
Procedure:

1: **# Initialization:**
2: $W = \{x_1, \ldots, x_w\}$, $updatestep = 0.05w$ and $\mathcal{M} = \phi$
3: Compute $\hat{h}_{i_1} = \left(\dfrac{R(K)}{\mu_2^2(K)R(\hat{f}_i'')w} \right)^{\frac{1}{5}}$ (Sect. 3.2)
4: **for** $k = 0$ to $U^1 - 1$ **do**   (Sect. 3.2)
5:   **for** $l = 0$ to $U^2 - 1$ **do**
6:    Put $s = lU^1 + k$ and $m_s = (u_k^1, u_l^2)$
7:    Compute $\hat{f}(m_s)$ using Eq. (3)
8:    Put $M_s = (m_s, \hat{f}(m_s))$
9:    $\mathcal{M} = \mathcal{M} \cup \{M_s\}$
10:   **end for**
11: **end for**
12: **while** a new sample x_t arrives in the stream **do**
13:   **if** $(1 \leq t \leq w)$ **then**
14:    Compute $\tilde{f}(x_t)$ using Eq. (7)
15:   **else**
16:    **# Update the resampling model:**   (Sect. 3.3)
17:    Remove x_{t-w} from W and add x_t to W
18:    **for** each dimension i **do**
19:     Update $\hat{\sigma}_{i_t}$ using x_t, x_{t-w}
20:     Compute $\tilde{h}_{i_t} = c\hat{\sigma}_{i_t} w^{-1/5}$
21:     Update $\hat{f}_i''(u_k^i)$
22:    **end for**
23.   **for** each dimension i **do**
24:     Compute $R(\hat{f}_i'')$, $\hat{h}_{i_t} = \left(\dfrac{R(K)}{\mu_2^2(K)R(\hat{f}_i'')w} \right)^{\frac{1}{5}}$
25:     $\forall s \, (0 \leq s \leq U^1U^2 - 1)$, update $\hat{f}(m_s)$ by Eq. (11)
     # Update the adaptive resampling:
26:     **if** $mod(t, updatestep) = 0$ **then**
27:      Compute $\bar{f}_i'' = \frac{1}{U^i} \sum_{j=0}^{U^i-1} |\hat{f}_i''(u_j^i)|$
28:      **if** $max(|\hat{f}_i''(u_k^i)|, |\hat{f}_i''(u_{k-1}^i)|) > \bar{f}_i''$ **then**
29:       $u_{temp}^i = (u_k^i - u_{k-1}^i)/2$
30:       Compute $\hat{f}_i''(u_{temp}^i)$
31:       Insert u_{temp}^i in $\mathcal{U}^i$
32:      **end if**
33:      $\Theta_i = max(|\hat{f}_i''(u_{l-1}^i)|, |\hat{f}_i''(u_l^i)|, |\hat{f}_i''(u_{l+1}^i)|)$
34:      **if** $\Theta_i < 0.05\bar{f}_i''$ **then**
35:       Merge $[u_{l-1}^i, u_l^i]$ and $[u_l^i, u_{l+1}^i]$
36:      **end if**
37:     **end if**
38:    **end for**
39:    Compute $\tilde{f}(x_t)$ using Eq. (7)
40:   **end if**
41: **end while**

$\max\{|\hat{f}_i''(u_j^i)|, |\hat{f}_i''(u_{j+1}^i)|\} > \bar{f}_i''$, where $\bar{f}_i'' = \frac{1}{U^i}\sum_{j=0}^{U^i-1}|\hat{f}_i''(u_j^i)|$ is the average of the second derivative absolute values. In this way, more resampling points are inserted in areas which are the boundary between dense and sparse areas or are dense with high peak values in density function. Two intervals $[u_{l-1}^i, u_l^i]$ and $[u_l^i, u_{l+1}^i]$ are merged to reduce the number of resampling points when the density function is close to linear, which means $|\hat{f}_i''(u_{l-1}^i)|, |\hat{f}_i''(u_l^i)|$ and $|\hat{f}_i''(u_{l+1}^i)|$ are close to zero (less than $0.05\bar{f}_i''$). In sparse regions, the PDF values are close to zero and the function is almost linear so the intervals are also merged to reduce the number of resampling points.

When updating the densities of resampling points in the model $\mathcal{M}$, the evolution of the data distribution should be considered. Here, a sliding window strategy is used to catch the evolution over time. The window size w is an application-dependent parameter and can be set based on the arrival rate of the data samples and the time interval during which we need to estimate the dynamic density. The window size also controls the robustness of KDE-Track against noisy data where an isolated outlier will increase the height of the PDF curve by maximum $1/w$, which will not be noticed. However, when a new pattern arrives, the new points will replace points in the sliding window from the old pattern and their contribution will be observed on the shape of the density function after receiving a reasonable number of data points from the new pattern. Let n_t denote the number of points that have been received until time t. Due to the difference between w and n_t, there are two different scenarios when updating the model $\mathcal{M}$, more specifically, updating the bandwidths and density $\hat{f}(\boldsymbol{m}_s)$.

When $n_t \leq w$ The received points cannot fill the whole window. The pilot bandwidth value at time t is calculated using all n_t points by the formula $\tilde{h}_{i_t} = c\hat{\sigma}_{i_t}n_t^{-1/5}$, where $\hat{\sigma}_{i_t}^2$ is the sample variance of the received data samples projected on the i-th axis calculated as $\hat{\sigma}_{i_t}^2 = \frac{1}{n_t-1}\left\{\sum_{j=1}^{n_t}x_{ij}^2 - \frac{1}{n_t}\left(\sum_{j=1}^{n_t}x_{ij}\right)^2\right\}, i \in \{1, 2\}$ [42], which can be updated with a constant time at each t. The pilot bandwidth is used to update the estimation of the second derivative of the data marginal distribution on dimension i. The roughness $R(f_i'')$ is then computed to estimate the bandwidth value on that dimension.

After receiving a point $\boldsymbol{x}_t$, the density at a resampling point $\boldsymbol{m}_s$ at time t is updated using *sample-point estimator* [43]

$$\hat{f}_t(\boldsymbol{m}_s) = \frac{n_t - 1}{n_t}\hat{f}_{t-1}(\boldsymbol{m}_s) + \frac{1}{n_t\hat{h}_{1_t}\hat{h}_{2_t}}K_{\hat{h}}\left(\boldsymbol{m}_s, \boldsymbol{x}_t\right), \tag{10}$$

where $K_{\hat{h}}$ is defined in Eq. (3). It is straightforward to show that the updated density $\hat{f}_t(\boldsymbol{m}_s)$ is a good approximation to the estimated density using all the n_t points. In particular, since $\hat{f}_t(\boldsymbol{x}) = \frac{1}{n_t}\sum_{j=1}^{t}\frac{1}{\hat{h}_{1_j}\hat{h}_{2_j}}K_{\hat{h}_j}\left(\boldsymbol{x}, \boldsymbol{x}_j\right) \geq 0$ and $\forall j$ the integration over x of $\frac{1}{\hat{h}_{1_j}\hat{h}_{2_j}} K_{\hat{h}_j}\left(x, x_j\right) = 1$, averaging the integrations of the two terms in (10) results in 1.

When $n_t > w$ In this case, the pilot bandwidth is calculated on the most recently received w points inside the window as follows: $\hat{h}_{i_t} = c\hat{\sigma}_{i_t} w^{-1/5}$, where the sample variance $\hat{\sigma}_{i_t}^2$ of the projected data on the i-th dimension can be easily updated by $\hat{\sigma}_{i_t}^2 = \frac{1}{w-1}(\sum_{j=t-w+1}^{t} x_{ij}^2 - \frac{1}{w}(\sum_{j=t-w+1}^{t} x_{ij})^2)$. The pilot bandwidth is used to update the estimation of the second derivative values and to compute the bandwidth $\hat{h}_{i_t}$, which is used to update the density function at the resampling points. The PDF values at the resampling points $\hat{f}_t(m_s)$ are updated by absorbing the new arrived point x_t and deleting the old point that was moved out from the window:

$$\hat{f}_t(m_s) = \hat{f}_{t-1}(m_s) + \frac{K_{\hat{h}_t}(m_s, x_t)}{w\hat{h}_{1_t}\hat{h}_{2_t}} - \frac{K_{\hat{h}_t}(m_s, x_{t-w})}{w\hat{h}_{1_{t-w}}\hat{h}_{2_{t-w}}}. \tag{11}$$

The probabilistic properties of updated density function $\hat{f}_t(x)$ can be proved as

1. $\hat{f}_t(x) \geq 0, \forall x$, due to the fact that $\hat{f}_t(x) = \frac{1}{w}\sum_{j=t-w+1}^{t} \frac{1}{\hat{h}_{1_j}\hat{h}_{2_j}} K_{\hat{h}_j}\left(x, x_j\right)$ is a summation of nonnegative terms.
2. the integration $\int_{-\infty}^{\infty} \hat{f}_t(x)dx_1 dx_2 = 1$. Since for any j the integration of $(\hat{h}_{1_j}\hat{h}_{2_j})^{-1} K_{\hat{h}_j}\left(x, x_j\right) = 1$, averaging w terms will also be 1.

Time and Space Complexity Analysis Based on the discussion earlier in this section, the time complexity of estimating the density for a new incoming data point is $O(U^1 + U^2)$, where U^1 and U^2 were given in the beginning of Sect. 3.2. They are independent of the number of points that have been received from the data stream. Updating the model when receiving a new point requires computing time linear to the total number of the resampling points $|\mathcal{M}|$, since all the function values at the resampling points are updated. The overall time complexity of processing each arriving point is linear to the model size, which is usually a limited small number. The time required for online density estimation of a data stream with n points is $O(n|\mathcal{M}|)$, which linearly increases with the number of received points from the stream.

Bandwidth selection requires maintaining a one-dimensional KDE-Track model on each dimension. Since $U^1 \times U^2 = |\mathcal{M}|$ and $U^1, U^2 \geq 1$, the total number of resampling points in both models of the one-dimensional KDE-Track is $U^1 + U^2 \leq |\mathcal{M} + 1|$, which will increase only the constant in the KDE-Track's time complexity formula. Thus, the KDE-Track time complexity is $O(n|\mathcal{M}|) = O(n \times U^1 \times U^2)$.

During the online density estimation process, KDE-Track keeps the resampling model $\mathcal{M}$ and the points in the sliding window in memory. Therefore, the memory usage is $|\mathcal{M}|w = U^1 \times U^2 \times w$. Note that the model size $|\mathcal{M}|$ changes upon the distribution variation in data streams due to merge/split operations in adaptive resampling.

Multidimensions Extending the two-dimensional KDE-Track to higher dimensions is straightforward. The same technique can be used for selecting the bandwidth using the marginal distribution of the data on each dimension. The KDE-Track model for estimating the density of d-dimensional data can be constructed as follows:

(1) Discretize the range of the data on the i-th dimension, with $1 \leq i \leq d$, using a set of points $\mathcal{U}^i$, (2) Define the set of resampling points as the cartesian product $\mathcal{U}^1 \times \mathcal{U}^2 \times \cdots \times \mathcal{U}^d$, and (3) Estimate the density function values at the set of resampling points and store them with their estimated density in the model $\mathcal{M}$.

The product kernel defined in Eq. (3) will be

$$K_h\left(\boldsymbol{x}, \boldsymbol{x}_j\right) = \prod_{i=1}^{d} \left\{ \frac{1}{h_i} K\left(\frac{x_i - x_{ji}}{h_i}\right) \right\}.$$

Some researchers avoid using the product kernel for the case of high-dimensional data and replace it with an orientation-invariant kernel function

$$K_h\left(\boldsymbol{x}, \boldsymbol{x}_j\right) = \frac{1}{h} K\left(\frac{\|\boldsymbol{x} - \boldsymbol{x}_j\|}{h}\right),$$

which may not be able to estimate densities with arbitrary shapes as it assumes equal variance values of the data on each dimension.

The KDE-Track's time complexity will remain linear in the size of the stream $O(n|\mathcal{M}|)$, but the constant in the complexity formula will increase according to the number of dimensions since $|\mathcal{M}| = |\mathcal{U}^1| \times |\mathcal{U}^2| \times \cdots \times |\mathcal{U}^d|$. Let $\overline{U} = \frac{1}{d} \sum_{i=1}^{d} |\mathcal{U}^i|$ then $|\mathcal{M}| = \overline{U}^d$ and the time complexity of KDE-Track will be $O(n\overline{U}^d)$.

The interpolation error for the case of d-dimensional data can be bounded as follows: let D_m^i be the maximum distance between the resampling points in dimension $i, i \in \{1, 2, \ldots, d\}$ and $D_m = \max\{D_m^i, 1 \leq i \leq d\}$. Then we can show using mathematical induction on d that

$$\tilde{f}(\boldsymbol{a}) = \hat{f}(\boldsymbol{a}) + \frac{D_m^2}{8} \left\{ \sum_{i=1}^{d} \hat{f}_{x_i x_i}(\boldsymbol{a}) \right\} + O_p(D_m^3).$$

The error is reducible by including more resampling points in certain regions. However, using KDE-Track for estimating high-dimensional density might become impractical as the data become sparse and the estimation error will be large. Besides, applications that rely on visualizing the density functions will not benefit from estimating the density for high-dimensional data as the density cannot be visualized in the case of data streams with more than three dimensions.

4 Density Estimation Performance Evaluation

This section evaluates the most popular density estimators, including (1) the traditional KDE [20] defined in Eq. (1); (2) the FFT-KDE [19, 20], which deploys FFT to convolve a very fine histogram of the data with a kernel function to produce a continuous density function; (3) the Cluster Kernels (CK) [10], which maintains a specific number of kernels by merging similar kernels; (4) SOMKE [12], which employs SOM to cluster the data into a specific number of clusters and uses the centroids of the clusters as the set of kernels; and (5) KDE-Track [18] presented in Sect. 3.2.

4.1 Estimation Accuracy on Synthetic Data

4.1.1 Datasets

The one-dimensional stream (S1D) was generated by extracting data samples from the fifteen densities suggested by Marron and Wand [44] and presented in Fig. 3. The stream is constructed by extracting 3×10^4 data samples from each density and concatenating the batches to get one stream of 4.5×10^5 data samples. The two-dimensional data stream (S2D) is generated by extracting data segments of size 10^5 from the seven densities presented in Fig. 4. The total size of the stream is 7×10^5

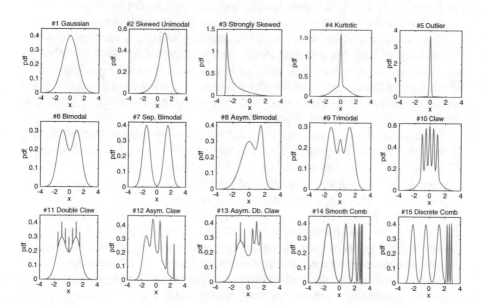

Fig. 3 The fifteen densities recommended by Marron and Wand [44] to evaluate univariate density estimators

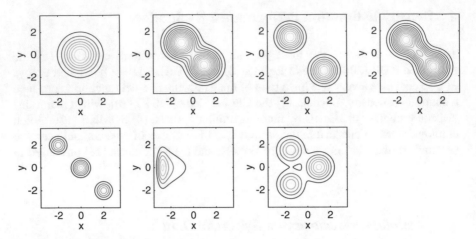

Fig. 4 The contours of the densities used to construct the 2D data stream that is used to evaluate the density estimators

data samples. These streams are selected because they contain challenging densities that are hard to be estimated accurately. Use of batches of the same size is to simplify the calculation of the true density only.

Estimation Accuracy Figure 5 shows the MAE and the l_∞ error incurred by the evaluated methods when estimating the density of S1D (subfigures a, b) and S2D (subfigures c, d). The MAE measures how the estimated density curve fits the curve of the true density, while the l_∞ measures the maximum variation between the true and the estimated curves. The error is computed by defining a set of checkpoints with the step of 1000. For each estimation method, at each checkpoint an evaluation set $E = \{e_1, \ldots, e_{1000}\}$ of 1000 samples is generated from the same distribution of the data in the sliding window. The true and estimated density values of the evaluation points are then compared to compute the MAE and the l_∞ error. The window size is set to 2×10^4 data samples.

The CK and FFT-KDE methods are not designed to capture the dynamic density of the data streams using the sliding window approach. To adapt these methods with sliding windows, we rebuild their model at each evaluation checkpoint by deleting the old model and creating a new model using the data samples in the current window. This adaptation preserves the estimation accuracy of the methods. However, the CK method is shown to be impractical for online density estimation due to its high computational cost, as we will show later. SOMKE is adapted for the case of sliding window by dividing the sliding window into batches of 1000 samples. At each evaluation checkpoint, the kernels that represent the removed batch out of the sliding window are deleted and replaced by the kernels that represent the most recent batch added to the sliding window. Notice that CK is not evaluated on S2D data as CK is proposed to estimate the density for univariate data only.

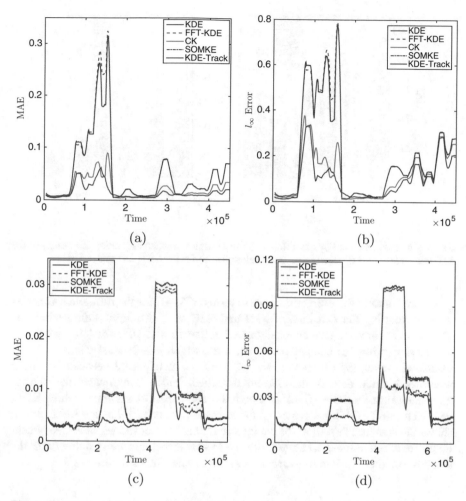

Fig. 5 The MAE and the l_∞ error incurred by the different density estimators when estimating the density of S1D and S2D streams. The window size is 2×10^4. (**a**) MAE of S1D, (**b**) l_∞ error of S1D, (**c**) MAE of S2D, (**d**) l_∞ error of S2D

Selecting the bandwidth values for each estimator is done using the same settings as in the references. All the baseline methods use the normal rule because of its efficiency, except the CK method which uses the Epanechnikov kernel function with a recommended constant $c = 1.06$. This setting enables CK to perform well when densities have high peaks and are multimodal. KDE-Track uses its own method for setting the bandwidths, that is, estimating the roughness of the second derivative $R(f'')$ and plugging it in Eq. (5), which increases its accuracy significantly. The results show that KDE-Track has the best performance (the smallest error). The high accuracy obtained by KDE-Track is mainly because of the accurate bandwidth selection method.

A. Qhatan et al.

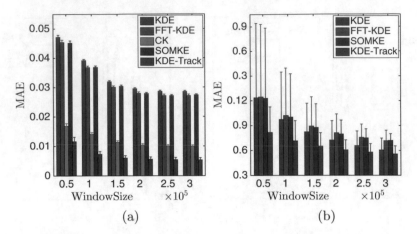

Fig. 6 The MAE incurred by the different density estimators when estimating the density of the S1D stream (**a**) and S2D stream (**b**). The window size varies from 5×10^4 to 3×10^5

Figure 6 shows the estimation error in terms of MAE for the different estimators when estimating the densities of S1D and S2D with different sliding windows. The sliding window's size changes from 5×10^4 to 3×10^5. For large windows, the density estimation becomes more accurate, which is reflected by smaller MAE values. However, the decrease in the MAE is not as expected because the larger windows include data from different densities, which complicates the density estimation process. KDE-Track is shown to have the most accurate results. KDE, FFT-KDE, and SOMKE have comparable results. In addition to the MAE, Fig. 6 shows the standard deviation for the sensitivity analysis of the window size, where KDE-Track is the most accurate (with the lowest error) and most stable (with the smallest standard deviation in error), especially in the S2D streams.

4.2 Computational Time Cost and Space Usage

Other important factors in the success of an online density estimator are its running time and space usage, as streaming data arrive fast and have unlimited size. Since we are estimating the dynamic density, which will be better represented by the most recent samples, all the methods are modified to use the sliding window technique. This technique requires storing the samples in the sliding window in the memory either for using them to estimate the density as in KDE or to update the density estimator's model as in CK, FFT-KDE, SOMKE, and KDE-Track. Hence, all the methods have comparable space complexity, which is linear in the window size.

The time complexity of KDE-Track, as discussed in Sect. 3.3, is $O(n|\mathcal{M}|)$. Estimating the density using KDE at any given data sample requires scanning the sliding window. Therefore, the time complexity of KDE when used for online density estimation is $O(nw)$, where w is the window size. The time complexity of CK is controlled by two main steps: (1) model reconstruction, which is performed at each evaluation checkpoint and has a complexity of $O(w)$; (2) density estimation at any sample of the evaluation points, which has a constant time complexity. Thus, using CK for online density estimation will have a complexity of $O(nw)$. However, the model's reconstruction step of CK is more expensive than the density estimation using all the kernels in KDE. It is expected that CK will be more timely efficient if the data is stationary and the model is updated online without reconstruction.

FFT-KDE also has two main steps: (1) model reconstruction, which involves updating the histogram after receiving a new data sample and convolving the histogram with kernel function; and (2) density estimation of the evaluation samples. The first step requires $O(B \log B)$, where B is the number of bins in the histogram, and the second step has a constant time complexity. The time complexity is thus $O(nB \log B)$.

The SOMKE model is built by training the SOM neurons with the current window which has a time complexity of $O(w)$. Estimating the density at the evaluation samples using the trained SOM neurons has a constant time complexity. The method's time complexity is then $O(nw)$, where the constant in the complexity formula is smaller than that for KDE and CK. Figure 7 shows the running time for using the density estimators for online density estimation of S1D and S2D streams. The results in the figure confirm our analysis.[3] KDE-Track and FFT-KDE are most efficient with very small running time, which is not affected by the size of w.

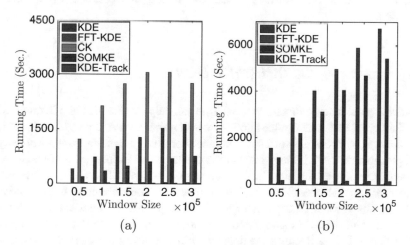

Fig. 7 The running time of the different density estimators when estimating the density of the S1D (**a**) and S2D (**b**) streams. The window size varies from 5×10^4 to 3×10^5

---

[3]All implementations were coded by C/C++ and run on Intel 2.5 GHz Dual-Core PC with 4 GB memory.

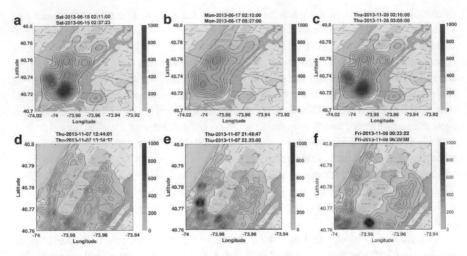

Fig. 8 The density estimated using New York Taxi trips data for different time intervals. (**a**) weekend, (**b**) working day, (**c**) Thanksgiving, (**d**) working hours, (**e**) night, (**f**) after midnight

5 Applications

In this section, we apply the estimated dynamic density to two different application problems: Taxi traffic real-time visualization and unsupervised online change detection.

5.1 Visualizing the Taxi Traffic Data

Visualizing the density function in real time can help service planners in monitoring the density of required services and forwarding more service providers to regions that demand more services at a specific time. For example, monitoring the density of taxi pickup data can tell the planners of taxi companies to forward more taxicabs to a specific region of the city. In this subsection, we visualize the dynamic traffic distribution in the New York Taxi trips dataset.[4] The dataset is freely available and contains records of trips that include pickup time, longitude and latitude of the pickup and drop-off location, etc. We are mainly interested in the pickup time and location. Figure 8 shows the density estimated using the pickup location with window size of 10^4, where the data records are sorted according to their pickup time. The first three subfigures show the pickup events occur in the early morning of a weekend day (subfigure a), of a regular working day (subfigure b), and of a national holiday (subfigure c). These subfigures show more pickup events during

---

[4]Available at: http://www.andresmh.com/nyctaxitrips/.

the weekends and holidays than during regular working days in the Greenwich and the east villages where there are many restaurants and nightclubs. The frequency of pickup events also increases during the weekends as it took less than 30 min to record 10^4 events in a weekend but more than 3 h in the early morning of a working day. More taxicabs are thus suggested in that region on similar events to satisfy the high demand.

Interesting patterns of community behavior can also be found in a regular working day. Figure 8d–f shows the pickup events on November 7 and 8, 2013, at different time intervals. The pickup events during the working hours (subfigure d) show close to uniform distribution within the area around the central park. Figure 8e shows high density at Lincoln center during the time interval 21:48–22:55, when a concert or other events may be over. After midnight, we can observe a small number of pickup events occurred as it took six hours to accumulate 10^4 events with more pickup events occurred around Trump and Freedom towers.

5.2 Online Change Detection

Change detection in data streams refers to the problem of finding time points, where for each point, there exists a significant change in the current data distribution. A typical window-based solution is to extract a fixed S_1 (reference window) from streaming samples and to update an S_2 (test window) with newly arriving samples [23, 28]. Changes are then detected by measuring the difference between the distributions in S_1 and S_2.

Modeling the data distribution and selecting a comparison criterion are essential for change detection in data streams. However, density estimation of multidimensional data is difficult. It becomes less accurate and more computationally expensive with increasing dimensionality. In this subsection, we introduce a framework which applies Principal Component Analysis (PCA) to project the multidimensional data from the stream on the principal components to obtain multiple 1D data streams. Density estimation, distribution comparison, and change-score calculations can then be conducted in parallel on those 1D data streams. Compared with projecting the data on the original coordinates (i.e., using the original variables), projecting on PCs has the following advantages: (1) it allows the detection of changes in data correlations, which cannot be detected in the original individual variables; (2) it guarantees that any changes in the original variables are reflected in PC projections; and (3) it reduces the computation cost by discarding trivial PCs. Proofs can be found in [33].

Change Detection Framework The framework is given in Algorithm 2, where D_M denotes any divergence metric for comparing two distributions.

Setting Windows Line 3 in Algorithm 2 sets the reference window S_1 to be the first w samples arriving after the change point t_c. Intuitively, when a data distribution shifts to a new one, the reference window should be updated to represent the new distribution. This update also enables the detection of further changes. Line 8 sets the test window S_2 as a collection of w samples after the reference window. This S_2 will slide along the data stream to include the newest w samples. The setting of this parameter is usually left to the user to give them the ability to monitor the long-/short-term changes, depending on their interests and the application sensitivity against changes [24].

Algorithm 2 CHANGEDETECTIONFRAMEWORK

Parameters: window size w, ξ, δ
Online flow in: streaming data $S = \{\mathbf{x}_1, \cdots, \mathbf{x}_t, \cdots\}$
Online output: time t when detecting a change
Procedure:
1: Initialize $t_c = 0$, $step = min(0.05w, 100)$
2: Initialize $\overline{Sc}, m, M$ to NULL.
3: Set reference window $S_1 = \{\mathbf{x}_{t_c+1}, \cdots, \mathbf{x}_{t_c+w}\}$
4: Extract principal components by applying PCA on S_1 to obtain $\mathbf{P}_1, \mathbf{P}_2, \cdots, \mathbf{P}_k$
5: Project S_1 on $\mathbf{P}_1, \mathbf{P}_2, \cdots, \mathbf{P}_k$ to obtain $\breve{S}_1$
6: $\forall i \ (1 \leq i \leq k)$ estimate $\hat{f}_i$ using data of the i-th component of $\breve{S}_1$
7: Clear S_1 and $\breve{S}_1$
8: Set test window $S_2 = \{\mathbf{x}_{t_c+w+1}, \cdots, \mathbf{x}_{t_c+2w}\}$
9: Project S_2 on $\mathbf{P}_1, \mathbf{P}_2, \cdots, \mathbf{P}_k$ to obtain $\breve{S}_2$
10: Clear S_2
11: Estimate $\hat{g}_i$ using data of the i-th component of $\breve{S}_2$
12: **while** a new sample $\mathbf{x}_t$ arrives in the stream **do**
13:     Project $\mathbf{x}_t$ on $\mathbf{P}_1, \mathbf{P}_2, \cdots, \mathbf{P}_k$ to obtain $\breve{\mathbf{x}}_t$
14:     Remove $\breve{\mathbf{x}}_{t-w}$ from $\breve{S}_2$
15:     $\forall i \ (1 \leq i \leq k)$ update $\hat{g}_i$ using $\breve{\mathbf{x}}_t^{(i)}$ and $\breve{\mathbf{x}}_{t-w}^{(i)}$
16:     **if** $mod(t, step) = 0$ **then**
17:         $curScore = \max_i \left(D_M \left(\hat{g}_i \| \hat{f}_i \right) \right)$
18:         **if** CHANGEFINDER($curScore, \overline{Sc}, m, M, \xi, \delta$) **then**
19:             Report a change at time t and set $t_c = t$
20:             Clear $\breve{S}_2$ and GOTO step 2
21:         **end if**
22:     **end if**
23: **end while**

Projecting the Data After receiving the first w data samples in reference window S_1, PCA is applied to extract the principal components from S_1. The first k PCs with the largest eigenvalues are selected if they account for 99.9% of data variance (i.e., $\sum_{i=1}^{k} \frac{\lambda_i}{\sum_{j=1}^{d} \lambda_j} \geq 0.999$). The data in the reference and test windows are then projected on these k components. On each component, projections of the reference and test windows are compared and a change-score value is recorded. The maximum value among the k change-score values is considered as the final change-score. Any new data sample is projected on the k components and the density functions of the projection of the test window are updated and compared with the reference densities.

Estimating Density Functions Because the change-score is used directly to trigger change alarms, PDFs for distribution comparison must be accurately and efficiently estimated. Here, KDE-Track is employed for estimating the density functions of the projected data of S_1 (line 6) and S_2 (line 11), as well as for updating the test densities when a new sample arrives (lines 13–15), due to its merits of both efficiency and accuracy. Note that the update at line 15 ensures that $\hat{g}_i$ is the current distribution of the i-th component in the data stream.

Computing the Change-Score Values Change-scores are computed by using a divergence metric on two density functions $\hat{f}_i$ and $\hat{g}_i$ (line 17), which are updated upon the arrival of each sample. However, it is not necessary to compute change-score at each time step, as the change of a distribution cannot be observed after a single data sample. Therefore, to reduce unnecessary repeated comparisons that may increase the execution time noticeably, we compute change-scores every $\min(0.05 * w, 100)$ samples (line 16). This setting of checkpoints complies with the monitoring requirements of users. Monitoring short-term changes with a small w needs frequent checkpoints while monitoring long-term changes by setting a large w allows bigger checkpoint intervals. The granularity can be adjusted by changing the 5% according to the users' needs.

Divergence metric $D_M(\hat{g}_i \| \hat{f}_i)$ is crucial for computing change-scores, and can be set differently. A widely used divergence metric, KL-divergence, is defined as

$$D_{\mathrm{KL}}\left(g\|f\right) = \int_x g(x) \log\left(\frac{g(x)}{f(x)}\right) dx. \tag{12}$$

It is a nonnegative (≥ 0) and nonsymmetric measure. It is 0 when the two distributions are completely identical, and becomes larger as the two distributions deviate from each other. The non-symmetry property of the D_{KL} complicates the procedure of setting the threshold for detecting changes in data streams. To overcome the problem of the KL-divergence, this framework provides two options on setting divergence metrics. The first divergence metric is a modified symmetric KL-divergence [45]

$$D_{\mathrm{MKL}}(\hat{g}\|\hat{f}) = \max\left(D_{\mathrm{KL}}\left(\hat{g}\|\hat{f}\right), D_{\mathrm{KL}}\left(\hat{f}\|\hat{g}\right)\right). \tag{13}$$

This divergence metric mimics powerful order selection tests developed in current statistics literature; see, for example, [46].

The second divergence metric is a measure of the intersection area under the curves of two density functions [47],

$$D_A\left(\hat{g}\|\hat{f}\right) = 1 - \int_x \min\left(\hat{f}(x), \hat{g}(x)\right) dx. \tag{14}$$

This D_A takes values in $[0, 1]$, where the value one means that the two distributions are completely different and zero means the two distributions are identical. This measure can also be computed using numerical integration techniques on the intersection area.

After computing the change-scores on all the PCs, the different change-score values are aggregated by taking the maximum over all values. This is necessary to maintain a single statistical quantity. The maximum is preferable for aggregating the change-score values as it allows for treating any changes happening at any component equally important. Also, when a change happens in a single PC, the change-score will not be affected by the small change-score values obtained from the other PCs.

Algorithm 3 CHANGEFINDER

Input: $curScore, \overline{Sc}, m, M, \xi, \delta$
Output: <True or False>: a change should be reported or not
1: Update $\overline{Sc}$ to include the $curScore$ in the average.
2: $newm = m + \overline{Sc} - curScore + \delta$
3: **if** $|newm| > M$ **then**
4:     $newM = newm$.
5: **end if**
6: $\tau_t = \xi * \overline{Sc}$
7: **if** $curScore > \tau_t$ **then**
8:     return True
9: **else**
10:     $M = newM, m = newm$
11:     return False
12: **end if**

Dynamic Threshold Settings Typical statistical tests for change detection start by considering the null hypothesis, which assumes that the data distribution is stationary. A change-score value is then calculated to determine the probability of rejecting the null hypothesis. The most popular technique to reject the null hypothesis is to specify a threshold and declare a change whenever the change-score becomes greater than the threshold. Most existing methods require a user-specified threshold [48], which has two main issues. First, the fixed threshold cannot be used to detect changes in different magnitudes. Second, the threshold is difficult to set, as it is sensitive to the divergence metric, window size, underlying distribution, and change types. In this framework, the threshold is dynamically adjusted in the detection process (Algorithm 3). More details of the setting can be found in [33].

Performance Evaluation The above presented change detection framework is evaluated together with two other change detection methods: *kdq-tree* [28] (using *kdq-tree* data structure to model data distribution) and PCA-SPLL [29] (using PCs with small variances for measuring distribution difference). Table 3 presents the results on datasets, which are generated following the same generation mechanism of [28] and contain the same types of changes: $M(\epsilon)$ means varying the mean value, $D(\epsilon)$ means varying the standard deviation, and $C(\epsilon)$ means varying the correlation. Each dataset contains 5×10^6 data samples with changes that occur every 5×10^4 data samples with a total of 99 change points. At each change point, a

Table 3 Evaluation results of Dasu's method (*kdq-tree*) with Average Absolute Difference (AAD) and KL-divergence metrics, Kuchneva's method (PCA-SPLL), CD-Area, and CD-MKL with KDE-Track as density estimators

Dataset	*kdq-tree* [28]		PCA-SPLL [29]	CD-Area [33]	CD-MKL [33]
	AAD	KL			
M(0.01)	30/15/1/54	3/7/0/89	10/16/5/73	**38/23/0/38**	34/31/1/34
M(0.02)	77/14/6/8	3/7/0/89	18/9/3/72	**87/12/0/0**	89/7/0/3
M(0.05)	98/1/4/0	12/21/0/66	66/11/9/42	**99/0/0/0**	95/0/4/4
D(0.01)	42/18/2/39	4/2/0/93	29/5/4/65	45/29/0/25	**54/25/0/20**
D(0.02)	98/1/9/0	12/7/0/80	85/0/2/14	**97/0/1/2**	94/1/2/4
D(0.05)	99/0/2/0	24/4/2/71	89/2/2/8	**99/0/0/0**	98/0/2/1
C(0.1)	67/13/2/19	8/4/1/87	55/4/2/40	68/19/0/12	**76/17/2/6**
C(0.15)	78/9/5/12	8/7/1/84	63/6/4/30	84/7/1/8	**85/9/0/5**
C(0.2)	96/3/10/0	21/5/3/73	75/3/3/21	**97/1/1/1**	93/2/0/4

The results are in the form TP/L/FP/FN with TP = true positives, L = late detections, FP = the false positives and FN = the false negatives. The best results are in bold

set of random numbers in the interval $[-\epsilon, -\epsilon/2] \cup [\epsilon/2, \epsilon]$ are generated and added to the distribution's parameters that will be changed. The parameter ϵ controls the magnitude of the change, where smaller values for ϵ make changes harder to detect and vice versa.

The presented framework using the maximum KL-divergence (Eq. (13)) is called CD-MKL, and the one using the Area metric (Eq. (14)) is called CD-Area. The performance of the different methods is measured according to the number of True Positives (TP), Late detections (L), False Positives (FP), and False Negatives (FN). By true positives, we mean the changes that were reported correctly before receiving $2w$ from the new distribution where w is the window size. Late detections are the changes reported after processing $2w$ data samples from the new distribution. False positives are changes reported by the method when there are no changes, and false negatives are the missed changes. The window size w is set to 10^4 for all methods.

Results The experimental results show that the presented framework outperforms *kdq-tree* and PCA-SPLL in terms of the number of correctly detected changes and less false positives. Also, the performance of the framework when using the Area metric (Eq. (14)) outperforms the performance of the framework when using the MKL (Eq. (13)) metrics in most of the evaluation datasets. PCA-SPLL [29] shows the worst performance, especially for detecting mean shifts.

The second set of experiments evaluates the detection accuracy in high-dimensional datasets. As changes become less observable when the data dimensionality increases, we used only datasets with a reasonable magnitude of change ($M(0.05)$, $D(0.05)$, and $C(0.2)$). For each type of change, we generated three datasets with 10, 20, and 30 dimensions. The changes in the data distribution affect only two dimensions (variables) and for more complicated change detection cases, we selected the variables that are affected by the change randomly at each change point.

Table 4 Evaluation results of *kdq-tree* with Average Absolute Difference (AAD) metric, PCA-SPLL, CD-MKL, and CD-Area for (a) changes in Gaussian high-dimensional data (1st 9 datasets), (b) changes in density shape from a list of non-Gaussian distributions (DistCh), and (c) changes in nonlinear dependent data streams (DEMC, DDC, and SWRL)

Dataset	*kdq-tree*(AAD) [28]	PCA-SPLL [29]	CD-MKL [33]	CD-Area [33]
M(10D)	96/1/9/2	63/10/10/26	**99/0/0/0**	**99/0/0/0**
M(20D)	75/7/3/17	32/11/1/56	96/1/2/2	**99/0/0/0**
M(30D)	59/8/4/32	33/13/2/53	97/0/1/2	**97/1/1/1**
D(10D)	93/3/3/3	70/1/1/28	**99/0/0/0**	**99/0/0/0**
D(20D)	87/3/7/9	59/0/0/40	**99/0/0/0**	**99/0/0/0**
D(30D)	78/4/4/17	56/1/4/42	**96/1/1/2**	96/0/0/3
C(10D)	64/7/10/28	67/1/0/31	95/0/5/4	**98/1/0/0**
C(20D)	29/12/12/58	49/0/0/50	**98/0/2/1**	95/2/2/2
C(30D)	11/4/5/84	48/0/0/51	97/0/2/2	**99/0/0/0**
DistCh	99/0/7/0	72/0/44/27	**99/0/0/0**	**99/0/0/0**
DEMC	84/6/2/9	24/15/28/60	84/6/3/9	**88/0/4/11**
DDC	96/2/3/1	19/18/29/62	94/1/3/4	**97/0/0/2**
SWRL	99/0/6/0	23/10/29/66	95/0/2/4	**99/0/2/0**

The best results are in bold

We report the results for the *kdq-tree* method with AAD-divergence only because KL-divergence shows very low accuracy for 2D data in Table 3. The results in Table 4 show that CD-Area and CD-MKL are not affected by data dimensionality. As all space partitioning methods, the *kdq-tree* method suffers from the curse of dimensionality as data dimensionality increases. The PCA-SPLL has low accuracy again.

The third set of experiments evaluates the detection accuracy in two special cases: (1) changes in the *density shape* of data streams, and (2) changes in data streams with *nonlinear dependencies*. Four datasets were generated for evaluation, where each dataset contains 100 batches with batch size of 5×10^4, resulting in a total of 99 changes. DistCh is generated by changing the data distribution in consecutive batches, where a distribution is randomly drawn from a list of preset distributions including standard Normal, highly skewed Normal, bimodal Normal, trimodal Normal, Gamma and Laplace distributions. In each batch, the mean, variance, and correlation are kept constant. The remaining three datasets evaluate the detection accuracy on *nonlinear dependent* data streams. The first dataset includes a Disc with EMpty Circle in the middle (DEMC). Changes in consecutive batches are introduced by randomly altering the radius of the empty circle without affecting the mean, variance, or correlation values. The second dataset includes Disc with Dense Circle in the middle (DDC). Nonlinear changes in the distribution are introduced by altering the radius of the dense circle in the center of the data points. The third dataset is a *Swiss roll* (SWRL) dataset with changes designed by altering the distance between any two consecutive contours of the *Swiss roll*.

The results presented in Table 4 show that both CD-MKL and CD-Area outperform other methods in DisCh, and that CD-Area obtained better results in all nonlinear dependent data streams. While the experimental results are promising, the proposed method is restricted by PCA's limitation to handling only linearly dependent data streams. Change detection in nonlinear dependent data streams will be left for future study.

Computational Cost Analysis The time complexity of the presented framework and the *kdq-tree* method is linear w.r.t. the size of the data stream. However, the constant in the complexity formula controls the efficiency of the evaluated methods. The cost of the PCA-based framework depends on three main subroutines: (1) The PCA routine for extracting PCs, which is called only when a change is reported. It has a complexity of $O(d^2 \times w \times R_c)$, where d is the data dimensionality, w is the window size, and R_c is the number of reported changes; (2) The incremental density update by KDE-Track, which requires a constant time at the arrival of a new data sample from the stream; and (3) Computing the divergence metric, which is done incrementally and costs a constant time upon the arrival of a new data sample. Therefore, the computational cost of the PCA-based framework is affected more by data dimensionality, as the PCA routine has quadratic complexity w.r.t. dimensionality d, and maintaining the density and computing the change-score may have to be done on more PCs when d is higher. The window size has a relatively less effect on the running time of our framework as it affects only the PCA routine.

The running time of the *kdq-tree* method increases fast w.r.t. window size and data dimensionality, mainly due to the expensive bootstrap sampling routine for computing the threshold to report changes. The overall complexity is $O(kdw \log(\frac{1}{v} R_c))$, where R_c is the number of reported changes.

The results in Fig. 9 confirm our analysis. We see a slight increase in the running time of CD-Area and CD-MKL when increasing the window size (Fig. 9 Left),

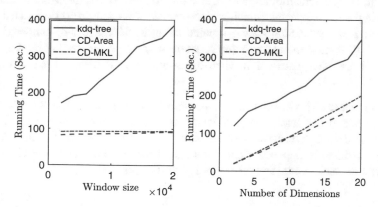

Fig. 9 Running time of the evaluated methods for different window sizes (left) and different number of dimensions (right). The stream size is 5×10^6 data samples. Changes occur every 5×10^4 data samples, and the number of changes is 99

while the runtime increases almost linearly w.r.t. data dimensionality (Fig. 9 Right). The running time of *kdq-tree* increases linearly with the window size and data dimensionality.

Last, it is worth mentioning that *kdq-tree* requires the data to be normalized (into [0, 1]) before running to ensure that growth of the tree growing is not stopped prematurely by the stopping condition of the minimum cell width. However, it is unrealistic to normalize data streams beforehand, as data arrives continuously. The PCA-framework does not require normalization and is thus superior to *kdq-tree* in processing a real data stream.

6 Summary and Future Work

Summary In this chapter, we studied the problem of estimating the dynamic density that comes with data streams. We presented KDE-Track [18] that can timely track the evolving distribution and accurately estimate the probability density function of evolving data streams. KDE-Track models the data distribution as a set of resampling points with their estimated PDF values, which are incrementally updated upon the arrival of new data samples from the stream. The effectiveness and efficiency of KDE-Track have been analytically studied and experimentally demonstrated on both synthetic and real-world data.

KDE-Track provides any time available density function, which has been utilized in two main applications. First, the estimated density is used for visualizing and monitoring the changes in New York Taxi pickup data. KDE-Track allows for visualizing the dynamic density estimated from the data in real time after receiving any data sample.

Second, KDE-Track is used to develop a framework for detecting abrupt changes in linearly dependent multidimensional data streams. The framework is based on projecting data on selected principal components. On each projection, densities in reference and test windows are estimated using KDE-Track and compared using effective divergence metrics to compute a change-score value that are used to report change in the data distribution when the score value increases for a reasonable period.

Possible Future Work First, more applications that benefit from KDE-Track as an accurate and computationally efficient density estimator can be studied. Second, the change detection framework depends mainly on the PCA, which can reflect the changes in linearly dependent data streams. However, more complex changes in nonlinear dependent data streams should be further studied.

References

1. Zhou, A., Cai, Z., Wei, L., Qian, W.: M-kernel merging: towards density estimation over data streams. In: DASFAA (2003)
2. Subramaniam, S., Palpanas, T., Papadopoulos, D., Kalogeraki, V., Gunopulos, D.: Online outlier detection in sensor data using non-parametric models. In: VLDB (2006)
3. Schaller, B.: A regression model of the number of taxicabs in U.S. cities. J. Public Transp. **8**, 63–78 (2005)
4. Zhou, Z., Matteson, D.: Predicting ambulance demand: a spatio-temporal kernel approach. In: KDD (2015)
5. Wu, F., Li, Z., Lee, W., Wang, H., Huang, Z.: Semantic annotation of mobility data using social media. In: WWW (2015)
6. Scott, D.: Multivariate Density Estimation: Theory, Practice, and Visualization. Wiley, New York (1992)
7. Qahtan, A., Zhang, X., Wang, S.: Efficient estimation of dynamic density functions with an application to outlier detection. In: CIKM (2012)
8. Babcock, B., Babu, S., Datar, M., Motwani, R., Widom, J.: Models and issues in data stream systems. In: ACM SIGMOD-SIGACT-SIGART (2002)
9. Zhang, X., Furtlehner, C., Germain-Renaud, C., Sebag, M.: Data stream clustering with affinity propagation. IEEE Trans. Knowl. Data Eng. **26**, 1644–1656 (2014)
10. Heinz, C., Seeger, B.: Cluster kernels: Resource-aware kernel density estimators over streaming data. IEEE Trans. Knowl. Data Eng. **20**, 880–893 (2008)
11. Boedihardjo, A.P., Lu, C., Chen, F.: A framework for estimating complex probability density structures in data streams. In: CIKM (2008)
12. Cao, Y., He, H., Man, H.: SOMKE: Kernel density estimation over data streams by sequences of self-organizing maps. IEEE Trans. Neural Netw. Learn. Syst. **23**, 1254–1268 (2012)
13. Zheng, Y., Jestes, J., Phillips, J., Li, F.: Quality and efficiency in kernel density estimates for large data. In: SIGMOD (2013)
14. Procopiuc, C., Procopiuc, O.: Density estimation for spatial data streams. In: SSTD (2005)
15. Gary, A., Moore, A.: Nonparametric density estimation: toward computational tractability. In: SDM (2003)
16. Lin, C., Wu, J., Yen, C.: A note on kernel polygons. Biometrika **93**, 228–234 (2006)
17. Hart, T., Zandbergen, P.: Kernel density estimation and hotspot mapping: examining the influence of interpolation method, grid cell size, and bandwidth on crime forecasting. Policing Int. J. Police Strateg. Manag. **37**, 305–323 (2014)
18. Qahtan, A., Wang, S., Zhang, X.: Kde-track: an efficient dynamic density estimator for data streams. IEEE Trans. Knowl. Data Eng. **29**, 642–655 (2017)
19. Wand, M.: Fast computation of multivariate kernel estimators. J. Comput. Graph. Stat. **3**, 433–445 (1994)
20. Silverman, B.: Density Estimation for Statistics and Data Analysis. Chapman and Hall, London (1986)
21. Yamanishi, K., Takeuchi, J., Williams, G., Milne, P.: On-line unsupervised outlier detection using finite mixtures with discounting learning algorithms. Data Min. Knowl. Disc. **8**, 275–300 (2004)
22. Ke, Y., Sukthankar, R., Hebert, M.: Event detection in crowded videos. In: ICCV (2007)
23. Kifer, D., Ben-David, S., Gehrke, J.: Detecting change in data streams. In: VLDB (2004)
24. Aggarwal, C.C.: A framework for diagnosing changes in evolving data streams. In: SIGMOD (2003)
25. Guralnik, V., Srivastava, J.: Event detection from time series data. In: KDD (1999)
26. Kawahara, Y., Sugiyama, M.: Change-point detection in time-series data by direct density-ratio estimation. In: SDM (2009)
27. Bifet, A., Gavaldà, R.: Learning from time-changing data with adaptive windowing. In: SDM (2007)

28. Dasu, T., Krishnan, S., Venkatasubramanian, S., Yi, K.: An information-theoretic approach to detecting changes in multi-dimensional data streams. In: Symposium on the Interface of Statistics, Computing Science, and Applications (2006)
29. Kuncheva, L.I., Faithfull, W.J.: PCA feature extraction for change detection in multidimensional unlabeled data. IEEE Trans. Neural Netw. Learn. Syst. **25**, 69–80 (2014)
30. Song, X., Wu, M., Jermaine, C., Ranka, S.: Statistical change detection for multi-dimensional data. In: KDD (2007)
31. Liu, S., Yamada, M., Collier, N., Sugiyama, M.: Change-point detection in time-series data by relative density-ratio estimation. In: International Conference on Structural, Syntactic, and Statistical Pattern Recognition, pp. 363–372 (2012)
32. Takeuchi, J., Yamanishi, K.: A unifying framework for detecting outliers and change points from time series. IEEE Trans. Knowl. Data Eng. **18**, 482–492 (2006)
33. Qahtan, A.A., Alharbi, B., Wang, S., Zhang, X.: A PCA-Based change detection framework for multidimensional data streams. In: SIGKDD, pp. 935–944 (2015)
34. Epanechnikov, V.A.: Non-parametric estimation of a multivariate probability density. Theory Probab. Appl. **14**, 153–158 (1969)
35. Turlach, B.: Bandwidth selection in kernel density estimation: a review. CORE and Institut de Statistique, vol. 19, pp. 1–33 (1993)
36. Scott, D., Terrell, G.: Biased and unbiased cross-validation in density estimation. J. Am. Stat. Assoc. **82**, 1131–1146 (1987)
37. Hall, P., Sheather, S., Jones, M., Marron, J.: On optimal data-based bandwidth selection in kernel density estimation. Biometrika **78**, 263–269 (1992)
38. Hall, P., Marron, J.: Estimation of integrated squared density derivatives. Stat. Probab. Lett. **6**, 109–115 (1987)
39. Jones, M.: The roles of ISE and mise in density estimation. Stat. Probab. Lett. **12**, 51–56 (1991)
40. Shimazaki, H., Shinomoto, S.: Kernel bandwidth optimization in spike rate estimation. J. Comput. Neurosci. **29**, 171–182 (2010)
41. Zheng, Y., Phillips, J.: l_∞ error and bandwidth selection for kernel density estimates of large data. In: KDD (2015)
42. Chan, T., Golub, G., LeVeque, R.: Algorithms for computing the sample variance: Analysis and recommendations. Am. Stat. **37**, 242–247 (1983)
43. Sain, R.: Multivariate locally adaptive density estimation. Comput. Stat. Data Anal. **39**, 165–186 (2002)
44. Marron, J., Wand, M.: Exact mean integrated squared error. Ann. Stat. **20**, 712–736 (1992)
45. Liu, D., Sun, D., Qiu, Z.: Feature selection for fusion of speaker verification via maximum kullback-leibler distance. In: ICSP (2010)
46. Jin, L., Wang, S., Wang, H.: A new nonparametric stationarity test of time series in time domain. J. R. Stat. Soc. Ser. B **77**, 893–922 (2015)
47. Cha, S.: Comprehensive survey on distance/similarity measures between probability density functions. Int. J. Math. Models Methods Appl. Sci. **1**, 300–307 (2007)
48. Dai, X.L., Khorram, S.: Remotely sensed change detection based on artificial neural networks. Photogramm. Eng. Remote. Sens. **65**, 1179–1186 (1999)

Incremental SVM Learning: Review

Isah Abdullahi Lawal

Abstract The aim of this paper is to present a review of methods for incremental Support Vector Machines (SVM) learning and their adaptation for data stream classification in evolving environments. We formalize a taxonomy of these methods based on their characteristics and the type of solution they provide. We discuss the strength and weakness of the various learning methods and also highlight some applications involving data stream, where incremental SVM learning has been used.

1 Introduction

Data streams from high-rate sources, such as wireless sensor network, surveillance camera network, and the Internet, require a timely and meaningful classification to facilitate their interpretation [19, 40]. The classification of the data stream can be performed with *supervised* [49], *unsupervised* [23], and *semi-supervised* [51] learning methods, when the data are labeled, unlabeled, or a mixture of both, respectively. Ideally, it is desirable to be able to consider all the samples in the data stream at once during the learning of the classification model (i.e., the classifier) in order to get the best estimation of the class distribution. However, when the number of samples becomes too large, the learning process can get computationally intractable [13]. In fact, the amount of memory required by the learning algorithms increases prohibitively with an increase in the number of samples in the data streams [24]. Moreover, sometimes the data stream exhibits a phenomenon referred to as *concept drift*, where the underlying data distribution changes over time [39, 47], making the classifier built with old samples become inconsistent with new samples [59]. Thus if the learning algorithms do not contemplate and adapt to the change in the data distribution, it will be impossible to properly learn a good classifier. Recently, researchers have proposed incremental learning techniques

I. A. Lawal (✉)
Department of Computer Engineering and Networks, Jouf University, Sakaka, Saudi Arabia
e-mail: ialawal@ju.edu.sa

© Springer International Publishing AG, part of Springer Nature 2019
M. Sayed-Mouchaweh (ed.), *Learning from Data Streams in Evolving Environments*, Studies in Big Data 41, https://doi.org/10.1007/978-3-319-89803-2_12

which allow the classifier to be built incrementally as the data arrives using the limited computing resources available and guaranteeing the effectiveness of the classifier over time [17, 57, 60]. In order to handle large-scale data stream, the learning process is constrained to have a restricted number of samples in memory. This is sometimes achieved by using fixed-window-size techniques which involve learning of new samples as they arrived while reinforcing relevant information learned from the previously seen samples. Moreover, some of the incremental learning methods incorporate unlearning ability into their framework to allow the removal of *knowledge* associated with old samples that are considered no longer relevant in the classifier development process [18, 27].

The aim of this contribution is to provide a review of the methods for incremental Support Vector Machines (SVM) learning. Though some reviews have been reported for incremental and online learning in general for data streams [10, 16, 40], but to the best of our knowledge, the articles that specifically discuss the various incremental learning methods for SVM are very limited, despite the extensive use of SVM for data classification. In fact, one survey article by Zhou et al. [61] came to our attention while submitting this paper for review. The article briefly discusses a set of papers ($\approx$25 references) that adapt SVM for real-time acquisition system, by categorizing them into *Primal* and *Dual* online algorithms based on the two SVM standard formulations. Unlike [61], we provide a comprehensive review of the incremental learning SVM methods by formalizing a taxonomy of the methods based on their characteristics and the type of solution they provide. Moreover, we compare the strength and weakness of the various methods and also highlight some of their applications in real world. Our paper is organized as follows, Sect. 2 provides a brief overview of the classical SVM algorithm, while Sect. 3 discusses the incremental SVM learning methods. Section 4 summarizes the discussion and compares the methods, while Sect. 5 discusses some of the applications and Sect. 6 concludes the paper.

2   SVM for Classification

In this section, we describe the classical SVM algorithm for a binary classification problem. Let $F^t = \{(x_1^t, y_1^t), \ldots, (x_N^t, y_N^t)\}$ be a set of data samples of size N gathered at a generic instant of time t, with $x_i^t \in \mathbb{R}^d$, where d and $y_i^t \in \mathcal{Y} = \{-1, 1\}$ are the dimension and label of x_i^t, respectively. The goal is to exploit F^t, in order to learn a decision function f^t (i.e., the classifier) that will generalize well on future unseen samples. If the samples in F^t are linearly separable, then a linear SVM can be used and f^t is defined as

$$f^t(x^t) = sign\left[{w^t}^T x^t + b^t\right] \tag{1}$$

where $b^t \in \mathbb{R}$ the bias and $\boldsymbol{w}^t \in \mathbb{R}^d$ the weights of the samples in F^t at t are the parameters of f^t that define the optimal separation of $\boldsymbol{x}^t \in F^t$ into two classes. The values of b^t and $\boldsymbol{w}^t$ are obtained by solving the following soft margin SVM primal quadratic programing problem [7]:

$$\min_{\boldsymbol{w}^t, b^t, \xi^t} \quad \frac{1}{2} \left\| \boldsymbol{w}^t \right\|^2 + C \sum_{i=1}^{N} \xi_i^t \tag{2}$$

$$y_i^t \left[\left(\boldsymbol{w}^{t^T} \boldsymbol{x}_i^t \right) + b^t \right] \geq 1 - \xi_i^t \quad \forall i \in \{1, \ldots, N\}$$

$$\xi_i^t \geq 0 \quad \forall i \in \{1, \ldots, N\},$$

where ξ_i^t are slack variables for penalizing misclassification errors and C is a hyper-parameter of the SVM known as the regularization parameter [7]. The problem of Eq. (2) can be solved directly in the primal formulation [5, 33], but it is easier when it is transformed into its Wolfe dual form [50]. Therefore, the problem is reformulated into a Lagrangian problem $L()$ first, by introducing two sets of Lagrange multipliers, $\alpha^t \in \mathbb{R}^N$ and $\mu^t \in \mathbb{R}^N$, one for each of the two constraints in the equation, respectively, as

$$L(\boldsymbol{w}^t, b^t, \xi^t) = \frac{1}{2} \left\| \boldsymbol{w}^t \right\|^2 + C \sum_{i=1}^{N} \xi_i^t$$

$$+ \sum_{i=1}^{N} \alpha_i^t \left(y_i^t \left[\left(\boldsymbol{w}^{t^T} \boldsymbol{x}_i^t \right) + b^t \right] - 1 + \xi_i^t \right) + \sum_{i=1}^{N} \mu_i^t \xi_i^t, \tag{3}$$

then the derivatives of Eq. (3) with respect to $\boldsymbol{w}^t$, b^t, and ξ^t are derived in order to obtain the Karush-Kuhn-Tucker (KKT) optimality conditions for the Lagrangian problem [7]

$$\frac{\partial L}{\partial \boldsymbol{w}_i^t} = 0 \rightarrow \boldsymbol{w}_i^t = \sum_{i=1}^{N} \alpha_i^t y_i^t \boldsymbol{x}_i^t, \quad i = 1, \ldots, N \tag{4}$$

$$\frac{\partial L}{\partial b^t} = 0 \rightarrow \sum_{i=1}^{N} \alpha_i^t y_i^t = 0 \tag{5}$$

$$\frac{\partial L}{\partial \xi_i^t} = 0 \rightarrow C - \alpha_i^t - \mu_i^t = 0, \quad i = 1, \ldots, N \tag{6}$$

$$\alpha_i^t \left(y_i^t \left[\left(\boldsymbol{w}^{t^T} \boldsymbol{x}_i^t \right) + b \right] - 1 + \xi_i^t \right) = 0, \quad \forall i = 1, \ldots, N \tag{7}$$

$$\mu_i^t \xi_i^t = 0, \quad li = 1, \ldots, N \tag{8}$$

$$\left(C - \alpha_i^t \right) \xi_i^t = 0, \quad i = 1, \ldots, N \tag{9}$$

$$\alpha_i^t \mu_i^t \xi_i^t = 0, \quad i = 1, \ldots, N. \tag{10}$$

Lastly, the values of w^t, b^t, and ξ^t obtained from the above are substituted in Eq. (2) in order to derive its final Wolfe dual form as

$$\min_{\alpha^t} \quad \frac{1}{2} \sum_{i=1}^{N} \sum_{j=1}^{N} \alpha_i^t \alpha_j^t Q_{ij}^t - \sum_{i=1}^{N} \alpha_i^t \tag{11}$$

$$0 \leq \alpha_i^t \leq C \quad \forall i \in \{1, \ldots, N\}$$

$$\sum_{i=1}^{N} y_i^t \alpha_i^t = 0,$$

where $Q_{ij} = y_i^t y_j^t K(x_i^t, x_j^t)$, and $K(.,.)$ is a kernel function which allows nonlinear mappings of $x^t \in F^t$ when they are not linearly separable in the *input space* [7]. The solution of Eq. (11), that is, $\alpha_i^t \ \forall i$, is obtained by using the Quadratic Programing (QP) solvers already developed in the literature [21]. Finally, the new expression for f^t is obtained by substituting w^t with $\alpha_i^t \ \forall i$ in Eq. (1) as

$$f^t(x^t) = sign \left[\sum_{i=1}^{N} y_i^t \alpha_i^t K(x^t, x_i^t) + b^t \right]. \tag{12}$$

The decision boundaries of the classifier f^t are defined by few patterns $S^t \in F^t$ known as the *support vectors*, that is, x_i^t with corresponding $0 < \alpha_i^t \leq C$ [3]. These support vectors can be categorized into two sets: The set of margin support vectors S_m^t with $y_i^t f^t(x_i^t) = 1$ and $\alpha_i^t \in [0, C]$; the set of bounded support vectors S_b^t with $y_i^t f^t(x_i^t) < 1$ and $\alpha_i^t = C$.

3   Incremental SVM Learning

At time $t + 1$, when a new set of samples $F^{t+1} = \left\{ (x_1^{t+1}, y_1^{t+1}), \ldots \right\}$ are gathered, f^t can be updated to f^{t+1} using incremental learning without having to recompute the classifier from the scratch (see Fig. 1). Depending on the application, the new samples can occur instantly in time, such as in time series data or in batches after long time intervals, such as research data and Web log records. Different SVM incremental learning methods have been reported in the literature to accommodate the ways in which the data streams are generated [54, 62]. These methods can be categorized into two groups: online and semi-online. The online methods process the data stream one sample at a time and ensure that the KKT conditions are

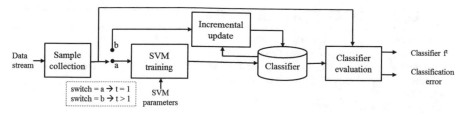

Fig. 1 Illustration of the incremental SVM learning procedure

maintained on all previously seen samples while updating the classifier [4]. The semi-online methods, on the other hand, process the samples in batches and while updating the classifier, they discard previously seen samples except those identified as support vectors [48]. In the following sections, we identify the pioneering work in both groups and explain the algorithm behind them. We also highlight their advantages and drawbacks.

3.1   Online Incremental SVM Learning Methods

Cauwenberghs and Poggio [4] proposed the first online recursive algorithm for incremental SVM learning. The method was later adapted to other variants of kernel machines [12, 15, 33]. When a new stream of data (x_1^{t+1}, y_1^{t+1}) is received, the algorithm computes the value of $y_1^{t+1} f^t(x_1^{t+1})$ in order to check whether the new sample has the potential to improve the classifier. If $y_1^{t+1} f^t(x_1^{t+1}) \leq 1$, it means the samples are useful. Thus, the algorithm initializes a coefficient α_1^{t+1} to 0 for the new sample and then perturbs the SVM by gradually increasing the value of the coefficient until the optimal SVM solution is obtained [4]. During the perturbation of the SVM, the other $\alpha_i^t \ \forall i \in S_m^t$ and S_b^t are also adjusted in order to maintain the KKT optimality conditions for all previously seen samples. Algorithm 1 describes the main steps of the Cawenbergs and Poggio's algorithm. The incremental SVM learning is also reversible whereby the patterns in the SVM solution can be unlearned one after another. To unlearn a pattern, the coefficient of the corresponding sample is gradually decremented to zero, while coefficients of the other samples are readjusted in order to maintain the KKT conditions [4]. In a later implementation of the Cawenbergh and Poggio's algorithm [9], the hyper-parameters (i.e., regularization and kernel parameters) of the SVM can be updated as well during the incremental learning process. Updating the hyper-parameters is very important because it guarantees the effectiveness of the classifier by enabling the selection of the best decision boundaries in cases where the underlying distribution of the data stream changes over time [1]. The Matlab code for this incremental

---

Algorithm 1 The Cawenbergs and Poggio's algorithm [4]

Input: (x_1^{t+1}, y_1^{t+1}): new sample at time $t+1$, f^t: decision function at t
Output: f^{t+1} : updated decision function at $t+1$
Definitions: α_i^t: coefficient of the ith sample at t, S_m^t: margin support vector set at t, S_b^t: bounded support vector set at t, Q: kernel matrix

1: *Begin:*
2: Compute $z = y_1^{t+1} f^t(x_1^{t+1})$,
3: If $z > 1$, then, $f^{t+1} \leftarrow f^t$, and go to step 10
4: Else,
5: Initilaize $\alpha_1^{t+1} \leftarrow 0$, for x_1^{t+1}
6: Compute Q_{i1} for $\forall\, x_i^t \in S_m^t$
7: Increment α_1^{t+1} to its largest value while maintaining the KKT optimality conditions on all previously seen samples
8: Check if one of the following conditions occurs:
   I. If $y_1^{t+1} f^t(x_1^{t+1}) = 1$, then $S_m^t \leftarrow x_1^{t+1}$
   II. Else if $\alpha_1^{t+1} = C$, then $S_b^t \leftarrow x_1^{t+1}$
   III. Else if any $x_i^t \in S_b^t$ become part of S_m^t, due to the change in their corresponding α_i^t, then update S_m^t and S_b^t accordingly
9: $f^{t+1} \leftarrow f^t$, $S_m^{t+1} \leftarrow S_m^t$ and $S_b^{t+1} \leftarrow S_b^t$
10: *End*

SVM learning is available online.[1] The Cawenberghs and Poggio's algorithm, however, suffers from the following limitations. First, it is designed for 2-class problems and cannot be applied directly to a multi-class data stream. It is possible to employ conventional methods such as One-Vs-One and One-Vs-All [45] to allow the algorithm cope with the multi-class problem, but this will significantly increase the computational cost of the learning process [14]. Second, being a supervised learning method, the algorithm requires the labeling of all the input training samples. However, getting enough labeled data is expensive especially in large-scale data stream applications [51].Thus, the algorithm is not directly applicable to data stream problems with purely unlabeled data or a mixture of labeled and unlabeled data.

To address the first problem, Boukharouba et al. [2] reformulated the problem of Eq. (2) as follows:

$$\min_{w^t, b^t, \xi^t} \quad \frac{1}{2}\sum_{i=1}^{K}\sum_{j=i+1}^{K}\left\|w_i^t - w_j^t\right\|^2 + \frac{1}{2}\sum_{i=1}^{K}\left\|w_i^t\right\|^2 + C\sum_{i=1}^{N}\sum_{j=i+1}^{K}\sum_{x_k^t \in C_{ij}^t}\xi_k^t \quad (13)$$

$$\forall x_k^t \in C_{ij}^t$$

$$y_k^t\left[\left(\left(w_i^t - w_j^t\right)^T x_k^t\right) + \left(b_i^t - b_j^t\right)\right] \geq 1 - \xi_k^t,$$

$$\xi_k^t \geq 0 \quad \forall i \in \{1, \ldots, K\}, \ \forall j \in \{i+1, \ldots, K\},$$

---

[1] https://github.com/diehl/Incremental-SVM-Learning-in-MATLAB.

where $K (K > 2)$ is the number of classes in the data stream, with i and j corresponding to the index of any two different classes. $C_{ij}^t = (C_i^t \cup C_j^t)$ and $y_k^t = 1$ if $x_k^t \in C_i^t$, and $y_k^t = -1$, if $x_k^t \in C_j^t$. Then proceeded to derive the solution of Eq. (13), after which the same learning procedure described in Algorithm 1 is applied in order to allow the method to simultaneously discriminate between K number of classes in the data stream.[2]

The second problem of the Cawenbergh and Poggio's algorithm is addressed by Chen et al. [6]. The method allows the incremental learning from partially labeled samples by using the principle of transduction [25, 55]. The method starts by using only the labeled samples gathered in the past to develop the initial decision function f^t and to determine the margins of the separating hyperplane of the SVM. At $t > 1$ when a new unlabeled sample x_1^{t+1} is received, it is classified and assigned a label using the current f^t. Also, its distance from the margins of the separating hyperplane of the current SVM is estimated. If the distance falls outside the margin of the SVM, then the sample contains no new information that will improve the decision boundaries of the SVM, thus the f^t remains unchanged and the sample is discarded. On the other hand, if the distance falls within the margins of SVM, then the sample contains useful information that should be incorporated in the SVM model. Thus, f^t is adjusted accordingly by updating the parameter of the SVM (as described in Algorithm 1) while maintaining the KKT condition of all the previously seen samples in the SVM. The adjustment of the decision boundaries can cause the labels of some of the previously labeled samples to change. In that case, all the affected samples are removed from the current SVM solution by decremental learning [4] and declared as unlabeled samples. These unlabeled samples are then subjected to an error correction process whereby they are relabeled, and used to retrain the SVM. Algorithm 2 summarizes the steps involved in this method.

Another online SVM learning method is that proposed by Rai et al. [43]. The method uses the concept of Minimum Enclosing Ball (MEB) to learn and/or update the SVM. It starts by initializing the learning process using the first sample (x_1^t, y_1^t) generated at $t = 1$ as the center c_i^t of the first MEB with radius $r_i^t = 0$. When a new sample (x_1^{t+1}, y_1^{t+1}) is received at $t + 1$, the algorithm checks whether the current MEB can enclose the new sample. If so, the sample is discarded because it does not contain new information that will improve the SVM. Otherwise, the sample is considered as a potential support vector for the SVM and it is placed in a set known as the *core set* S^t. Then create a new MEB with c_i^{t+1} and r_i^{t+1} which can cope with the new changes introduced by the addition of the new sample. The c_i^{t+1} is shown to lie on the straight line joining c_i^t and x_1^{t+1} [43]. Thus, the shortest distance δ^t between c_i^t and c_i^{t+1} is computed as $||c_i^{t+1} - c_i^t|| = \delta^t$, where $\delta^t = \frac{1}{2}(||x_1^{t+1} - c_i^t|| - r_i^t)$. The radius r_i^{t+1} of the new MEB is then obtained as [43]

---

[2]For brevity and clarity, the derivation of the solution of Eq. (13) is not shown in this paper, the interested reader is referred to [2] for a detail explanation.

Algorithm 2 The Chen's SVM algorithm [6]

Input: $F^{t+1} = (x_1^{t+1})$: new unlabeled sample at time $t + 1$, f^t: decision function at t, $\lambda^t = \{(x_1^t, y_1^t), \ldots, (x_k^t, y_k^t)\}$: set of k previously labeled samples
Output: f^{t+1} : updated decision function
1: *Begin:*
2: Classify x_1^{t+1} and assign a label to it, $y_1^{t+1} = \text{sign}(f^t(x_1^{t+1}))$
3: Check *if* $|f^t(x_1^{t+1})| < 1$, *then*
4: Update f^{t+1} with (x_1^{t+1}, y_1^{t+1}) using incremental learning (see Algorithm 1)
5: Compute $y*_i^t = \text{sign}(f^{t+1}(x_i^t)), \forall x_i^t \in \lambda^t$
6: Check *if* $y_i^t \neq y*_i^t \forall y_i^t \in \lambda^t$, *then*
7: Unlearn the defaulting (x_i^t, y_i^t) from f^{t+1} using decremental learning [4], and mark then as unlabeled samples
8: *end if*
9: *else,* $\lambda^{t+1} = \left\{ \lambda^t \cup (x_1^{t+1}, y_1^{t+1}) \right\}$
10: *end if*
11: Return f^{t+1}, λ^{t+1}
12: *End*

$$r_i^{t+1} = r_i^t + \delta^t, . \tag{14}$$

The set of MEBs created over time is represented as L. As more samples are received and can be enclosed by any of the MEBs, some of the MEBs are merged together. The algorithm is implemented using the L_2-SVM formulation and Algorithm 3 describes the steps of the algorithm. The difference between L_2-SVM formulation and that shown in Eq. (2) is that L_2-SVM uses $C \sum_{i=1}^{N} \xi_i^{2^t}$ instead of $C \sum_{i=1}^{N} \xi_i^t$ as it is a penalty term. As shown in Algorithm 3, the steps that update the weight vector w^t and margin R^t of the SVM correspond to the update of the MEB center and radius, respectively, as discussed above. The advantage of this algorithm is that its space and time complexity is relatively small. This is because only w^t and R^t need to be stored and/or updated over time. Moreover, the number of updates required is limited by the size of the S and L which are generally small [43].

3.2   Semi Online Incremental SVM Learning Methods

The pioneer work in this group of incremental SVM learning is the one presented by Syed et al. [48]. The algorithm processes the stream of data in batches of fixed sample size. The first batch of data $F^t = \{(x_1^t, y_1^t), \ldots, (x_N^t, y_N^t)\}$ gathered at time $t = 1$ is used to learn the f^t as described in Sect. 2, then the algorithm discards all the previously seen samples except those of the margin support vectors set S_m^t. At time $t + 1$, when a new batch of data $F^{t+1} = \left\{(x_1^{t+1}, y_1^{t+1}), \ldots, (x_N^{t+1}, y_N^{t+1})\right\}$ is gathered, the algorithm combined the new data with the previously stored support vectors in order to create a new set of training data $\tilde{F}^{t+1} = \left\{F^{t+1} \cup S_m^t \cup S_b^t\right\}$.

Algorithm 3 The Rai's SVM algorithm [43]

Input: $F^{t+1} = (x_1^{t+1}, y_1^{t+1}),$: new sample at time $t + 1$, w^t: weight vector at t, R^t: margin of SVM at t, C: regularization parameter, ξ^t: slack variable at t, S^t: size of the support vector set at t

Output: f^{t+1}: updated decision function, R^{t+1}: updated SVM margin at $t + 1$, S^{t+1}: updated size of the support vector set, ξ^{t+1}: slack variable at $t + 1$

1: *Begin:*
2: Compute the distance of the new sample x_1^{t+1} to the center of the current MEB $\delta = \sqrt{||w^t - y_1^{t+1}x_1^{t+1}||^2 + \xi^{2^t} + \frac{1}{C}}$
3: Check *if* $\delta \geq R^t$, *then*
4: Compute $w^{t+1} = w^t + \frac{1}{2}(1 - \frac{R^t}{\delta})(y_1^{t+1}x_1^{t+1} - w^t)$
5: Compute $R^{t+1} = R^t + \frac{1}{2}(\delta - R^t)$
6: Compute $\xi^{2^{t+1}} = \xi^{2^t}[1 - \frac{1}{2}(1 - \frac{R^{t+1}}{\delta})]^2 + [\frac{1}{2}(1 - \frac{R^{t+1}}{\delta})]^2$
7: Compute $S^{t+1} = S^t + 1$
8: Compute f^{t+1} using w^{t+1} as shown in Eq. (1)
9: *end if*
10: Return f^{t+1}, R^{t+1}, ξ^{t+1} and S^{t+1}
11: *End*

The method then uses the samples in $\tilde{F}^{t+1}$ to retrain the SVM as described in Sect. 2. This process is repeated every time new stream of data is collected (see Algorithm 4). Although the method is simple and efficient because only a small amount of data are processed at each time of the incremental learning, however, it suffers from three major issues. First, the algorithm needs to store all the support vectors that were previously learned, which is not sustainable for lifelong learning. This is because the number of support vectors will grow over time, and this will increase the space and time requirement of the learning process. Second, the method also assumes that the properties of the batches of data do not vary over time; thus, the decision boundaries of the classifier learnt using the first batch of data would remain the same all the time. This is however not always true, because, in practice, the distribution of data can change and the decision boundaries of the classifier need to be adjusted accordingly in order to cope with the changes [46]. Third, the method discards all input samples which are not support vectors during incremental learning. However, the deleted samples may become support vectors later, especially when new classes or new distributions data arrived. Thus, if the discarded samples do not appear again in the new set of training data, then the current decision boundaries of the SVM will become obsolete and the general performance of the classifier will decline over time [22].

One way to address the first issue of the Syed's algorithm is to analyze the support vectors obtained during each phase of the incremental learning process in order to check whether they are linearly dependent in the feature space [41, 44]. If they are, then it is better to preserve only the support vectors which cannot be expressed as a function of the others in the feature space. This will reduce the number of support vectors that need to be stored [41]. Moreover, the second issue can be

---

Algorithm 4 The Syed's algorithm [48]

Input: $F^{t+1} = \left\{ (x_1^{t+1}, y_1^{t+1}), \ldots, (x_N^{t+1}, y_N^{t+1}) \right\}$: new samples at time $t + 1$, S_m^t: margin support vector set at t, S_b^t: bounded support vector set at t, f^t: decision function at t

Output: f^{t+1} : updated decision function, S_m^{t+1}: margin support vector set at $t + 1$, S_b^{t+1}: bounded support vector set at $t + 1$

Definitions: α_i^{t+1}: coefficient of the ith sample at $t + 1$

1: *Begin:*
2: Compute $\tilde{F}^{t+1} = \left\{ F^{t+1} \cup S_m^t \cup S_b^t \right\}$,
3: Using $\tilde{F}^{t+1}$ as input, solve for the problem of Eq. (11) to obtain α_i^{t+1} $\forall i$
4: Compute f^{t+1} by solving the problem of Eq. (12)
5: Compute $S_m^{t+1} = \left\{ S_m^t \cup x_i^{t+1} : y_i^{t+1} f^{t+1}(x_i^{t+1}) = 1 \quad \& \quad \alpha_i^{t+1} > 0 \right\}$
6: Compute $S_b^{t+1} = \left\{ S_b^t \cup x_i^{t+1} : y_i^{t+1} f^{t+1}(x_i^{t+1}) < 1 \quad \& \quad \alpha_i^{t+1} = C \right\}$
7: Discard all x_i^{t+1} for which $\alpha_i^{t+1} = 0$
8: Return f^{t+1}, S_m^{t+1} and S_b^{t+1}
9: *End*

---

addressed by incorporating an optimization technique such as Dynamic Particle Swarm Optimization (DPSO) [26] to allow the incremental selection of the best SVM hyper-parameters when a new batch of data is being learned. This will ensure that the classifier adapts to change in the distribution of the data stream over time. Finally, the third issue can be addressed, for example, by retaining the previously seen samples and then using a Least Recently Used (LRU) scheme [53] to discard the oldest samples that have not been used after a given period of time. This will minimize the risk of deleting potential support vectors among the previously seen data samples.

Katagiri and Abe [28] presented a method for solving the problem identified above. They showed that training samples that are likely to become support vectors in the incremental learning future steps would typically be lying near the separating hyperplane of SVM solution. Thus, they created a minimum volume hypersphere for each class of the training data, and then a smaller concentric hyper-cone with a vertex at the center of the hypersphere. After each incremental learning step, the algorithm deletes only the samples that are trapped inside the smaller hyper-cone, while that of those that exist near the boundary of the hypersphere are retained as potential candidates for support vectors in the next step of the incremental learning.

Domeniconi et al. [11] extends the Syed's algorithm in order to cope with situations where the characteristics of the data change over time, such that SVM built with old samples are no longer suitable for predicting the class of future samples. The method processes the stream of data in batches of fixed sample size N, and at any time t it retains the decision functions, that is, $f_1^t, \ldots, f_k^t$ built from the last $1, \ldots, k$ batches of data, respectively, in memory. f_k^t is the oldest decision function and f_1^t is the most recent decision function which is representative of the current distribution of the data stream in memory. At time $t + 1$ when new samples are received, the algorithm discards f_k^t and updates the remaining k-1 decision functions (i.e., $f_2^t, \ldots, f_k^t$) incrementally using the new batch of data. Meanwhile, only f_1^t, is built from scratch using the received batch of data. The memory overhead of this method is limited because at any point in time it needs to store only k decision functions and N samples in memory.

The idea of using prototype (compress data) for incremental SVM learning in order to be able to handle large-scale data stream applications effectively has been studied in the literature [52, 60]. The prototypes are generated according to the properties of the data stream and are used to learn the support vectors of the SVM. Because the size of the prototypes set is much smaller compared to the original data, the incremental learning can be completed very fast. The work of Xing et al. [56] is an example of this approach. The method incorporates a Learning Prototype Network (LPN) which consists of an interconnection of representative prototypes arranged in layers known as the Prototype Support Layer (PLS). The LPN learns the prototypes P_i of each class in the data stream, then uses them to train SVM incrementally. Each P_i is associated with a 3-tuple: $\left\{ W_{P_i} T_{P_i}, M_{P_i} \right\}$ which represent the weight vector, the similarity threshold, and the accumulated number of input data for each prototype, respectively. For a binary classification problem, two sets of prototypes P_- and P_+ are created one for each of the two classes, using the first sample from each of the class (x_1^t, y_1^t), (x_2^t, y_2^t) received at $t = 1$. The prototypes are obtained as

$$P_i^t = \left\{ W_{p_i}^t = x_i, T_{p_i}^t = ||x_1^t - x_2^t||, M_{p_i}^t = 1 \right\}, i = 1, 2. \qquad (15)$$

Algorithm 5 The Xing's algorithm [56]

Input: $F^{t+1} = \left\{ (x_1^{t+1}, y_1^{t+1}), \ldots, (x_N^{t+1}, y_N^{t+1}) \right\}$: new samples at $t + 1$, P_+^t: set of prototype for positive class samples at t, P_-^t: set of prototype for negative class samples at t,
Output: f^{t+1}: updated classifier at $t + 1$
Definitions: α_i^{t+1}: coefficient of the ith sample at $t + 1$
1: *Begin:*
2: *For* i=1:N,
3: *if* $y_i^{t+1} = 1$,
4: Update $P_+^t \rightarrow P_+^{t+1}$ using the LPN procedure
5: *else,*
6: Update $P_-^t \rightarrow P_-^{t+1}$ using the LPN procedure
7: *end if*
8: *end For*
9: Compute $P^{t+1} = \left\{ P_+^{t+1} \cup P_-^{t+1} \right\}$,
10: Using P^{t+1} as input, solve for the problem of Eq. (11) to obtain α_i^{t+1} $\forall i$
11: Compute f^{t+1} by solving the problem of Eq. (12)
12: Return f^{t+1}, P_+^{t+1} and P_-^{t+1}
13: *End*

At $t + 1$ when a new sample(x_i^{t+1}, y_i^{t+1}) is received, a similarity factor $z = ||x_i^{t+1} - W_{p_i}^t||$ is computed using the new sample and the weight vector $W_{p_i}^t$ of all the P_i^t in the current set of prototypes. If the value of z is greater than the similarity threshold $T_{p_i}^t$, then a new prototype is created as

$$P_i^{t+1} = \left\{ W_{p_i}^t = x_i^{t+1}, T_{p_i}^t = +\infty, M_{p_i}^t = 1 \right\}. \tag{16}$$

Otherwise, the current set of prototypes is sufficient to cover the information contained in the new sample and the elements of current P_i^t are updated as follows:

$$M_{p_i}^{t+1} = M_{p_i}^t + 1 \tag{17}$$

$$W_{p_i}^{t+1} = W_{p_i}^t + \frac{1}{M_{p_i}^{t+1}} \left(x_i^{t+1} - M_{p_i}^t \right). \tag{18}$$

Then the algorithm updates the connections between the set of prototypes in the LPN using the *Hebbian learning rule* [30] as well. Over time, the concept of connection obsolescence is considered in the LPN, that is, the connections whose *age* is larger than a predefined threshold are removed. Once update of the prototypes is completed, the algorithm proceeds with the SVM training. Algorithm 5 summarizes the procedure of the proposed method.

4 Discussion and Comparison

Table 1 compares the property of some of the incremental SVM learning methods discussed in the previous sections. The methods discussed in Sect. 3.1 (e.g., [2, 4, 6, 9, 43]) modify the SVM optimization procedure to allow learning one sample at a time. The advantages of these methods include providing an exact SVM solution by maintaining the KKT optimality condition for all the previously seen samples in the data streams, and also providing an efficient means of unlearning of non-informative samples from the SVM solution. This particular feature is good for systems that require long life learning. However, the preservation and use of all previously seen

Table 1 Comparison of the main properties of some of the incremental SVM learning methods

Properties	[48]	[11]	[41]	[28]	[56]	[4]	[9]	[2]	[6]	[43]	[27]
Allows learning a sample at a time						✓	✓	✓	✓	✓	✓
Handles concept drift problem		✓	✓		✓		✓			✓	
Allows learning with unlabeled data									✓		
Handles multi-class problem					✓			✓			
Suitable for large-scale data	✓	✓	✓	✓	✓	✓				✓	
Suitable for longlife learning			✓	✓	✓	✓	✓	✓	✓	✓	✓

samples increase the computational and memory overhead of the methods over time
[31]. This overhead can be reduced by discarding some previously seen samples and
keeping only the support vectors in memory at the end of each incremental learning
process. But then this raises the question of how to choose which samples or support
vectors to keep or discard during the learning process. The methods discussed in
Sect. 3.2 (e.g., [11, 28, 41, 48, 54, 56]), on the other hand, use the conventional
quadratic programming solvers to solve the SVM learning problem in batch mode
and only keep the support vectors in memory. These methods are efficient and
suitable for large-scale data stream problems. But the indiscriminate removal of
nonsupport vectors samples at the end of each incremental learning phase can affect
the performance of the classifier over time. This is because some of the discarded
samples can become support vectors in the future. A summary of the peculiarity
of each group of the SVM incremental learning previously discussed in this paper
is presented in Table 2. It can be seen that each group of the learning methods
has its own advantage and drawback in terms of generalization power, memory, or
computational cost. The choice of which algorithm to use in any application strongly
depends on the requirements of the application.

5 Applications of Incremental SVM Learning

Incremental SVM learning algorithms have been used in many data stream applications [32, 38, 42, 58]. In this section, we identify four application areas and briefly
discuss them:

Document Stream Classification In this application, data appear as the flow of
documents, and the information and content included in the documents can evolve
over time [20]. There are two major addressable issues in this type of application: (1)
huge amount of data is involved and (2) the data evolve over time. Incremental SVM
learning had been used to realize an adaptive classifier for this kind of application
in the past [20, 37, 38]. For example, Ngo Ho et al. [38] applied incremental SVM
for the classification of streams of scanned documents according to their contents,
in order to facilitate the efficient and adaptive digitization process of the documents.
Similarly, Naqa et al. [37] applied SVM incremental learning for the purpose of
supporting medical diagnoses through a computationally efficient mammogram
image retrieval system.

Scene Characterization In this application, video data are generated as huge
streams of features representing the characteristic of objects in a scene under
surveillance [32]. The properties of this stream of features can change over time due
to the dynamic nature of the scene. For example, the appearance of an object can
change due to the illumination change in the scene or change in object orientation
when they move across the scene, etc. Incremental SVM learning has been used to
successfully classify scenes [34], objects [29], people [35], or flow of moving people
[32] in a highly dynamic video scenes for intelligent video surveillance system.

Table 2 Summary of the advantage and disadvantage of the SVM learning methods discussed in Sect. 3

Methods	Mode of training	Space complexity	Time complexity	Advantage	Disadvantage
Non-incremental SVM in **section 2.0** (e.g., [7])	Batch with the entire training set	**	**	Creates a robust SVM model because of its global view of the entire data space	(1) Model update is computationally expensive. (2) Not suitable for online applications. (3) Not suitable for large-scale data stream application
Incremental SVM in **section 3.1** (e.g., [4, 9])	Online with one sample at a time	$O(S^2)$	Quadratic in the number of samples learned	(1) Creates robust model. (2) Handles problem of concept drift. (3) Provides exact SVM solution. (4) Good for lifelong learning	(1) Requires more memory space to store the previously seen samples. (2) Not suitable for large-scale data stream application
Incremental SVM in **section 3.2** (e.g., [41, 48])	Batch with subset of the training set	**	**	(1) Less memory requirements. (2) Less computational cost. (3) Good for large-scale data stream	(1) Provides a framework for approximate SVM solution. (2) Not suitable for situations where the distribution of the data changes frequently.

S is the size of the support vector set

** The value depends on the choice of conventional QP solver used

Fault Diagnosis In this type of application, the acquisition of diagnostic data from sensors occurs continuously over time and with possible strong concept drift presence. Incremental SVM learning has been employed to build an adaptive classifier for automatic diagnosis of fault in systems equipped with sensors. For example, fault detection and isolation of Heating, Ventilation, and Air Conditioning (HVAC) system [8] and Diesel engine fault diagnosis [58].

Robotics and Autonomous System Autonomous systems are systems that continuously updated itself with the information gathered from its environment, in order to able to perform a given task with a high degree of autonomy. The updating process is open-ended which requires an adaptive and efficient computational model that is inherently incremental. Incremental SVM learning has been used in this type of application as well, which allows the system adaptation to be performed incrementally and also facilitates achieving a controlled-memory growth, reduced computational complexity, and better system performance [36, 42].

6 Conclusion

The continuous growth of real-world applications involving massive and evolving data streams has inevitably made incremental learning an important research issue. Different proposals have been made to efficiently address the scalability and adaptability of the learning process. In this paper, we have surveyed various incremental learning methods for Support Vector Machines (SVMs). A comparison of the methods in terms of their strengths and weaknesses is summarized in a tabular form to ease readability. Moreover, we identified and discussed some areas where incremental SVM learning has been successfully applied. Though each work reviewed in this paper presents an improvement either in terms of memory requirements, computational complexity or both in the learning process, however, the incremental SVM learning problem is still very much open to further research. Future research could be directed toward incorporating an online model selection method (which is, by the way, a very important part of the SVM model development process) in the incremental learning framework in order to guarantee the effectiveness of the SVM for data stream applications.

References

1. Anguita, D., Ghio, A., Lawal, I.A., Oneto, L.: A heuristic approach to model selection for online support vector machines. In: Proceedings of the International Workshop on Advances in Regularization, Optimization, Kernel Methods and Support Vector Machines: Theory and Application, pp. 77–78 (2013)
2. Boukharouba, K., Bako, L., Lecoeuche, S.: Incremental and decremental multi-category classification by support vector machines. In: Proceedings of the International Conference on Machine Learning and Applications, pp. 294–300 (2009)

3. Burges, C.J.C.: A tutorial on support vector machines for pattern recognition. Data Min. Knowl. Disc. **2**(2), 121–167 (1998)
4. Cauwenberghs, G., Poggio, T.: Incremental and decremental support vector machine learning. In: Proceedings of the International Conference on Advances in Neural Information Processing Systems, pp. 409–415 (2000)
5. Chapelle, O.: Training a support vector machine in the primal. Neural Comput. **19**(5), 1155–1178 (2007)
6. Chen, M.S., Ho, T.Y., Huang, D.Y.: Online transductive support vector machines for classification. In: Proceedings of the International Conference on Information Security and Intelligent Control, pp. 258–261 (2012)
7. Cortes, C., Vapnik, V.: Supportvector networks. Mach. Learn. **20**, 273–297 (1995)
8. Dehestani, D., Eftekhari, F., Guo, Y., Ling, S., Su, S., Nguyen, H.: Online support vector machine application for model based fault detection and isolation of HVAC system. Int. J. Mach. Learn. Comput. **1**(1), 66–72 (2011)
9. Diehl, C., Cauwenberghs, G.: SVM incremental learning, adaptation and optimization. In: Proceedings of the International Joint Conference on Neural Networks, pp. 2685–2690 (2003)
10. Diethe, T., Girolami, M.: Online learning with multiple kernels: a review. Neural Comput. **25**(3), 567–625 (2013)
11. Domeniconi, C., Gunopulos, D.: Incremental support vector machine construction. In: Proceedings of the IEEE International Conference on Data Mining, pp. 589–592 (2001)
12. Duan, H., Shao, X., Hou, W., He, G., Zeng, Q.: An incremental learning algorithm for lagrangian support vector machines. Pattern Recogn. Lett. **30**(15), 1384–1391 (2009)
13. Fan, H., Song, Q., Yang, X., Xu, Z.: Kernel online learning algorithm with state feedbacks. Knowl.-Based Syst. **89**, 173–180 (2015)
14. Galmeanu, H., Andonie, R.: A multi-class incremental and decremental svm approach using adaptive directed acyclic graphs. In: Proceedings of the International Conference on Adaptive and Intelligent Systems, pp. 114–119 (2009)
15. Gâlmeanu, H., Sasu, L.M., Andonie, R.: Incremental and decremental svm for regression. Int. J. Comput. Commun. Control **11**(6), 755–775 (2016)
16. Gama, J.: A survey on learning from data streams: current and future trends. Prog. Artif. Intell. **1**(1), 45–55 (2012)
17. Gepperth, A., Hammer, B.: Incremental learning algorithms and applications. In: Proceedings of the European Sympoisum on Artificial Neural Networks, pp. 357–368 (2016)
18. Guo, J.: An improved incremental training approach for large scaled dataset based on support vector machine. In: Proceedings of the International Conference on Big Data Computing Applications and Technologies, pp. 149–157 (2016)
19. He, H., Chen, S., Li, K., Xu, X.: Incremental learning from stream data. IEEE Trans. Neural Netw. **22**(12), 1901–1914 (2011)
20. Ho, A.K.N., Ragot, N., Ramel, J.Y., Eglin, V., Sidere, N.: Document classification in a non-stationary environment: a one-class svm approach. In: Proceedings of the International Conference on Document Analysis and Recognition, pp. 616–620 (2013)
21. Hsieh, C.J., Si, S., Dhillon, I.: A divide-and-conquer solver for kernel support vector machines. In: Proceedings of the International Conference on Machine Learning, pp. 566–574 (2014)
22. Ikeda, K., Yamasaki, T.: Incremental support vector machines and their geometrical analyses. Neurocomputing **70**(13–15), 2528–2533 (2007)
23. JinHyuk, H., Sung-Bue, C.: Incremental support vector machine for unlabeled data classification. In: Proceedings of the International Conference on Neural Information Processing, pp. 1403–1407 (2002)
24. Joachims, T.: Making large-scale support vector machine learning practical. In: Schölkopf, B., Burges, C.J.C., Smola, A.J. (eds.) Advances in Kernel Methods, pp. 169–184. MIT Press, Cambridge, MA (1999)
25. Joachims, T.: Transductive inference for text classification using support vector machines. In: International Conference on Machine Learning, pp. 200–209 (1999)

26. Kapp, M.N., Sabourin, R., Maupin, P.: Adaptive incremental learning with an ensemble of support vector machines. In: Proceedings of the International Conference on Pattern Recognition, pp. 4048–4051 (2010)
27. Karasuyama, M., Takeuchi, I.: Multiple incremental decremental learning of support vector machines. IEEE Trans. Neural Netw. **21**(7), 1048–1059 (2010)
28. Katagiri, S., Abe, S.: Incremental training of support vector machines using hyperspheres. Pattern Recogn. Lett. **27**(13), 1495–1507 (2006)
29. Kembhavi, A., Siddiquie, B., Miezianko, R., McCloskey, S., Davis, L.S.: Incremental multiple kernel learning for object recognition. In: Proceedings of the International Conference on Computer Vision, pp. 638–645 (2009)
30. Kohonen, T.: Self-organized formation of topologically correct feature maps. In: Anderson, J.A., Rosenfeld, E. (eds.) Neurocomputing: Foundations of Research, pp. 509–521. MIT Press, Cambridge, MA (1988)
31. Laskov, P., Gehl, C., Kruger, S., Muller, K.R.: Incremental support vector learning: analysis, implementation and applications. J. Mach. Learn. Res. **7**, 1909–1936 (2006)
32. Lawal, I.A., Poiesi, F., Anguita, D., Cavallaro, A.: Support vector motion clustering. IEEE Trans. Circuits Syst. Video Technol. **27**(11), 2395–2408 (2017)
33. Liang, Z., Li, Y.: Incremental support vector machine learning in the primal and applications. Neurocomputing **72**, 2249–2258 (2009)
34. Lin, H., Deng, J.D., Woodford, B.J.: Anomaly detection in crowd scenes via online adaptive one-class support vector machines. In: Proceedings of the International Conference on Image Processing, pp. 2434–2438 (2015)
35. Lu, Y., Boukharouba, K., Boonært, J., Fleury, A., Lecuche, S.: Application of an incremental svm algorithm for on-line human recognition from video surveillance using texture and color features. Neurocomputing **126**, 132–140 (2014)
36. Luo, J., Pronobis, A., Caputo, B., Jensfelt, P.: Incremental learning for place recognition in dynamic environments. In: Proceedings of the IEEE International Conference on Intelligent Robots and Systems, pp. 721–728 (2007)
37. Naqa, I., Yang, Y., Galatsanos, N., Wernick, M.: Relevance feedback based on incremental learning for mammogram retrieval. In: Proceedings of the International Conference on Image Processing, pp. 729–732 (2003)
38. Ngo Ho, A.K., Eglin, V., Ragot, N., Ramel, J.Y.: Multi one-class incremental svm for document stream digitization. In: Proceedings of the International Workshop on Document Analysis Systems, pp. 5–6 (2016)
39. Nguyen, H.M., Cooper, E.W., Kamei, K.: Online learning from imbalanced data streams. In: Proceedings of the International Conference of Soft Computing and Pattern Recognition, pp. 347–352 (2011)
40. Nguyen, H.L., Woon, Y.K., Ng, W.K.: A survey on data stream clustering and classification. Knowl. Inf. Syst. **45**(3), 535–569 (2015)
41. Orabona, F., Castellini, C., Caputo, B., Jie, L., Sandini, G.: On-line independent support vector machines. Pattern Recogn. **43**(4), 1402–1412 (2010)
42. Pronobis, A., Jie, L., Caputo, B.: The more you learn, the less you store: memory-controlled incremental SVM for visual place recognition. Image Vis. Comput. **28**(7), 1080–1097 (2010)
43. Rai, P., Daumé, H., Venkatasubramanian, S.: Streamed learning: one-pass svms. In: Proceedings of the International Jont Conference on Artifical Intelligence, pp. 1211–1216 (2009)
44. Ralaivola, L., d'Alché Buc, F.: Incremental support vector machine learning: a local approach. In: Proceedings of the International Conference on Artificial Neural Networks, pp. 322–329. Springer, Berlin (2001)
45. Rifkin, R., Klautau, A.: In defense of one-vs-all classification. J. Mach. Learn. Res. **5**, 101–141 (2004)
46. Ruping, S.: Incremental learning with support vector machines. In: Proceedings of the International Conference on Data Mining, pp. 641–642 (2001)
47. Sayed-Mouchaweh, M.: Learning from Data Streams in Dynamic Environments. Springer International Publishing, Berlin (2016)

48. Syed, N.A., Huan, S., Kah, L., Sung, K.: Incremental learning with support vector machines. In: Proceedings of the 16th International Joint Conference on Artificial Intelligence, pp. 161–168 (1999)
49. Tsai, C.H., Lin, C.Y., Lin, C.J.: Incremental and decremental training for linear classification. In: Proceedings of the International Conference on Knowledge Discovery and Data Mining, pp. 343–352 (2014)
50. Vapnik, V.N.: The Nature Of Statistical Learning Theory. Springer, New York (1995)
51. Wang, J., Yang, D., Jiang, W., Zhou, J.: Semisupervised incremental support vector machine learning based on neighborhood kernel estimation. IEEE Trans. Syst. Man Cybern. Syst. **PP**(99), 1–11 (2017)
52. Wu, C., Wang, X., Bai, D., Zhang, H.: Fast SVM incremental learning based on the convex hulls algorithm. In: Proceedings of the International Conference on Computational Intelligence and Security, vol. 1, pp. 249–252 (2008)
53. Xiao, R., Wang, J., Zhang, F.: An approach to incremental SVM learning algorithm. In: Proceedings of the International Conference on Tools with Artificial Intelligence, pp. 268–273 (2000)
54. Xie, W., Uhlmann, S., Kiranyaz, S., Gabbouj, M.: Incremental learning with support vector data description. In: Proceedings of the International Conference on Pattern Recognition, pp. 3904–3909 (2014)
55. Xihuang, Z., Wenbo, X.: The implementation of online transductive support vector machine. In: Li, D., Wang, B. (eds.) Artificial Intelligence Applications and Innovations. AIAI 2005. IFIP—The International Federation for Information Processing, vol. 187, pp. 231–238. Springer, Boston (2005)
56. Xing, Y., Shen, F., Luo, C., Zhao, J.: L3-svm: a lifelong learning method for svm. In: Proceedings of the International Joint Conference on Neural Networks, pp. 1–8 (2015)
57. Xu, S., Wang, J.: A fast incremental extreme learning machine algorithm for data streams classification. Expert Syst. Appl. **65**, 332–344 (2016)
58. Yin, G., Zhang, Y.T., Li, Z.N., Ren, G.Q., Fan, H.B.: Online fault diagnosis method based on incremental support vector data description and extreme learning machine with incremental output structure. Neurocomputing **128**, 224–231 (2014)
59. Zang, W., Zhang, P., Zhou, C., Guo, L.: Comparative study between incremental and ensemble learning on data streams: case study. J. Big Data **1**(1), 1–16 (2014)
60. Zheng, J., Shen, F., Fan, H., Zhao, J.: An online incremental learning support vector machine for large-scale data. Neural Comput. Appl. **22**(5), 1023–1035 (2013)
61. Zhou, X., Zhang, X., Wang, B.: Online support vector machine: a survey. In: Kim, J., Geem, Z. (eds.) Harmony Search Algorithm. Advances in Intelligent Systems and Computing, vol. 382, pp. 269–278. Springer, Berlin (2016)
62. Zhu, Z., Zhu, X., Guo, Y.F., Xue, X.: Transfer incremental learning for pattern classification. In: Proceedings of the 19th ACM International Conference on Information and Knowledge Management, pp. 1709–1712 (2010)

On Social Network-Based Algorithms for Data Stream Clustering

Jean Paul Barddal, Heitor Murilo Gomes, and Fabrício Enembreck

Abstract Extracting useful patterns from data is a challenging task that has been extensively investigated by both machine learning researchers and practitioners for many decades. This task becomes even more problematic when data is presented as a potentially unbounded sequence, the so-called data streams. Albeit most of the research on data stream mining focuses on supervised learning, the assumption that labels are available for learning is unverifiable in most streaming scenarios. Thus, several data stream clustering algorithms were proposed in the last decades to extract meaningful patterns from streams. In this study, we present three recent data stream clustering algorithms based on insights from social networks' theory that exhibit competitive results against the state of the art. The main distinctive characteristics of these algorithms are the following: (1) they do not rely on a hyper-parameter to define the number of clusters to be found; and (2) they do not require batch processing during the offline steps. These algorithms are detailed and compared against existing works on the area, showing their efficiency in clustering quality, processing time, and memory usage.

1 Introduction

Learning from data streams is a fast-growing research topic due to the ubiquity of data generation and gathering in several real-world situations. Examples of data streams include sensor networks, wearable sensors, computer network traffic and video surveillance, to name a few. Given their ephemeral nature, data stream sources

J. P. Barddal (✉) · F. Enembreck
Graduate Program in Informatics (PPGIa), Pontifícia Universidade Católica do Paraná, Curitiba, Brazil
e-mail: jean.barddal@ppgia.pucpr.br; fabricio@ppgia.pucpr.br

H. M. Gomes
Institut Mines-Télécom, Department of Computer Science and Networks (INFRES), Université Paris-Saclay, Paris, France
e-mail: heitor.gomes@telecom-paristech.fr

© Springer International Publishing AG, part of Springer Nature 2019
M. Sayed-Mouchaweh (ed.), *Learning from Data Streams in Evolving Environments*, Studies in Big Data 41, https://doi.org/10.1007/978-3-319-89803-2_13

are expected to present changes in their data distribution, thus giving rise to both concept drifts and evolutions [13].

By far, the most common approach for learning from data streams is classification. However, the task of classification works under the assumption that labeled data is frequently—or periodically—input to the learner so that it can update its predictive model accordingly. Labeling is a costly process [23] that is nearly impossible in most streaming scenarios, since halting the processing of a stream to gather labeled data may result in data loss. For the above reasons, researchers have put an honorable amount of effort into extracting patterns from data streams using unsupervised techniques [2, 9, 17, 20].

In this paper, we sum up three data stream clustering algorithms [5–7] tailored on concepts stemmed from social networks theory. Social networks theory has been successfully applied in many research fields due to its formal description of structural variables. In contrast to classical graph theory, social networks theory provides insights on how networks evolve through time, thus overlapping with drifting data streams. The common ground of these algorithms is the layering of micro-clusters in a graph, which is connected and updated given a variation of the Scale-free Network model and homophily [3].

Individually, each of these algorithms has shown interesting results in clustering quality and computational resources usage. However, they were not benchmarked against each another in a common evaluation scenario. This gap disallows the identification of which algorithm is the most efficient and efficient across different scenarios and why each of its traits is being beneficial. Furthermore, in this study, we compare these proposals against several data stream clustering algorithms in an environment that encompasses both synthetic and real-world data.

This paper is divided as follows. Section 2 formalizes data stream clustering, while Sect. 3 presents related work. Section 4 then sums up three social network-based data stream clustering proposals: CNDenStream [5], SNCStream [6] and SNCStream+ [7], which are later evaluated in Sect. 5. Finally, Sect. 6 concludes this paper and provides insights on future work for social network-based data stream clustering algorithms.

2   Data Stream Clustering

Extracting useful knowledge from data streams is a challenge per se. In contrast to traditional batch machine learning schemes, data stream mining must iteratively process a potentially unbounded incoming data sequence [13]. As a consequence, instances should be processed right after their arrival (single-pass processing) or in limited size batches (chunks) of data [24]. More importantly, due to the temporal aspect of streams, their underlying generation process is expected to be ephemeral, thus giving rise to concept drifts and evolutions [13].

Let $S = [\mathbf{x}_t]_{t=0}^{t \to \infty}$ denote a data stream providing instances $\mathbf{x}_t$ rapidly and intermittently, where $\mathbf{x}_t$ is a d-dimensional data object which arrives at a timestamp

t. The task of data stream clustering can be described as the act of grouping streaming data in a set of meaningful clusters $K = \{k_1, k_2, \ldots, k_n\}$ [4]. The rationale behind most clustering techniques is that instances within a cluster are more similar to each other when compared to instances in other clusters [13].

Data stream clustering algorithms must be capable of dealing with *concept drifts* and *evolutions*. Concept drifts occur whenever the data distribution $P[\mathbf{x}]$ changes [13], while concept evolutions refer to the appearance or disappearance of clusters, that is, changes in cardinality of K [23]. Ideally, clustering algorithms must: (i) detect concept drifts and adapt its clusters accordingly; (ii) detect concept evolutions and create/delete clusters independently from user intervention; (iii) discern between seeds of new clusters and noisy data; and finally (iv) not rely on a multitude of parameters.

When combining items (ii) and (iv), we must also consider that defining optimal values for parameters very often depends on the incoming data. Thus, if a concept drift or evolution occurs, the clustering algorithm parameters' values become outdated and waiting for user intervention seems too optimistic and unrealistic. A canonical example is a parameter that defines the ground-truth number of clusters to be found n. Any algorithm, for example, k-means [22], that demands a predefined value of n is thus unable to cope with concept evolutions with no user intervention.

3 Related Work

A variety of data stream clustering algorithms was developed in the last years and its majority [2, 9, 17, 20] process incoming instances intercalating online and offline steps.

During the online step, algorithms incrementally update specific data structures to deal with the evolving nature of data streams and time/space constraints. To efficiently represent instances, the feature vector structure and its variants are widely adopted. A feature vector is a triplet $CF = \langle LS, SS, N \rangle$, where LS stands for the sum of the objects summarized, SS is the squared sum of these objects and N is the number of objects [24]. An instance $\mathbf{x}_i$ can increment a feature vector CF_j as follows: $LS_j \leftarrow LS_j + \mathbf{x}_i$, $SS_j \leftarrow SS_j + \mathbf{x}_i^2$ and $N_j \leftarrow N_j + 1$. Also, two feature vectors CF_i and CF_j can be merged in a third CF_l as follows: $LS_l \leftarrow LS_i + LS_j$, $SS_l \leftarrow SS_i + SS_j$ and $N_l \leftarrow N_i + N_j$.

Due to the ephemeral characteristics of data streams, a common approach to assigning higher importance to recently retrieved instances and to "forget" older concepts is windowing. In the clustering context, two techniques are widely used: landmark and damped windows [14]. Landmark windows process streams in disjoint batches of data, such that each of these is often sized following a user-given parameter or a timespan (hourly, daily, weekly, and so on). Conversely, damped windows associate weights to each instance, which decay with time to provide higher importance to more recent data compared to those in the past.

During the offline step, often triggered by user requests, traditional batch clustering algorithms such as k-means [22] and DBSCAN [12] are applied in batch mode. These algorithms must be adapted to work with data structures, for example, feature vectors, instead of instances themselves.

In this section, we present existing work on data stream clustering that was used to compare the proposed methods. We forfeit from providing a complete survey on the topic as this has been recently reported in [24]. All of the following algorithms will be used during the empirical analysis, later presented in Sect. 5. The criteria for selecting these clustering algorithms are as follows: (1) code availability for experiment reproducibility, (2) interesting results in different data domains, and (3) number of citations.

3.1 CluStream

During the online step, CluStream breaks the stream into chunks of data, whose size is defined by a horizon parameter H [2]. During each landmark, q feature vectors are created and updated to represent the instances obtained from the stream until a new landmark is reached.

During the offline step, the original CluStream uses an adaptation of the k-means algorithm [22] to obtain clusters based on the q feature vectors computed during the online step.

3.2 ClusTree

ClusTree [20] maintains feature vectors in a hierarchical structure provided by an R-Tree [15]. ClusTree creates a hierarchy of feature vectors at different granularity levels. Accordingly to user-given thresholds, it is determined whether an instance should be merged with an existing feature vector. In the negative case, a new feature vector is created and added to the R-Tree. ClusTree is also capable of handling noisy data by using outlier-buffers.

During the offline step, algorithms such as k-means [22] and DBSCAN [12] are used to find final clusters, where feature vectors' centers are treated as centroids.

3.3 DenStream

DenStream [9] extends the DBSCAN [12] algorithm to allow density-based clustering for data streams. DenStream heavily relies on the definition of core microcluster. A core object is an object that its ϵ-neighborhood has at least ψ neighbors while a dense area is the union of all ϵ-neighborhoods of all core objects. Therefore, a core micro-cluster is a CF that is defined as $CMC(w, c, r, t_c, t_u)$ to a group of near instances $\mathbf{x}_i, \ldots, \mathbf{x}_n$ where w stands for its weight, c its center, r its radius, t_c the timestamp of its creation and the timestamp of its last update (increment or addition) t_u; where $w \geq \psi$, $r \leq \epsilon$, $f(\cdot, \lambda)$ is a exponential decay function with decaying factor λ and $d(\cdot, \cdot)$ is an Euclidean distance. Core micro-clusters are classified based on their weights w: if $w \geq \beta\psi$, it is a potential micro-cluster (PMC), otherwise, it is considered an outlier micro-cluster (OMC).

The online step of DenStream updates both PMCs and OMCs if the merging of an instance in its closest micro-cluster results in a structure with a radius less than ϵ. On the other hand, new OMCs are created to represent instances that were not aggregated.

Finally, the offline step of DenStream applies DBSCAN to find final clusters upon the potential micro-clusters maintained during the online step.

3.4 HAStream

To automatically detect clusters of distinct densities, HAStream [17] performs hierarchical density-based clustering that automatically adapts its density thresholds according to the arrival of data.

During the online step, instances are processed using any feature vector model such as provided by CluStream [2], ClusTree [20] or DenStream [9]. Following the implementation used by the authors in the original paper, the implementation used during the evaluation also follows the online step used in DenStream.

During the offline step, HAStream generates the final clusters using a hierarchical density-based procedure. Since returning every possible hierarchical cluster forces the evaluator to define the correct amount of clusters n, HAStream attempts different dendrogram pruning options and returns the n clusters which maximize a concept called cluster stability [17].

4 Social Network-Based Approaches

This section details three social network-based data stream clustering algorithms. CNDenStream [5], SNCStream [6] and SNCStream+ [7] are based on the hypothesis that intra-cluster data are related due to diminished dissimilarity while inter-cluster data are not related, due to higher dissimilarity. To create and keep track of high-quality clusters, these algorithms model clusters and their dynamics as a social network, in which nodes are either instances or micro-clusters, edges represent connections between these nodes and subgroups in the network represent clusters.

Social networks theory has been applied in many research fields, from computer science to sociology, due to its formal description of structural variables based on graph theory. However, in contrast to Graph Theory, Social Network Analysis focuses on more subjective topics such as an individual behavior in society, and how networks are formed and updated over time. This section starts with an overview of social networks which justify some of the core decisions of the clustering algorithms which will be presented in the following sections.

4.1 Background on Social Networks Theory

Even though Social Network Analysis focuses on subjective topics, its building blocks (nodes and edges) are usually represented computationally as a graph $G = (V, E, W)$ where V is the set of nodes, E the set of edges between nodes and W is the set of tuples which associates to each edge in E a weight. This notation will be used throughout this paper to denote both social networks and to explain the fundamentals of all the discussed algorithms.

Different social network construction models were developed over the years, with the objective of modeling both generation and evolution of networks, where we emphasize: random [11], small-world [26], and scale-free [3].

The Random generation model is based on the hypothesis that the existence of a connection between a pair of nodes is given by a probability p. The Small-world model incorporates attributes of both random and regular networks. Consequently, this topology presents a high clustering coefficient, inherited from the regular networks (which are those where all nodes have the same degree), and a small average path length, as random networks [26].

The objective of the Scale-free model is to represent the dynamics of real networks, where connections between nodes can be replaced with time such as the World Wide Web and Cellular Networks [3]. Thus, authors in [3] developed generation and evolution elements. A Scale-free network starts with a small network size n. For every time unit t, a new node is added to the network establishing ω connections with already existing nodes. Also, on the Scale-free model, when choosing the nodes which this new node will establish its connections, it is assumed a probability $\prod(\omega_i) = \frac{\omega_i}{\sum_j \omega_j}$, where ω_i represents the degree of a node v_i. This process is called "preferential attachment", where nodes with higher degrees tend to establish even more connections.

As a consequence of the preferential attachment process, Scale-free networks are "dominated" by a few vertices denominated *hubs*. Thus, Scale–free network degree distribution follows the distribution $p(k) \sim k^{-\xi}$ where $p(k)$ is the probability of a random node being attached to ω other nodes and $\{\xi \in \mathbb{R} \mid 2 \leq \lambda \leq 3\}$ for many real networks [3]. Also, at each time unit t, besides the addition of new nodes, a rewiring process exists. The rewiring component is based on the homophily definition, where nodes tend to eradicate connections with dissimilar nodes, replacing them by new connections with more similar ones with probability θ. Since homophily tends to establish connections between similar nodes, this process is useful for the task of clustering.

4.2 CNDenStream

CNDenStream generates and updates a complex network $G = (V, E, W)$ and an outlier micro-cluster buffer $\mathscr{B}$. To keep track of clusters in the stream without the need of batch processing during the offline step, CNDenStream uses homophily-based insertion and rewiring procedures inspired by complex networks theory.

Initially, CNDenStream stores the first N instances retrieved from S in a burnout buffer to an initial DBSCAN run, thus finding initial potential and outlier micro-clusters. While outlier micro-clusters are stored in an outlier buffer $\mathscr{B}$, potential micro-clusters PMC_i are added to the network G, where each potential micro-cluster establishes with the ω closest possible neighbors (considering Euclidean distances) currently in V as described in lines 1–7 of Algorithm 1.

Afterward, all PMC_i are added to V, and edges and their corresponding weights are added to its correspondent sets E and W. Figure 1 presents the insertion of 4 potential micro-clusters, namely PMC_1 to PMC_4. The insertion procedure can connect the last added node to the ω-most similar nodes in G. Nevertheless, the same cannot be said for the other nodes currently in G. In Fig. 1d one can see that after the addition of PMC_4, PMC_1 should be connected to PMC_4 instead of PMC_3, since $d(PMC_1, PMC_4) < d(PMC_1, PMC_3)$, where $d(\cdot, \cdot)$ is an Euclidean distance given by Eq. (1) and v iterates over all dimensions of a pair of instances.

Algorithm 1 Insertion procedure. **Input:** a micro-cluster PMC_i, the network $G = (V, E, W)$ and the amount of connections each micro-cluster will establish at its insertion in the network ω.

1: *neighbors* $\leftarrow$ the ω closest micro-clusters in V to PMC_i
2: $V \leftarrow V \cup \{PMC_i\}$
3: **for all** $PMC_j \in neighbors$ **do**
4:     *newEdge* $\leftarrow \langle PMC_i, PMC_j \rangle$
5:     $E \leftarrow E \cup \{ newEdge \}$
6:     $W \leftarrow W \cup \{\langle newEdge, d(PMC_i, PMC_j)\rangle\}$
7: **end for**
8: {rewiring procedure}
9: **for all** $PMC_j \in V$ **do**
10:     $d_j \leftarrow \deg(PMC_j)$
11:     *newNeighbors* $\leftarrow$ the d_j closest micro-clusters in V to PMC_j
12:     Remove all edges connecting PMC_j from E and its correspondent weights from W
13:     **for all** $PMC_k \in newNeighbors$ **do**
14:         {Establishes new edges}
15:         *newEdge* $\leftarrow \langle PMC_j, PMC_k \rangle$
16:         $E \leftarrow E \cup \{newEdge\}$
17:         $W \leftarrow W \cup \{\langle newEdge, d(PMC_j, PMC_k)\rangle\}$
18:     **end for**
19: **end for**

$$d(\mathbf{x}_i, \mathbf{x}_j) = \sqrt{\sum_{k=i}^{d} \left(\mathbf{x}_i[v] - \mathbf{x}_j[v]\right)^2} \qquad (1)$$

After the addition of a PMC_i to the network, all nodes $PMC_j \in V$ such that $i \neq j$ perform rewirings based on homophily, such that each PMC_j replaces its edges with higher dissimilarities w by edges to its closest neighbors, that is, edges with lower dissimilarity. For every PMC_j, the Euclidean distances for all of its 2-hop neighbors are then computed. A 2-hop neighborhood is assumed since potential closest nodes are likely to be neighbors (2-hop) of the current neighbors (1-hop). This 2-hop neighborhood is an approximation to prevent distance computation between all nodes, which would be computationally costly. With the results of these Euclidean distances, PMC_j replaces edges by the most dissimilar instances with some similar ones, yet maintaining its degree ω_j. Lines 9–19 of Algorithm 1 detail this process.

To exemplify how the rewiring procedure works, we refer back to the addition of nodes presented in Fig. 1, where PMC_1 is capable of connecting itself with higher similar nodes. Therefore, Fig. 2 presents the rewiring of node PMC_1. First, Euclidean distances between PMC_1 and its 2-hop neighborhood are computed and compared to its current neighbors (1-hop). In Fig. 2b one can see that $d(PMC_1, PMC_4) < d(PMC_1, PMC_3)$. Consequently, PMC_1, to maintain its degree $\omega_1 = 2$, must eliminate its current most dissimilar edge to replace it with a similar one. Figure 2c depicts the substitution of the edge between PMC_1 and PMC_3 by a new one connecting PMC_1 and PMC_4.

Due to the rewiring process, communities of potential micro-clusters tend to appear since some intra-clusters edges between similar micro-clusters grow, while those linking dissimilar micro-clusters (interclusters) shrink. For instance, Fig. 3

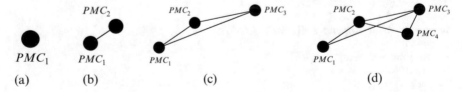

Fig. 1 Insertion example using $\omega = 2$. Adapted from [5]. (**a**) PMC_1. (**b**) PMC_2. (**c**) PMC_3. (**d**) PMC_4

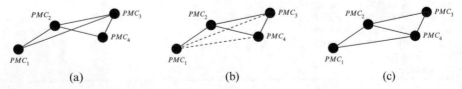

Fig. 2 Example of node PMC_1 rewiring. Adapted from [5]. (**a**) Node PMC_1 rewiring. (**b**) Euclidean distance comparison. (**c**) Node PMC_1 rewired

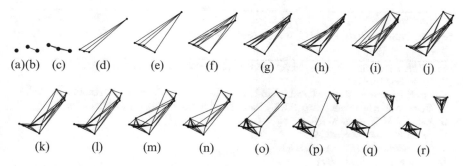

Fig. 3 Insertion of potential micro-clusters obtained from the insertion of micro-clusters during the RBF$_2$ experiment. Adapted from [5].

presents the evolution of a network as instances arrive, where rewirings enlarge the amount of intra-cluster edges while decreasing intercluster connections. This procedure is repeated until, in Fig. 3r, two clusters emerge.

After the DBSCAN execution and the initial network is built, all arriving instances x_i are processed according to an adaptation of the DenStream algorithm. First, CNDenStream finds the potential micro-cluster in V which minimizes the dissimilarity with x_i: PMC_i. Afterward, CNDenStream verifies whether the addition of x_i with PMC_i results in a micro-cluster with a radius below ϵ. If this condition holds, x_i is then merged with PMC_i. Otherwise, this process is repeated within the outlier micro-clusters. If x_i is not merged with any potential or outlier micro-clusters, it is used to instantiate a new outlier micro-cluster, which is then added to $\mathcal{B}$.

When an outlier micro-cluster OMC_j is promoted to a potential micro-cluster, that is, $w(OMC_j) \geq \beta\psi$, it is removed from the outlier buffer $\mathcal{B}$ and is inserted in the network G.

As in DenStream, both potential and outlier micro-clusters' weights decay over time according to an exponential decay function presented in Eq. (2), where $\Delta t = t_i - t_u$ is the difference between the current timestamp and the instant of the last update of each core-micro-cluster, N is the amount of instances summarized by such micro-cluster and λ is the parameter for the exponential function.

$$w(CMC_i) = 2N^{-\lambda\Delta t} \tag{2}$$

Algorithm 2 Rewiring procedure for removal. **Input:** a micro-cluster PMC_j to rewire, a micro-cluster PMC_i which is about to be removed from G and the network $G = (V, E, W)$.

1: $\{PMC_j$ is rewired to its d_j closest micro-clusters with the exception of $PMC_i.\}$
2: $d_j \leftarrow \deg(PMC_j)$
3: $newNeighbors \leftarrow$ the d_j closest micro-clusters in V to PMC_j with the exception of PMC_i
4: Remove all edges connecting PMC_j from E and its correspondent weights from W
5: **for all** $PMC_k \in newNeighbors$ **do**
6:     {Establishes new edges}
7:     $newEdge \leftarrow \langle PMC_k, PMC_j \rangle$
8:     $E \leftarrow E \cup \{newEdge\}$
9:     $W \leftarrow W \cup \{\langle newEdge, d(PMC_k, PMC_j)\rangle\}$
10: **end for**

When the weight $w(PMC_i)$ of a micro-cluster PMC_i is below $\beta\psi$, it is removed from the network. Whenever removal occurs, all neighbors PMC_j of PMC_i are allowed to rewire to maintain their degree ω_j. This procedure for a PMC_j node is described by Algorithm 2.

Following the concerns stated by the authors of DenStream [9], verifying all potential micro-clusters weights according to the arrival of each instance $\mathbf{x}_i$ can be too computationally costly. Therefore, CNDenStream also encompasses a periodic verification performed based on a clean-up window size T_p. Equation 3 presents the computation of the clean-up window size T_p [9].

$$T_p = \left\lceil \frac{1}{\lambda} \log \frac{\beta\psi}{\beta\psi - 1} \right\rceil \tag{3}$$

4.3 SNCStream

The Social Network Clustering Stream (SNCStream) [6] is built on the same hypothesis of CNDenStream, that is, that intra-cluster data are related due to high similarity, and intercluster are not related due to low similarity. SNCStream extends CNDenStream to overcome one of its major limitations: the initial DBSCAN run.

SNCStream is divided in three phases: *initial network construction*, *network transformation* and *network evolution*.

During the first phase, the initial network construction, SNCStream gathers the initial N instances to build its initial network. The construction of the network follows the same procedure adopted by CNDenStream. At the arrival of an instance $\mathbf{x}_i$, such that $i < N$ holds, SNCStream computes Euclidean distances between $\mathbf{x}_i$ and all vertices $v_j \in V$ of the network. One important difference between CNDenStream and SNCStream is that in the latter, v_j represent instances until this point. Later, SNCStream then establishes edges between $\mathbf{x}_i$ and the ω closest neighbors of x_i. Later, the rewiring procedure is performed, allowing all other vertices in the network to locally optimize their distances to neighbors.

After the processing of this burnout window, SNCStream proceeds with the network transformation step, where all vertices in the network are replaced, giving place to potential micro-clusters. Although simple, this transformation allows the network to scale up to higher amounts of incoming instances. Maintaining a network where its vertices represent instances would be computationally prohibitive since the number of instances in the stream is potentially unbounded, and the number of distance computations and rewirings would be cumbersome. In practice, all instances x_i in the network are replaced by potential micro-clusters PMC with $LS = x_i$, $SS = (x_i)^2$ and $N = 1$. All edges and weights are maintained intact. Additionally, the outlier micro-cluster buffer $\mathscr{B}$ is initialized as an empty set.

The final phase of SNCStream follows the same procedure adopted by CNDen-Stream, that is, it tries to merge upcoming instances within existing micro-clusters. Weights' fading and verification are also computed as in DenStream according to Eqs. (2) and 3, respectively.

The main advantage of SNCStream when compared to CNDenStream is that SNCStream can provide clustering results regardless of how many instances were observed. For instance, if a clustering request occurs to CNDenStream before its initial DBSCAN run, no clustering results would be provided, a fact that does not hold for SNCStream. Additionally, since SNCStream does not depend on DBSCAN, the size of the burnout window in SNCStream can be decreased. In [6] authors show that smaller burnout windows, for example, $W = 100$, are sufficient for a variety of scenarios and can improve clustering quality results significantly.

4.4 SNCStream+

The core of both CNDenStream and SNCStream is the rewiring procedure. Additionally, both are capable of processing gigantic amounts of data by adopting specific density-based statistical summaries, that is, core micro-clusters. After the introduction of [6], several questions were raised over the impact of some of SNCStream's parameters.

In [7], authors introduced SNCStream+. SNCStream+ introduces three enhancements to its original version, including: (1) an optimized rewiring procedure, (2) a new micro-cluster weight window size, and (3) the adoption of a fractional distance metric for high-dimensional streams.

The original rewiring procedure used in CNDenStream and SNCStream performs linear access to all existing vertices in the network, and most of the times, unnecessarily. This is mainly because most of the vertices in the network are unlikely to be affected from the location of a vertex insertion. Instead of performing linear access to all vertices in the network, SNCStream+ adopts a propagation scheme to trigger rewirings in the network. Starting from the neighbors of the newly added vertex, rewirings are triggered only by the neighbors who replaced at least one

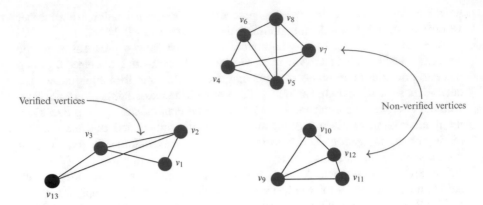

Fig. 4 Example of rewiring after the insertion of vertex v_{13}, where verified vertices are displayed in blue and non-verified in red. Adapted from [7]

of their edges. In practice, this propagation rewiring procedure decreases the number of rewirings in the network, while converging to the original rewiring procedure [6]. In Fig. 4, we present an example of a rewiring procedure after the insertion of v_{13} to the network. In this case, only blue vertices were tested for possible rewirings, while red nodes were not due to the lack of connections connecting subnetworks.

SNCStream verifies micro-clusters' weights according to a clean-up window size T_p, originally used in [9]. Intuitively, if these verifications take too long to take place, micro-clusters that do not represent the current stream concept will join the network and jeopardize the quality of final clusters. Conversely, performing this verification too often may jeopardize algorithms' processing time. In [7], authors showed that assuming $T_p = 1$ is beneficial not only to promptly remove micro-clusters with insufficient weight but also in processing time. Smaller values are preferred since that infrequent weight verification allows micro-clusters to stay longer in memory, and consequently, the number of distance computations performed during the arrival of each instance also increases. Experiments conducted in [7] showed that with the increase of T_p, the number of distance computations exponentially increases due to the greater amount of micro-clusters in memory. Thus, adopting $T_p = 1$ is beneficial since it allows prompt removal of insufficiently weighted micro-clusters from memory, and fewer micro-clusters lead to a smaller amount of distance computations.

Finally, the experiments provided in [6] showed that with the increase of the stream dimensionality, the clustering quality of SNCStream decreased. This phenomenon is known as "curse of dimensionality" [9], where Euclidean distances fail to represent the dissimilarity between points in high-dimensional spaces, enforcing algorithms to fall in its vastitude. In most high-dimensional applications, the choice of the distance metric is concealed and the computation of dissimilarities is rather heuristical [1]. There is very little work in the literature providing guidance on choosing the correct distance metric to calculate dissimilarity between two instances. By far, the most widely used distance metric is the L_p norm. The L_p

distance between two instances $\mathbf{x}_i$ and $\mathbf{x}_j$ can be computed according to Eq. (4), where v iterates over all dimensions of the pair of instances.

$$d_{L_p}(\mathbf{x}_i, \mathbf{x}_j) = \left[\sum_{v=1}^{d} | \mathbf{x}_i[v] - \mathbf{x}_j[v] |^p \right]^{\frac{1}{p}} \tag{4}$$

In [1], authors discuss about different values of p, enlightening that $p = 1$ (Manhattan distance) and $p = 2$ (Euclidean distance) are theoretically more efficient than $p \geq 3$ for high-dimensional problems. Additionally, authors provide proofs that L_p rapidly decays with the increase of d. Encouraged by this trend, authors examined the behavior of fractional distance metrics, where $0 \leq p \leq 1$. After several studies, $p = 0.3$ was pointed out as an interesting value for several domains. Following these results, SNCStream+ adopted the fractional $L_{.3}$ distance metric, which showed improvements in clustering quality with the increase of streams' dimensionality [7].

5 Evaluation

Even though CNDenStream [5], SNCStream [6] and SNCStream+ [7] original papers presented and compared their performances, they were not compared against one another and related work. In this section, we merge the experimental protocols used in these three papers and apply it to extensively evaluate them against each other and related work.

5.1 Evaluation Procedure

In the following experiments, clustering quality is calculated using the Cluster Mapping Measure (CMM), a metric developed aiming the evolving characteristics of data streams [21]. In contrast to batch clustering evaluation metrics (e.g., Purity, Precision, Recall), CMM is an external metric that accounts for nonassociated, misassociated instances and noisy data inclusion. It is also important to emphasize that CMM considers recently retrieved instances with more weight than older ones by using an exponential decay function. CMM is bounded in the [0; 1] interval, 1 being the representation of a perfect clustering given a ground-truth set of clusters. For more details on CMM's computation and variants, the reader is referred to [21].

Processing time is computed as the time (in seconds) that each algorithm used of CPU and memory usage is given in RAM-Hours, where each RAM-Hour equals to 1 GB of RAM used in a processing hour. All experiments presented in this paper were performed on an Intel Xeon CPU E5649 @ 2.53 GHz ×8 based computer running CentOS with 16 GB of memory using Massive Online Analysis (MOA) framework [8].

To determine whether a significant statistical difference between algorithms exists in any evaluation metric, a combination of Friedman's and Nemenyi's [10] nonparametric hypothesis tests with a 95% confidence level was used. The results of these tests are graphically reported with critical difference (CD) plots.

5.2 Parametrization

All algorithms parameters were set according to their original papers. CluStream parameters are: horizon $H = 1000$ and $q = 1000$ [2]. ClusTree parameters are: a horizon $H = 1000$ and a maximum tree height of 8 [20]. DenStream parameters are: $\psi = 1$, $\mathcal{N} = 1000$, $\lambda = 0.25$, $\epsilon = .02$, $\beta = 0.2$ and an offline multiplier $\eta = 2$[9]. HAStream uses a DenStream-like online step, therefore adopts the same density parameters [17].

Both CluStream and ClusTree were evaluated twice by changing their offline step approach to find final clusters, that is, the usage of an informed k-means and the adoption of DBSCAN. In the first case, whenever a clustering request is performed, k-means received as input the ground-truth amount of clusters to be found n, which can be seen as an optimistic clustering approach since this number is often unknown in advance. On the other hand, the parameters for DBSCAN were the same as in DenStream.

In Table 1 we present the parameters adopted by CNDenStream, SNCStream, and SNCStream+. The main differences amongst them are: (1) CNDenStream requires a larger burnout window N, (2) the fractional distance metric adopted by SNCStream+, and (3) SNCStream also assumes $T_p = 1$ instead of using Eq. (3). Again, we emphasize that neither of the social network-based algorithms require the ground-truth amount of clusters to be found n. CNDenStream, SNC-Stream, and SNCStream+ can be downloaded from https://sites.google.com/site/

Table 1 CNDenStream, SNCStream, and SNCStream+ parameters

Parameter	CNDenStream [5]	SNCStream [6]	SNCStream+ [7]
ω	4	4	4
ψ	1	1	1
N	1000	100	100
λ	0.25	0.25	0.25
ϵ	0.02	0.02	0.02
β	0.2	0.2	0.2
Distance metric	Euclidean L_2	Euclidean L_2	Fractional $L_{.3}$
T_p	Eq. (3)	Eq. (3)	1

moasocialbasedalgorithms/home as a plugin for the Massive Online Analysis (MOA) framework [8].

5.3 Synthetic Data

To evaluate whether a learning algorithm can work in different scenarios, it is necessary to assess its performance in different data domains. Synthetic data stream generators are important and often used due to their flexibility since they offer a precise definition of drifts types and location during the streams. To synthesize data streams, the Radial Basis Function (RBF) generator was used. This generator creates a user-given amount of drifting centroids, which are defined by a label, center, weight, and standard deviation given by a Gaussian distribution. Additionally, ground-truth clusters appear and disappear during the stream, thus giving rise to concept evolutions. We adopt the RBF_d notation to refer to an experiment that uses the RBF generator with a dimensionality d. In all RBF-based experiments, streams were created with 100,000 instances where the ground-truth amount of ground-truth clusters varied in the [2; 8] interval and an appearance/disappearance of a centroid randomly occurs every 3000 instances.

5.4 Real-World Datasets

Besides the usage of data generators, the performance of algorithms should also be evaluated along publicly available real datasets. Even though it is reasonably difficult to justify if and when drifts and evolutions occur, it is still important to verify how proposals act on different data domains. The following paragraphs briefly describe these datasets.

Airlines (AIR) This dataset contains 539,383 instances and 8 attributes and represents all flight arrivals and departures from US airports, from October 1987 until April 2008 [18].

Electricity (ELEC) This dataset was created by the Australian New South Wales Electricity Market and stores energy prices obtained every 5 min [16]. In this problem, energy prices are not fixed as they vary according to market supply and demand. The Electricity dataset consists of 45,312 instances and 8 attributes.

Forest Covertype (COV) This dataset models the forest covertype prediction problem based on cartographic variables [19]. This dataset consists of 900 m^2 cells obtained from US Forest Service Region 2 Resource Information System and contains 581,012 instances with 54 attributes.

KDD98 This dataset[1] contains data about 95,412 donations described by 56 attributes. Donators groups are known to be nonstationary [20], thus is a challenge for clustering algorithms. Additionally, this dataset is mostly constituted of categorical features, which are either ignored or naively converted to numeric by clustering algorithms.

KDD99 This dataset is composed of raw TCP dump data for a LAN. Each connection has 42 attributes, and most of the 4,898,431 instances are labeled as conventional connections. Also, this dataset is known to present appearances of clusters over time, representing previously unknown types of cyber-attacks [2].

Body Posture and Movements (BPaM) This dataset consists of 165,632 instances collected on 8 h of activities of 4 healthy subjects. The original goal is to classify whether the subject is sitting down, standing up, standing, walking, or sitting based on 18 attributes [25].

5.5   Results

In this section we evaluate CNDenStream, SNCStream, and SNCStream+ against each other and algorithms presented in Sect. 3 using both synthetic and real data.

Table 2 presents the results obtained for the Clustering Mapping Measure (CMM) quality metric. CMM results show that social network-based clustering algorithms present higher clustering quality values, even surpassing informed k-

Table 2 Average CMM obtained during experiments

	CMM								
	CluStream		ClusTree		Den	HA	CNDen	SNC	SNC
Experiment	k-Means	DBSCAN	k-Means	DBSCAN	Stream	Stream	Stream	Stream	Stream+
RBF$_2$	0.91	0.83	0.88	0.88	0.94	0.88	0.96	0.97	**0.99**
RBF$_5$	**0.99**	0.67	0.98	0.8	0.84	0.86	0.93	0.95	**0.99**
RBF$_{20}$	0.97	0.47	0.99	0.88	0.85	0.91	0.92	0.93	**0.99**
RBF$_{50}$	0.76	0.48	0.99	0.44	0.85	0.81	0.93	0.91	**0.99**
Airlines	0.93	0.67	0.76	0.94	0.50	0.70	0.87	**0.98**	**0.98**
Electricity	0.75	0.41	0.76	0.44	0.61	0.43	0.88	0.89	**0.98**
Forest cover-type	0.76	0.41	0.75	0.43	0.38	0.37	0.81	**0.97**	**0.97**
KDD'98	0.37	0.37	0.37	0.37	0.37	0.37	0.37	0.38	**0.42**
KDD'99	0.40	0.50	0.44	0.40	0.69	0.77	0.84	0.90	**0.95**
BPaM	0.88	0.61	0.87	0.69	0.74	0.74	0.90	0.91	**0.99**

---

[1] http://kdd.ics.uci.edu/databases/kddcup98/kddcup98.html.

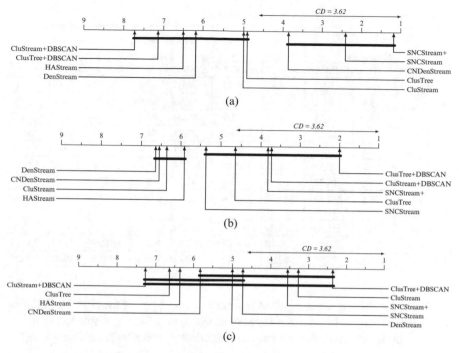

Fig. 5 Nemenyi test results. (**a**) CMM. (**b**) CPU Time. (**c**) *RAM-Hours*

means approaches. Results also highlight the efficiency of SNCStream+, showing high clustering quality results regardless of the dimensionality and domain of the stream. We emphasize the results obtained in the KDD'98, which is known for not having a clear data cohesion and several categorical attributes. Thus, clustering becomes a highly challenging task [6].

Significant differences amongst algorithms were found after performing Friedman and Nemenyi tests, in which the average ranks differ by at least a Critical Difference (*CD*). In Fig. 5a we report the results obtained, where {SNCStream+, SNCStream, CNDenStream} ≻ {ClusTree, CluStream, DenStream, ClusTree + DBSCAN, CluStream + DBSCAN} with a 95% confidence level.

In Table 3 we present the results obtained for processing time, in seconds. The results obtained show that ClusTree, on average, presents the best efficiency. This occurs due to the hierarchical framing of micro-clusters in an R-Tree. This allows quicker processing of incoming instances since the number of computations is logarithmic with the number of micro-clusters stored, instead of linear, as it occurs in other proposals. In Fig. 5b we report the critical differences obtained for processing time comparison, where {ClusTree + DBSCAN, CluStream + DBSCAN, SNCStream+, ClusTree, SNCStream} ≻ {HAStream, CluStream, CNDenStream, DenStream}. Even though the social network-based algorithms possess higher asymptotic complexities, both SNCStream+ and SNCStream are practical as their processing times are comparable to both ClusTree and CluStream.

Table 3 Processing time (s) obtained during experiments

Experiment	CPU Time (s)								
	CluStream k-Means	DBSCAN	ClusTree k-Means	DBSCAN	Den Stream	HA Stream	CNDen Stream	SNC Stream	SNC Stream+
RBF$_2$	12.59	9.31	7.82	**7.47**	36.08	19.98	32.91	30.81	26.33
RBF$_5$	13.67	9.02	10.39	**6.74**	28.88	18.91	19.50	18.25	15.60
RBF$_{20}$	35.18	15.13	27.72	18.99	38.00	34.87	17.89	16.74	**14.31**
RBF$_{50}$	71.62	24.59	56.27	**18.60**	23.71	64.49	25.99	24.53	20.79
Two moon	2.31	2.70	2.28	**1.05**	4.77	7.32	3.60	3.40	2.88
Airlines	262.18	179.65	161.38	**108.12**	240.76	108.44	156.16	144.92	124.93
Electricity	29.17	15.32	17.24	9.98	32.35	**9.22**	47.73	44.29	38.18
Forest covertype	523.65	334.43	335.90	218.14	328.34	291.91	161.63	148.70	**129.30**
KDD'98	1774.32	1255.20	1783.24	1261.51	**1104.74**	1823.21	1485.15	1390.10	1188.12
KDD'99	751.71	280.30	319.10	**196.67**	657.26	594.50	936.39	883.95	749.11
BPaM	114.29	77.06	**72.46**	53.90	179.46	176.39	125.26	119.25	100.21

Finally, in Table 4, we present the memory consumption of the algorithms during the experiments and Fig. 5c. For memory consumption, we were able to verify that there is no significant difference between algorithms with a 95% confidence level.

6 Conclusion

In this study, we discussed the problem of data stream clustering and reviewed three recent social network-based algorithms. These were discussed, compared and thoroughly evaluated in both synthetic and real-world datasets that included concept drifts and evolutions. Results showed the efficiency of social network-based methods when compared to baselines and state-of-the-art clustering algorithms regarding clustering quality, processing time and memory usage. Additionally, none of the proposed methods require batch processing during the offline step to finding clusters, nor demand the number of clusters to be found n, which are two important and expected traits of data stream clustering algorithms.

Trends for future works involving data stream clustering techniques encompass the adoption of these and other proposals in semi-supervised learning settings. In semi-supervised learning, instances' labels are periodically made available to learners. Therefore, learners must maintain clusters that represent the current data distribution to update their classification models during unlabeled periods of the stream and use these clusters to detect concept drifts and evolutions based on unlabeled data. In opposition to existing semi-supervised learning schemes, the adoption of social network-based clustering algorithms would be beneficial since they do not rely on a user-given amount of clusters to be found and have shown high clustering quality in a variety of scenarios.

Table 4 *RAM-Hours* obtained during experiments

| Experiment | RAM-Hours (GB-Hour) | | | | | | | | |
| | CluStream | | ClusTree | | DenStream | HAStream | CNDenStream | SNCStream | SNCStream+ |
	k-Means	DBSCAN	k-Means	DBSCAN					
RBF_2	$\mathbf{7.40 \times 10^{-8}}$	1.74×10^{-6}	3.81×10^{-6}	4.55×10^{-8}	1.04×10^{-6}	1.01×10^{-6}	2.91×10^{-6}	2.47×10^{-6}	2.33×10^{-6}
RBF_5	9.73×10^{-8}	4.56×10^{-6}	3.40×10^{-6}	$\mathbf{7.34 \times 10^{-8}}$	9.17×10^{-7}	9.59×10^{-7}	1.15×10^{-6}	1.02×10^{-6}	9.23×10^{-7}
RBF_{20}	4.69×10^{-7}	8.76×10^{-6}	6.24×10^{-6}	$\mathbf{3.68 \times 10^{-7}}$	2.43×10^{-6}	1.77×10^{-6}	9.71×10^{-7}	8.47×10^{-7}	7.77×10^{-7}
RBF_{50}	1.84×10^{-6}	3.70×10^{-5}	2.80×10^{-5}	$\mathbf{1.45 \times 10^{-6}}$	7.93×10^{-6}	3.27×10^{-6}	2.02×10^{-6}	1.77×10^{-6}	1.61×10^{-6}
Two moon	$\mathbf{3.22 \times 10^{-9}}$	3.45×10^{-9}	3.99×10^{-8}	4.18×10^{-8}	3.62×10^{-7}	5.54×10^{-8}	8.67×10^{-8}	6.72×10^{-8}	6.34×10^{-8}
Airlines	1.65×10^{-6}	8.32×10^{-5}	5.01×10^{-5}	$\mathbf{1.01 \times 10^{-6}}$	7.06×10^{-6}	3.41×10^{-5}	8.67×10^{-6}	7.35×10^{-6}	6.93×10^{-6}
Electricity	2.32×10^{-7}	7.88×10^{-6}	5.13×10^{-6}	$\mathbf{1.36 \times 10^{-7}}$	1.09×10^{-6}	3.25×10^{-6}	2.40×10^{-6}	2.04×10^{-6}	1.92×10^{-6}
Forest covertype	4.81×10^{-5}	2.27×10^{-4}	1.48×10^{-4}	3.07×10^{-6}	1.37×10^{-5}	1.27×10^{-5}	1.39×10^{-5}	1.20×10^{-5}	$\mathbf{1.11 \times 10^{-5}}$
KDD'98	3.44×10^{-5}	1.38×10^{-4}	8.47×10^{-5}	9.15×10^{-5}	3.89×10^{-5}	8.03×10^{-4}	1.07×10^{-4}	2.91×10^{-5}	$\mathbf{2.69 \times 10^{-5}}$
KDD'99	1.44×10^{-5}	2.33×10^{-5}	6.08×10^{-6}	1.63×10^{-4}	5.69×10^{-5}	1.45×10^{-3}	$\mathbf{4.74 \times 10^{-6}}$	4.13×10^{-5}	3.79×10^{-5}
BPaM	7.66×10^{-5}	3.23×10^{-5}	2.26×10^{-5}	$\mathbf{4.82 \times 10^{-7}}$	6.22×10^{-6}	1.24×10^{-4}	1.84×10^{-5}	1.62×10^{-5}	1.47×10^{-5}

Acknowledgements This research was financially supported by the *Coordenação de Aperfeiçoa-mento de Pessoal de Nível Superior* (CAPES) through the *Programa de Suporte à Pòs-Graduação de Instituições de Ensino Particulares* (PROSUP) program and Fundação Araucária.

References

1. Aggarwal, C.C., Hinneburg, A., Keim, D.A.: On the surprising behavior of distance metrics in high dimensional space. In: International Conference on Database Theory 2001, pp. 420–434. Springer, Berlin (2001). https://doi.org/10.1007/3-540-44503-X_27
2. Aggarwal, C.C., Han, J., Wang, J., Yu, P.S.: A framework for clustering evolving data streams. In: Proceedings of the 29th International Conference on Very Large Data Bases - Volume 29, VLDB Endowment, VLDB '03, pp. 81–92 (2003)
3. Albert, R., Barabási, A.L.: Statistical mechanics of complex networks. In: Reviews of Modern Physics, pp. 139–148. The American Physical Society (2002)
4. Amini, A., Wah, T.Y.: On density-based data streams clustering algorithms: a survey. J. Comput. Sci. Technol. **29**(1), 116–141 (2014). https://doi.org/1.1007/s11390-014-1416-y
5. Barddal, J.P., Gomes, H.M., Enembreck, F.: A complex network-based anytime data stream clustering algorithm. In: Neural Information Processing - 22nd International Conference, ICONIP 2015, Istanbul, Turkey, November 9–12, 2015, Proceedings, Part I, pp. 615–622 (2015). https://doi.org/10.1007/978-3-319-26532-2_68
6. Barddal, J.P., Gomes, H.M., Enembreck, F.: SNCStream: a social network-based data stream clustering algorithm. In: Proceedings of the 30th Annual ACM Symposium on Applied Computing (SAC). ACM, New York (2015)
7. Barddal, J.P., Gomes, H.M., Enembreck, F., Barthès, J.P.: SNCStream+: extending a high quality true anytime data stream clustering algorithm. Inf. Syst. (2016). https://doi.org/10.1016/j.is.2016.06.007
8. Bifet, A., Holmes, G., Kirkby, R., Pfahringer, B.: Moa: Massive online analysis. J. Mach. Learn. Res. **11**, 1601–1604 (2010)
9. Cao, F., Ester, M., Qian, W., Zhou, A.: Density-based clustering over an evolving data stream with noise. In: Proceedings of the 2006 SIAM International Conference on Data Mining, pp 328–339 (2006)
10. Corder, G., Foreman, D.: Nonparametric Statistics for Non-Statisticians: A Step-by-Step Approach. Wiley, London (2011)
11. Erdos, P., Rényi, A.: On the evolution of random graphs. In: Publication of the Mathematical Institute of the Hungarian Academy of Sciences, pp. 17–61 (1960)
12. Ester, M., Kriegel, H.P., Sander, J., Xu, X.: A density-based algorithm for discovering clusters in large spatial databases with noise. In: Simoudis, E., Han, J., Fayyad, U.M. (eds.) KDD-96 Proceedings, pp. 226–231. AAAI Press, Menlo Park (1996)
13. Gama, J., Zliobaite, I., Bifet, A., Pechenizkiy, M., Bouchachia, A.: A survey on concept drift adaptation. ACM Comput. Surv. **46**(4), 1–37 (2014). https://doi.org/1.1145/2523813
14. Gomes, H.M., Barddal, J.P., Enembreck, F., Bifet, A.: A survey on ensemble learning for data stream classification. ACM Comput. Surv. **50**(2), 1–36 (2017). https://doi.org/10.1145/3054925
15. Guttman, A.: R-trees: a dynamic index structure for spatial searching. In: Proceedings of the 1984 ACM SIGMOD International Conference on Management of Data SIGMOD'84, pp. 47–57. ACM, New York (1984). https://doi.org/1.1145/602259.602266
16. Harries, M., Wales, N.S.: Splice-2 comparative evaluation: Electricity pricing (1999)
17. Hassani, M., Spaus, P., Seidl, T.: Adaptive multiple-resolution stream clustering. In: Perner, P. (ed.) Machine Learning and Data Mining in Pattern Recognition. Lecture Notes in Computer Science, vol. 8556, pp. 134–148. Springer International Publishing, Berlin (2014)

18. Ikonomovska, E., Gama, J., Zenko, B., Dzeroski, S.: Speeding-up hoeffding-based regression trees with options. In: Proceedings of the 28th International Conference on International Conference on Machine Learning, pp. 537–544 (2011)
19. Kosina, P., Gama, J.: Very fast decision rules for multi-class problems. In: Proceedings of the 27th Annual ACM Symposium on Applied Computing, SAC'12, pp. 795–800. ACM, New York (2012). https://doi.org/1.1145/2245276.2245431
20. Kranen, P., Assent, I., Baldauf, C., Seidl, T.: The clustree: indexing micro-clusters for anytime stream mining. Knowl. Inf. Syst. **29**(2), 249–272 (2011)
21. Kremer, H., Kranen, P., Jansen, T., Seidl, T., Bifet, A., Holmes, G., Pfahringer, B.: An effective evaluation measure for clustering on evolving data streams. In: Proceedings of the 17th ACM Conference on Knowledge Discovery and Data Mining (SIGKDD 2011), San Diego, CA, pp. 868–876. ACM, New York (2011)
22. Lloyd, S.: Least squares quantization in PCM. IEEE Trans. Inf. Theory **28**(2), 129–137 (1982)
23. Masud, M.M., Gao, J., Khan, L., Han, J., Thuraisingham, B.M.: Classification and novel class detection in concept-drifting data streams under time constraints. IEEE Trans. Knowl. Data Eng. **23**(6), 859–874 (2011)
24. Silva, J.A., Faria, E.R., Barros, R.C., Hruschka, E.R., Carvalho, A.C.P.L.F.D., Gama, J.: Data stream clustering: a survey. ACM Comput. Surv. **46**(1), 1–31 (2013). https://doi.org/1.1145/2522968.2522981
25. Ugulino, W., Cardador, D., Vega, K., Velloso, E., Milidia, R., Fuks, H.: Wearable computing: accelerometers' data classification of body postures and movements. In: Advances in Artificial Intelligence - SBIA 2012. Lecture Notes in Computer Science, pp. 52–61. Springer, Berlin (2012)
26. Watts, D.J., Strogatz, S.H.: Collective dynamics of small-world networks. Nature **393**(6684), 440–442 (1998)

Printed in the United States
By Bookmasters